Freeform Vector Diffraction Theory

Freeform Vector Diffraction Theory traces the conceptual journey from classical scalar diffraction to a rigorous and generalized framework for computing the vector diffraction patterns of arbitrary freeform wavefronts and complex, spatially varying polarization states.

Readers will gain a unified theoretical perspective that enables accurate prediction of electromagnetic field behavior in high-numerical-aperture systems and modern non-conventional optics. The book delivers these fundamental insights through detailed mathematical and geometrical toolkits, including vector calculus and Zernike polynomials, which facilitate rigorous derivations of the generalized Richards-Wolf integrals in both cylindrical and Cartesian coordinates. It also explores key practical applications in fields such as super-resolution microscopy, advanced lithography, and optical trapping. Through in-depth analysis of apodization functions and the response of radially, azimuthally, and linearly polarized light, readers will learn to precisely characterize complex optical systems affected by significant aberrations.

This work is an indispensable resource for leading researchers, optical design engineers, and graduate students in optics and photonics who seek to accurately model non-stigmatic and aberrated systems. The text clearly sets itself apart from other treatises on the subject by moving beyond the classical assumptions of spherical and ideal wavefronts typical of traditional theories. Instead, it offers a robust analytical framework for freeform surfaces and complex vectorial beams encountered in today's technology. By masterfully integrating concepts such as the Malus-Dupin theorem and the angular spectrum representation, the book equips professionals with the essential tools to analyze the vector nature of light and optimize optical instruments in an era where structured light continues to reveal profound new insights into the electromagnetic nature of radiation.

Rafael G. González-Acuña studied Industrial Physics Engineering at the Tecnologico de Monterrey and received a master's degree in Optomechatronics at the Optics Research Center, A.C. He studied PhD at the Tecnologico de Monterrey. He has 45 papers in indexes journals. Rafael was awarded the 2019 Optical Design and Engineering Scholarship by SPIE and is the co-author of three books. He is currently a senior lens designer at Huawei Technologies.

Series in Optics and Optoelectronics

Fourier Optics in Image Processing
Neil Collings

Holography
Principles and Applications
Raymond K. Kostuk

An Introduction to Quantum Optics, Second Edition
Photon and Biphoton Physics
Yanhua Shih

Polarized Light and the Mueller Matrix Approach, Second Edition
José J. Gil, Razvigor Ossikovski

Introduction to Holography, Second Edition
Vincent Toal

Optical Coherence Tomography in Dentistry
Scientific Developments to Clinical Applications
Anderson S. L. Gomes, Denise M. Zezell, Cláudia C. B. O. Mota, John M. Girkin

Holography: Principles and Applications, Second Edition
Raymond K. Kostuk

CAD-Based Optical Design with Quadoa
Rafael G. González-Acuña

Anisotropy of Metamaterials: Beyond Conventional Paradigms
Vladimir Mityushev, Tatjana Gric, Radoslaw Kycia and Natalia Rylko

Optical Science and Engineering
Elias Glytsis

Advanced Optical Techniques in Biosciences
Nirmal Mazumder, Yury Kistenev and Igor Lednev

Fundamental Optical Models
William A. Challener

Optical Activity: Basic Properties
Ulrich Zurcher

Field-Emission Vacuum Nanoelectronics
A Bird's Eye View of Theory and Practical Applications
Vinod Kumar Khanna

Freeform Vector Diffraction Theory
Rafael G. González-Acuña

For more information about this series, please visit:
https://www.crcpress.com/Series-in-Optics-and-Optoelectronics/book-series/TFOPTICSOPT

Freeform Vector Diffraction Theory

Rafael G. González-Acuña

CRC Press
Taylor & Francis Group
Boca Raton London New York

CRC Press is an imprint of the
Taylor & Francis Group, an **informa** business

Front cover image: Rafael G. González-Acuña

First edition published 2027
by CRC Press
2385 NW Executive Center Drive, Suite 320, Boca Raton FL 33431

and by CRC Press
4 Park Square, Milton Park, Abingdon, Oxon, OX14 4RN

CRC Press is an imprint of Taylor & Francis Group, LLC

ISBN: 978-1-041-20198-4 (hbk)
ISBN: 978-1-041-21295-9 (pbk)
ISBN: 978-1-003-72649-4 (ebk)

DOI: 10.1201/9781003726494

Typeset in CMR10 font
by KnowledgeWorks Global Ltd.

Publisher's note: This book has been prepared from camera-ready copy provided by the authors.

To Jesus Christ and the Virgin Mary.

Contents

Acknowledgments

To Almighty God, in whose glory this book is offered. Thank You, Lord, for the intellect, the passion, the faith, and the provision that enabled its completion. May I ever be Your faithful servant, deserving through grace the promises of Your Son, Jesus Christ. Without You, man does not own himself. You are the way, the truth, and the life. (John 14:6)

I want to thank my parents, my mom, Carmen Leticia Acuña Medellín, and my dad, Rogelio González Cantú, for encouraging and loving me so much.

To several professors and researchers in my path, who have advised me wisely: R. Barry Johnson, Simon Thibault, Reinhard Kettle, Julio C. Gutierrez-Vega, Bernardino Barrientos-García, Abundio Dávila-Álvarez, Mateusz Szatkowski, and Benjamin Perez-Garcia.

I want to thank my friend Héctor A. Chaparro-Romo, for all these years working in optics together.

I would also like to thank my friends and familiy: Rolando González Acuña, Rogelio González Acuña, Sonia Villarreal, María Isabel González Villarreal, José Miguel González Villarreal, Juan Pablo González Villarreal, Lilia Acuña Medellín, Leopoldo Fernández, Daniel Lomas Satarain, Eduardo de Villa, Jose Urbina, Mira Kankkonen, Lorena Patricia, Yoshio Catillejos, Israel Meléndez, Roberto Martínez, Fernando Tamayo, Roberto Vera, Miguel Rojas, Zane Sibley, Jussi Echeverría, Jesús Garibaldi, Luis Garza Soto, Jorge Sánchez, Lionel González, Adrián Lozano, Miguel Ianilli, Mauricio Reyes Arroyo, Guillermo Tamargo, Oona Niklas-Salminen, and Ghulam Murtaza.

Preface

Diffraction of light, one of the most fundamental and elegant phenomena in optical physics, has fascinated scientists since the early observations of shadows and slits, evolving into the precise manipulation of electromagnetic fields at nanoscale resolutions. This book traces a historical and conceptual journey that begins with the classical scalar theories—which, in their elegant simplicity, captured the essential wave nature of light—and progresses toward the rigorous vectorial descriptions that now enable accurate predictions of light behavior in high-numerical-aperture systems and under complex polarization conditions.

Scalar theories, from the seminal contributions of Kirchhoff, Huygens, Fresnel, and Fraunhofer, provided powerful tools for understanding propagation in paraxial regimes and image formation in conventional optical systems. Rooted in solutions to the scalar Helmholtz equation and principles such as the Huygens principle, these approximations proved remarkably successful in classical applications: from microscope resolution limits to lens design large aperture telescopes. However, when light is focused tightly—with high numerical apertures or significant aberrations—the vectorial nature of the electromagnetic field becomes unavoidable. Polarization is no longer a secondary attribute but a central player, giving rise to effects such as longitudinal field components at the focus, and non-paraxial structured light patterns.

In this context, the pioneering work of B. Richards and E. Wolf (1959) represented a turning point. Their vectorial diffraction integrals, derived from energy conservation, far-field approximations, and plane-wave angular spectrum representations, enabled precise description of the focused field in aplanatic high-aperture systems. The Richards-Wolf equations became the standard tool for analyzing tightly focused beams with arbitrary polarizations—linear, circular, radial, azimuthal, or vortex—and have found widespread application in confocal microscopy, optical trapping, electron acceleration, and the generation of optical vortex knots. Despite their robustness, the original Richards-Wolf integrals were limited to stigmatic, aberration-free systems with spherical wavefronts and circular pupils. The demand to model more realistic and advanced systems—aspheric lenses, freeform surfaces, arbitrary pupil shapes, and complex aberrations—has driven subsequent extensions and generalizations. A notable recent advancement is the contribution by Rafael G. González-Acuña and Simon Thibault (published in Physica Scripta in 2023), who introduced a generalized vector diffraction integral capable of computing diffraction patterns for aspheric and freeform converging wavefronts, with arbitrary pupil shapes and polarization states. Grounded in energy conservation, far-field approximations, and angular spectrum principles, this formulation reduces to the classical Richards-Wolf integrals under ideal conditions while significantly broadening applicability to modern non-conventional optics.

This book aims to chart that evolution: from the scalar roots of diffraction theory to the most advanced vectorial frameworks, with particular emphasis on the Richards-Wolf equations and their contemporary generalization. The chapters provide detailed derivations, numerical examples, validations against limiting cases, and practical applications, offering both a solid theoretical foundation useful for researchers and students in physical optics, photonics, and related fields.

It is my hope that this work will serve not only as a technical reference but also as inspiration for new generations to push the boundaries of vectorial diffraction in an era where structured light and optics continue to unveil profound insights into the electromagnetic nature of light.

Rafael G. González-Acuña
Tampere, Finland
March 2026

Rafael G. González-Acuña studied his PhD at the Tecnológico de Monterrey (2020). He has written 47 optics papers in indexed journals. He has been awarded several prizes, including the Romulo Garza Science Awards and the Recognition of Civic Merit, the highest distinction that the State of Nuevo León grants. SPIE has awarded Rafael the 2019 Optical Design and Engineering Scholarship, and he is the co-author of three books: Analytical Lens Design, Stigmatic Optics, and Optical Path Theory. He is known for analytically solving a 2,400-year-old unsolved problem by obtaining the general equation of the stigmatic lens. He currently works as a senior lens designer for Huawei Technologies.

Part I

Mathematical, physical, and optical background

1

Mathematical toolkit

This chapter presents a comprehensive mathematical toolkit essential for scalar and vector diffraction theory, bridging classical optics with advanced freeform wavefront formulations. Key topics include vector algebra in $\mathbb{R}^3$, Rodrigues' rotation formula for polarization transformations, parametric representations of curves and surfaces, normal vectors, and the Jacobian determinant for coordinate changes in diffraction integrals. Complex numbers are reviewed for phase-amplitude description, while the two-dimensional Fourier and Hankel transforms are emphasized as core tools for field propagation and radially symmetric systems. These foundations enable rigorous derivation and generalization of the Richards–Wolf equations to non-stigmatic, freeform wavefronts, providing a unified framework for subsequent vector diffraction analysis.

1.1 Introduction

Vector diffraction theory bridges electromagnetism, geometrical optics, and wave propagation to describe the vectorial nature of light diffraction. This theory has a robust mathematical foundation, which is indispensable for both conceptual clarity and practical applications [1]. This initial chapter serves as a comprehensive mathematical toolkit for this theory. The purpose of the toolkit is to show and prove several mathematical statements that are going to be used in the generalization of the stigmatic vector diffraction integral, the Richards-Wolf equations, to freeform wavefronts [2].

Therefore, this chapter may be unnecessary for math-skilled readers. Also it may seem a little bit unorganized, since it grasps several math theorems, concepts, here and there, and put them in a single chapter. But with this chapter, we will have in a single document the mathematical background from scalar diffraction theory and vector diffraction theory, to freeform vector

DOI: 10.1201/9781003726494-2

diffraction theory. In other words, this chapter works to achieve the ultimate goal of the book, which is to provide a source that describes the whole journey, which starts from the scalar that starts from the scalar diffraction theory, then studies the stigmatic vector diffraction theory (the Richards-Wolf equations), and finally finishes with the generalization of the Richards-Wolfs equations to freeform wavefronts.

This chapter equips readers with the essentials for the mentioned task, tools, vector calculus, parametric representation, the Jacobian matrix, Rodrigues' rotation formula, to navigate the intricate vectorial formulations and transformations inherent in diffraction phenomena. Vector calculus provides the core framework for analyzing electromagnetic fields, including gradients of scalar potentials, divergences of flux densities, and curls representing rotational aspects of wave vectors, which are fundamental to Maxwell's equations and diffraction integrals (the Richards-Wolf equations).

We start with reviewing vectors and their main properties. We put special attention to Rodrigues' rotation formula. Rodrigues' formula will be used, later in this book, to describe rotations of polarization vectors and coordinate frames. It will be essential for understanding polarization of the electric field along the freeform and aspherical wavefronts.

The parametric representation will be studied as well in this chapter. Since it enables the precise description of curved surfaces, apertures, and wavefronts encountered in optical systems, facilitating the parameterization of boundaries in integral equations like the Kirchhoff diffraction formula or, as we will see in Chapter 3. Or the generalization of the Richards-Wolfs equations to freeform wavefronts in Chapter 7.

The Jacobian matrix is crucial for handling coordinate transformations and change of variables in multidimensional integrals, ensuring accurate evaluations of diffraction patterns under varying geometries or media. In Chapters 7 and 8, the Jacobian matrix plays a major role in the coordinate transformations of the wavefront in spherical coordinates to Cartesian coordinates. This enables, in part, the generalization of the Richards-Wolfs equations to freeform wavefronts. Therefore, it is good to have a refresher of the Jacobian matrix.

Finally, the concepts behind the Fourier transform and the Hankel transform are reviewed. The review focuses on their main properties and application of the mentioned integrals transforms. These concepts would be applied in Chapters 3, 4, 7, and 8.

By consolidating these interconnected concepts into a dedicated toolkit chapter, this book not only refreshes prerequisite knowledge but also fosters a unified perspective, preparing the reader, whether students, researchers, or practitioners, to tackle these advanced topics, Richards-Wolf equations and their generalization onto freeform wavefronts. This foundational approach minimizes barriers to entry, promotes self-contained exposition, and underscores the mathematical elegance underlying vector diffraction theory.

1.2 Introduction to vectors in $\mathbb{R}^3$

This book is entitled *Freeform Vector Diffraction Theory.* Throughout the book, we are going to discuss all the words in the title. For the moment, we are going to focus on the word *vector.* A *vector* is an identity in two-dimensional Euclidean space ($\mathbb{R}^2$) or in three-dimensional Euclidean space ($\mathbb{R}^3$). It is a mathematical object with both magnitude and direction. Vectors are used to represent quantities such as displacement, velocity, and force in physics, engineering, and computer graphics. But the most important (for us) vectors are also used to describe the electric field **E**, which plays a major role in diffraction phenomena (as detailed in Parts 3 and 4 of the book). So it is good to have a good reminder of vectors' properties in $\mathbb{R}^2$ and $\mathbb{R}^3$, after all **E** is a vector in $\mathbb{R}^3$.

Geometrically, a vector in $\mathbb{R}^2$ and $\mathbb{R}^3$ is visualized as an arrow, where the length represents the magnitude and the orientation represents the direction [3]. See Fig. 1.1.

In two dimensions, a vector **v** is given by its components along the x, and y, as is displayed in *left*-side of Fig. 1.1,

$$\mathbf{v} = \begin{bmatrix} v_1 \\ v_2 \end{bmatrix} \quad \text{or} \quad \mathbf{v} = [v_1, v_2].$$

In $\mathbb{R}^3$, a vector **v** is represented by its components along the x, y, and z axes along the Cartesian coordinate system,

$$\mathbf{v} = \begin{bmatrix} v_1 \\ v_2 \\ v_3 \end{bmatrix} \quad \text{or} \quad \mathbf{v} = [v_1, v_2, v_3].$$

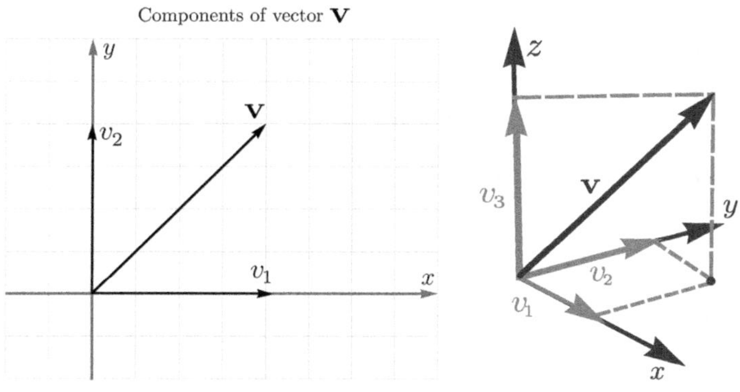

FIGURE 1.1
Diagram of the components of the vector **v** in $\mathbb{R}^2$ (*left*) and $\mathbb{R}^3$ (*right*).

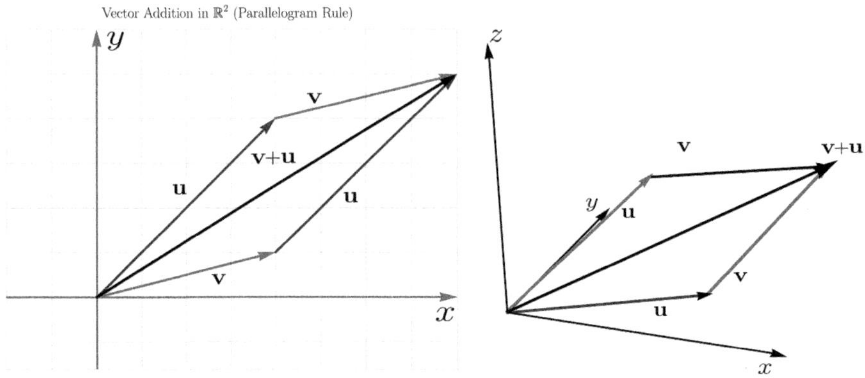

FIGURE 1.2
Additions of two vectors, $\mathbf{u}+\mathbf{v}$ in $\mathbb{R}^2$ (*left*) and $\mathbb{R}^3$ (*right*). In the parallelogram method, two vectors placed tail-to-tail form adjacent sides; their diagonal from the origin is the resultant sum.

1.2.1 Basic vector operations

Vectors in $\mathbb{R}^2$ have the following operations [3],

Addition: Given $\mathbf{u} = [u_1, u_2]$ and $\mathbf{v} = [v_1, v_2]$, the addition is

$$\mathbf{u} + \mathbf{v} = [u_1 + v_1, u_2 + v_2]. \tag{1.1}$$

Graphically the addition of two vectors can be represented by a parallelogram as displayed in Fig. 1.2. When two vectors are added using the parallelogram method, they are placed tail-to-tail at a common origin, forming two adjacent sides of a parallelogram. The diagonal of this parallelogram, drawn from the common origin to the opposite corner, represents the resultant vector, which is the sum of the two original vectors both in magnitude and direction.

Scalar Multiplication: Let c be a scalar, and a vector $\mathbf{v} = [v_1, v_2]$, the scalar multiple is as follows,

$$c\mathbf{v} = [cv_1, cv_2]. \tag{1.2}$$

Magnitude: The magnitude of a vector $\mathbf{v}$ in $\mathbb{R}^2$ is

$$\|\mathbf{v}\| = \sqrt{v_1^2 + v_2^2}. \tag{1.3}$$

The magnitude is always a positive real value. Therefore, $\|\mathbf{v}\| \in \mathbb{R}^+$.

Just like vectors in $\mathbb{R}^2$. Vectors in $\mathbb{R}^3$ can be manipulated using the following operations:

Addition: For two vectors $\mathbf{u} = [u_1, u_2, u_3]$ and $\mathbf{v} = [v_1, v_2, v_3]$, their sum is

$$\mathbf{u} + \mathbf{v} = [u_1 + v_1, u_2 + v_2, u_3 + v_3]. \tag{1.4}$$

Scalar Multiplication: For a scalar c and a vector $\mathbf{v} = [v_1, v_2, v_3]$, the scalar multiple is

$$c\mathbf{v} = [cv_1, cv_2, cv_3]. \tag{1.5}$$

Magnitude: The magnitude of a vector $\mathbf{v} = [v_1, v_2, v_3]$ is

$$\|\mathbf{v}\| = \sqrt{v_1^2 + v_2^2 + v_3^2}. \tag{1.6}$$

Like in two-dimensions, the magnitude is positive, thus, $\|\mathbf{v}\| \in \mathbb{R}^+$.

1.2.2 Dot product

The **dot product** of two vectors $\mathbf{u} = [u_1, u_2]$ and $\mathbf{v} = [v_1, v_2]$ in $\mathbb{R}^2$ is a scalar defined as [3],

$$\mathbf{u} \cdot \mathbf{v} = u_1v_1 + u_2v_2. \tag{1.7}$$

Geometrically, the dot product can also be expressed as

$$\mathbf{u} \cdot \mathbf{v} = \|\mathbf{u}\|\|\mathbf{v}\| \cos\theta, \tag{1.8}$$

where θ is the angle between the vectors. The geometrical interpretation of the dot product is the projection of vector $\mathbf{u}$ over $\mathbf{v}$.

In the same way the **dot product** of two vectors $\mathbf{u} = [u_1, u_2, u_3]$ and $\mathbf{v} = [v_1, v_2, v_3]$ in $\mathbb{R}^3$ is a scalar defined as,

$$\mathbf{u} \cdot \mathbf{v} = u_1v_1 + u_2v_2 + u_3v_3. \tag{1.9}$$

Representation of the dot product in $\mathbb{R}^2$ and $\mathbb{R}^3$ is displayed in Fig. 1.3. The dot product is commutative,

$$\mathbf{u} \cdot \mathbf{v} = \mathbf{v} \cdot \mathbf{u}. \tag{1.10}$$

Notice that the dot product is zero if the vectors are perpendicular ($\theta = 90°$, since $\cos 90° = 0$). So it is very useful to compute projections or the angle between vectors. This is going to be very useful for Chapter 7, when we apply it to describe the polarization and the freeform wavefront, since the propagation vector of the wavefront must be perpendicular to the polarization vector.

Example: Consider vectors $\mathbf{u} = [1, 2, 3]$ and $\mathbf{v} = [4, -1, 0]$. Compute the dot product,

$$\mathbf{u} \cdot \mathbf{v} = (1 \cdot 4) + (2 \cdot (-1)) + (3 \cdot 0) = 4 - 2 + 0 = 2.$$

To find the angle θ between them, compute their magnitudes:

$$\|\mathbf{u}\| = \sqrt{1^2 + 2^2 + 3^2} = \sqrt{1 + 4 + 9} = \sqrt{14},$$

$$\|\mathbf{v}\| = \sqrt{4^2 + (-1)^2 + 0^2} = \sqrt{16 + 1 + 0} = \sqrt{17}.$$

Using the dot product formula,

$$\cos\theta = \frac{\mathbf{u} \cdot \mathbf{v}}{\|\mathbf{u}\|\|\mathbf{v}\|} = \frac{2}{\sqrt{14} \cdot \sqrt{17}} \approx \frac{2}{7.727} \approx 0.129641.$$

Thus, $\theta \approx \arccos(0.129641)$.

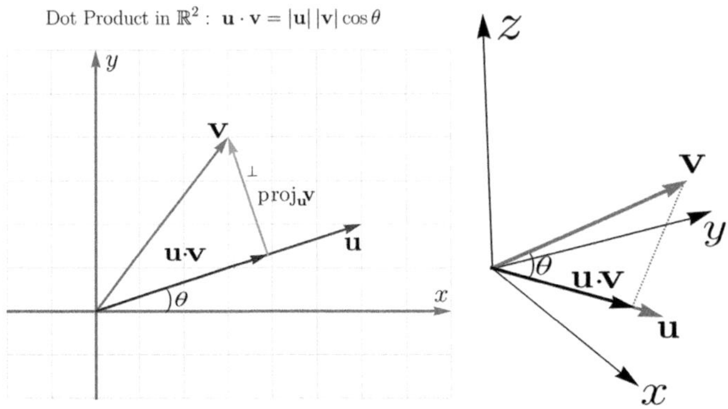

FIGURE 1.3
Dot product of two vectors, $\mathbf{u} \cdot \mathbf{v}$, $\mathbb{R}^2$ (*left*) and $\mathbb{R}^3$ (*right*). θ is the angle between the vectors. The geometrical interpretation of the dot product is the projection of vector $\mathbf{u}$ over $\mathbf{v}$.

1.2.3 Cross product

The **cross product** of two vectors $\mathbf{u} = [u_1, u_2, u_3]$ and $\mathbf{v} = [v_1, v_2, v_3]$ in $\mathbb{R}^3$ is a vector perpendicular to both $\mathbf{u}$ and $\mathbf{v}$, defined as [3],

$$\mathbf{u} \times \mathbf{v} = \begin{bmatrix} u_2v_3 - u_3v_2 \\ u_3v_1 - u_1v_3 \\ u_1v_2 - u_2v_1 \end{bmatrix}. \tag{1.11}$$

This can be computed using the determinant

$$\mathbf{u} \times \mathbf{v} = \begin{vmatrix} \hat{\mathbf{x}} & \hat{\mathbf{y}} & \hat{\mathbf{z}} \\ u_1 & u_2 & u_3 \\ v_1 & v_2 & v_3 \end{vmatrix}, \tag{1.12}$$

where $\hat{\mathbf{x}}$, $\hat{\mathbf{y}}$, and $\hat{\mathbf{z}}$ are the unit vectors along the x, y, and z axes.

The magnitude of the cross product is

$$\|\mathbf{u} \times \mathbf{v}\| = \|\mathbf{u}\|\|\mathbf{v}\| \sin\theta, \tag{1.13}$$

where θ is the angle between the vectors. The direction of $\mathbf{u} \times \mathbf{v}$ follows the right-hand rule, which is expressed in the matrix at Eq. (1.11). The resultant vector is perpendicular to $\mathbf{u}$ and $\mathbf{v}$ as it can be seen in Fig. 1.4.

Properties of the cross product include:

- Anti-commutative: $\mathbf{u} \times \mathbf{v} = -\mathbf{v} \times \mathbf{u}$.
- Zero if the vectors are parallel ($\theta = 0°$ or $180°$, since $\sin 0° = \sin 180° = 0$).
- Orthogonal to the plane formed by $\mathbf{u}$ and $\mathbf{v}$.

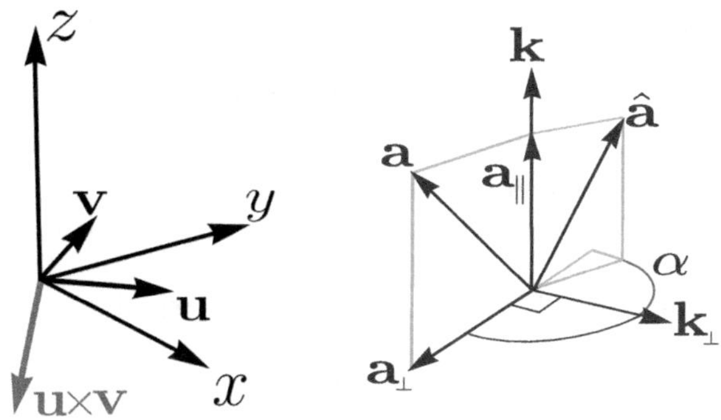

FIGURE 1.4
Left: Diagram of Rodrigues' rotation formula. Given a vector $\mathbf{a} \in \mathbb{R}^3$, an axis of rotation $\mathbf{k} \in \mathbb{R}^3$, and an rotation angle θ, this rotates vector to $\hat{\mathbf{a}}$. *Right:* The cross product of two vectors $\mathbf{u}$ and $\mathbf{v}$ in $\mathbb{R}^3$. The resultant vector $\mathbf{u} \times \mathbf{v}$ is perpendicular to $\mathbf{u}$ and $\mathbf{v}$.

Therefore, the cross product is usually used for vectors that are orthogonal to each other.

Example: Using the same vectors $\mathbf{u} = [1, 2, 3]$ and $\mathbf{v} = [4, -1, 0]$, compute the cross product,

$$\mathbf{u} \times \mathbf{v} = \begin{vmatrix} \hat{\mathbf{x}} & \hat{\mathbf{y}} & \hat{\mathbf{z}} \\ 1 & 2 & 3 \\ 4 & -1 & 0 \end{vmatrix}.$$

Calculate each component,

$$\hat{\mathbf{x}}\text{-component, } (2 \cdot 0) - (3 \cdot (-1)) = 0 + 3 = 3,$$

$$\hat{\mathbf{y}}\text{-component, } -((1 \cdot 0) - (3 \cdot 4)) = -(0 - 12) = 12,$$

$$\hat{\mathbf{z}}\text{-component, } (1 \cdot (-1)) - (2 \cdot 4) = -1 - 8 = -9.$$

Thus,

$$\mathbf{u} \times \mathbf{v} = [3, 12, -9].$$

To verify, the resulting vector should be perpendicular to both $\mathbf{u}$ and $\mathbf{v}$. Check using the dot product:

$$\mathbf{u} \cdot (\mathbf{u} \times \mathbf{v}) = (1 \cdot 3) + (2 \cdot 12) + (3 \cdot (-9)) = 3 + 24 - 27 = 0,$$

$$\mathbf{v} \cdot (\mathbf{u} \times \mathbf{v}) = (4 \cdot 3) + ((-1) \cdot 12) + (0 \cdot (-9)) = 12 - 12 + 0 = 0.$$

Since both dot products are zero, $\mathbf{u} \times \mathbf{v}$ is perpendicular to both vectors.

1.3 Rodrigues' rotation formula for vectors in $\mathbb{R}^3$

Rodrigues' rotation formula, named after Olinde Rodrigues, provides an efficient way to compute the rotation of a vector in three-dimensional space around a given axis by a specified angle [4]. For our purpose, it is very useful to use it for the rotation of the polarization vector along the propagation vector of the freeform wavefront. So, let's start by explaining the formula by providing a geometric proof.

Given a vector $\mathbf{a} \in \mathbb{R}^3$, a unit vector $\mathbf{k} \in \mathbb{R}^3$ (with $\|\mathbf{k}\| = 1$) representing the axis of rotation, and an angle α, the rotated vector $\hat{\mathbf{a}}$ according to Olinde Rodrigues is

$$\hat{\mathbf{a}} = \mathbf{a}\cos\alpha + (\mathbf{k} \times \mathbf{a})\sin\alpha + \mathbf{k}(\mathbf{k} \cdot \mathbf{a})(1 - \cos\alpha). \tag{1.14}$$

If $\mathbf{k}$ is not a unit vector, it should be normalized first.

This formula rotates $\mathbf{a}$ counterclockwise around $\mathbf{k}$ by α when viewed along the direction of $\mathbf{k}$. See *left* side of Fig. 1.4. The formula decomposes the rotation into three terms. The first term, $\mathbf{a}\cos\alpha$ scales the original vector by $\cos\alpha$, accounting for the projection in the rotated direction. The second term, $(\mathbf{k} \times \mathbf{a})\sin\alpha$, adds a component perpendicular to both $\mathbf{k}$ and $\mathbf{a}$, representing the rotational *twist*. Finally, the last term, $\mathbf{k}(\mathbf{k} \cdot \mathbf{a})(1 - \cos\alpha)$, adjusts the component parallel to $\mathbf{k}$, which remains unchanged but is *restored* since the first term scales it incorrectly.

Geometrically, the rotation leaves the component of $\mathbf{a}$ parallel to $\mathbf{k}$ unchanged, while rotating the perpendicular component in the plane orthogonal to $\mathbf{k}$. To derive Rodrigues' formula, decompose $\mathbf{a}$ into components parallel and perpendicular to the axis $\mathbf{k}$,

$$\mathbf{a} = \mathbf{a}_{\|} + \mathbf{a}_{\perp}, \tag{1.15}$$

where, $\mathbf{a}_{\|}$ is the parallel component of $\mathbf{a}$ with respect to $\mathbf{k}$ and $\mathbf{a}_{\perp}$ is the perpendicular component of $\mathbf{a}$ with respect to $\mathbf{k}$. Therefore,

$$\mathbf{a}_{\|} = (\mathbf{k} \cdot \mathbf{a})\mathbf{k}, \quad \mathbf{a}_{\perp} = \mathbf{a} - \mathbf{a}_{\|}. \tag{1.16}$$

Since rotation around $\mathbf{k}$ does not affect $\mathbf{a}_{\|}$, the rotated vector is:

$$\hat{\mathbf{a}} = \mathbf{a}_{\|} + \hat{\mathbf{a}}_{\perp}, \tag{1.17}$$

where $\hat{\mathbf{a}}_{\perp}$ is the rotation of $\mathbf{a}_{\perp}$ by α in the plane perpendicular to $\mathbf{k}$.

In the plane orthogonal to $\mathbf{k}$, the rotated vector $\hat{\mathbf{a}}_{\perp}$ can be expressed using the basis formed by $\mathbf{a}_{\perp}$ and a vector perpendicular to both $\mathbf{k}$ and $\mathbf{a}_{\perp}$. The vector perpendicular to both is $\mathbf{k} \times \mathbf{a}_{\perp} = \mathbf{k} \times \mathbf{a}$ (since $\mathbf{k} \times \mathbf{a}_{\|} = \mathbf{0}$).

The rotation in this plane is,

$$\hat{\mathbf{a}}_\perp = \mathbf{a}_\perp \cos\alpha + (\mathbf{k}\times\mathbf{a}_\perp)\frac{\sin\alpha}{\|\mathbf{a}_\perp\|}\cdot\|\mathbf{a}_\perp\| = \mathbf{a}_\perp\cos\alpha + (\mathbf{k}\times\mathbf{a}_\perp)\sin\alpha. \tag{1.18}$$

Notice that the unit vector in the direction of $\mathbf{k}\times\mathbf{a}_\perp$ is $(\mathbf{k}\times\mathbf{a}_\perp)/\|\mathbf{a}_\perp\|$, and multiplying by the length $\|\mathbf{a}_\perp\|$ preserves magnitude.

Substitute $\mathbf{a}_\perp = \mathbf{a} - \mathbf{a}_\| = \mathbf{a} - (\mathbf{k}\cdot\mathbf{a})\mathbf{k}$,

$$\hat{\mathbf{a}}_\perp = [\mathbf{a} - (\mathbf{k}\cdot\mathbf{a})\mathbf{k}]\cos\alpha + (\mathbf{k}\times\mathbf{a})\sin\alpha. \tag{1.19}$$

Now, add back the parallel component,

$$\hat{\mathbf{a}} = \mathbf{a}_\| + \hat{\mathbf{a}}_\perp = (\mathbf{k}\cdot\mathbf{a})\mathbf{k} + [\mathbf{a} - (\mathbf{k}\cdot\mathbf{a})\mathbf{k}]\cos\alpha + (\mathbf{k}\times\mathbf{a})\sin\alpha. \tag{1.20}$$

Expand and rearrange,

$$\hat{\mathbf{a}} = (\mathbf{k}\cdot\mathbf{a})\mathbf{k} + \mathbf{a}\cos\alpha - (\mathbf{k}\cdot\mathbf{a})\mathbf{k}\cos\alpha + (\mathbf{k}\times\mathbf{a})\sin\alpha. \tag{1.21}$$

Factor the terms involving $(\mathbf{k}\cdot\mathbf{a})\mathbf{k}$:

$$\hat{\mathbf{a}} = \mathbf{a}\cos\alpha + (\mathbf{k}\times\mathbf{a})\sin\alpha + (\mathbf{k}\cdot\mathbf{a})\mathbf{k}(1-\cos\alpha), \tag{1.22}$$

which is Rodrigues' formula.

This derivation relies on vector decomposition and the geometry of rotation in a plane. It assumes $\mathbf{k}$ is a unit vector; if not, the formula holds after normalization, as the projections and cross products scale appropriately.

Example: Rotation Around the z-axis. Consider rotating $\mathbf{a} = [1,0,0]$ around $\mathbf{k} = [0,0,1]$ by $\alpha = \frac{\pi}{2}$:

$$\begin{aligned}\hat{\mathbf{a}} &= [1,0,0]\cos\frac{\pi}{2} + ([0,0,1]\times[1,0,0])\sin\frac{\pi}{2}\\ &\quad + [0,0,1]([0,0,1]\cdot[1,0,0])(1-\cos\frac{\pi}{2})\\ &= [1,0,0]\cdot 0 + [0,1,0]\cdot 1 + [0,0,1]\cdot 0\cdot 1\\ &= [0,1,0],\end{aligned}$$

which is the expected 90-degree rotation in the xy-plane.

Rodrigues' rotation formula offers a compact, axis-angle representation of $\mathbb{R}^3$ rotations, avoiding the need for full rotation matrices in some applications. The proof highlights its geometric intuition through vector decomposition, making it accessible and verifiable.

1.4 Functions and mapping

The electric field $\mathbf{E}$ is a vector whose components can be complex (complex number will be reviewed soon) or real numbers, but it also depends on variables that can be real or complex, depending on the representation of it (later we are going to discuss the parametric representation). Therefore, it is positive for the reader to be reminded of the basic concepts of real vector functions and complex vector functions.

A real multivariable vector function $\mathbf{f} : \mathbb{R}^m \to \mathbb{R}^n$ maps a vector $\mathbf{x} = [x_1, x_2, \ldots, x_m] \in \mathbb{R}^m$ to a vector $\mathbf{f}(\mathbf{x}) = [f_1(\mathbf{x}), f_2(\mathbf{x}), \ldots, f_n(\mathbf{x})] \in \mathbb{R}^n$, where each component $f_i : \mathbb{R}^m \to \mathbb{R}$ is a real-valued function. These functions generalize scalar functions and single-variable vector functions by handling multiple inputs and outputs [5].

The set that is being mapped is called domain, and once this set is mapped is called codomain also called range or image. To illustrate this in Fig. 1.5, we set the mapping of a parabolic function. The function $f(x) = x^2$ maps each real number x to its square. Its domain is the entire set of real numbers, denoted $\mathbb{R}$, because any real value can be substituted for x and the operation x^2 is always defined. The codomain, traditionally taken as $\mathbb{R}$ for this function, is the set of all real numbers; however, the actual range (image) of the function is the set of non-negative real numbers $[0, +\infty)$, since the square of any real

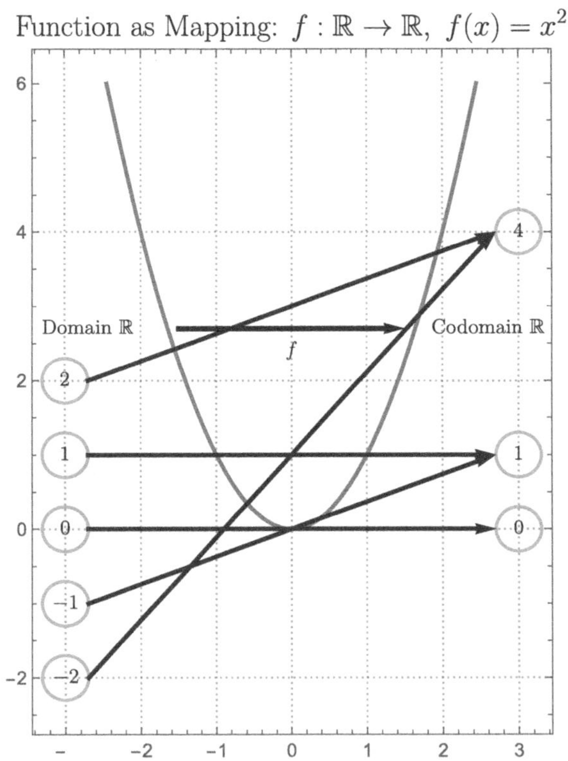

FIGURE 1.5
The function $f(x) = x^2$ has domain $\mathbb{R}$ (all real numbers) and codomain $\mathbb{R}$, though its range is $[0, +\infty)$. The column shows domain being mapped to the codomain by an arrow.

number is never negative. Thus, $f : \mathbb{R} \to \mathbb{R}^+$, where every input $x \in \mathbb{R}$ produces an output $f(x) \geq 0$.

Example: Consider $\mathbf{f} : \mathbb{R}^2 \to \mathbb{R}^2$ defined by $\mathbf{f}(x, y) = [x^2 + y, xy]$. For $\mathbf{x} = [1, 2]$,

$$\mathbf{f}(1, 2) = [1^2 + 2, 1 \cdot 2] = [3, 2]$$

This maps the vector $[1, 2]$ to the vector $[3, 2]$ in $\mathbb{R}^2$.

1.4.1 Basic operations

Operations on real multivariable vector functions are defined component-wise:

- **Addition**: For $\mathbf{f}, \mathbf{g} : \mathbb{R}^m \to \mathbb{R}^n$, $\mathbf{f} + \mathbf{g} = [f_1(\mathbf{x}) + g_1(\mathbf{x}), \ldots, f_n(\mathbf{x}) + g_n(\mathbf{x})]$.
- **Scalar Multiplication**: For a scalar $c \in \mathbb{R}$, $(c\mathbf{f})(\mathbf{x}) = [cf_1(\mathbf{x}), \ldots, cf_n(\mathbf{x})]$.
- **Dot Product**: For $\mathbf{f}, \mathbf{g} : \mathbb{R}^m \to \mathbb{R}^n$, the dot product is $\mathbf{f}(\mathbf{x}) \cdot \mathbf{g}(\mathbf{x}) = f_1(\mathbf{x})g_1(\mathbf{x}) + \cdots + f_n(\mathbf{x})g_n(\mathbf{x})$.

Example: Let $\mathbf{f}(x, y) = [x + y, y^2]$ and $\mathbf{g}(x, y) = [x, 2y]$. For $\mathbf{x} = [1, 1]$, compute the addition

$$\mathbf{f}(1, 1) = [1 + 1, 1^2] = [2, 1], \quad \mathbf{g}[1, 1] = [1, 2 \cdot 1] = [1, 2] \tag{1.23}$$

so,

$$(\mathbf{f} + \mathbf{g})(1, 1) = [2 + 1, 1 + 2] = [3, 3]. \tag{1.24}$$

Compute the dot product,

$$\mathbf{f}(1, 1) \cdot \mathbf{g}(1, 1) = 2 \cdot 1 + 1 \cdot 2 = 2 + 2 = 4. \tag{1.25}$$

1.4.2 Continuity

A vector function $\mathbf{f} : \mathbb{R}^m \to \mathbb{R}^n$ is continuous at a point $\mathbf{x}_0 \in \mathbb{R}^m$ if, for every $\epsilon > 0$, there exists a $\delta > 0$ such that $||\mathbf{x} - \mathbf{x}_0|| < \delta$ implies $||\mathbf{f}(\mathbf{x}) - \mathbf{f}(\mathbf{x}_0)|| < \epsilon$, where $|| \cdot ||$ is the Euclidean norm (similar to the magnitude of the vector). Equivalently, $\mathbf{f}$ is continuous if each component $f_i(\mathbf{x})$ is continuous [5].

Example: Consider $\mathbf{f}(x, y) = [x^2 - y, xy]$. The components $f_1(x, y) = x^2 - y$ and $f_2(x, y) = xy$ are polynomials, which are continuous everywhere in $\mathbb{R}^2$. Thus, $\mathbf{f}$ is continuous for all $(x, y) \in \mathbb{R}^2$. For $\mathbf{x}_0 = (1, 1)$:

$$\mathbf{f}(1, 1) = [1^2 - 1, 1 \cdot 1] = [0, 1]$$

As $\mathbf{x} \to (1, 1)$, $\mathbf{f}(\mathbf{x}) \to (0, 1)$, confirming continuity.

Example: The function $f(x) = (x - 1)^2 + 1$ is continuous at any point $x = a$ because the two-sided limit equals the value of the function at that point.

First, clearly $f(a) = (a-1)^2 + 1$. The left-hand limit is $\lim_{x \to a^-} f(x) = \lim_{x \to a^-}[(x-1)^2 + 1]$. Since $(x-1)^2 + 1$ is a polynomial, it is continuous from the left, so $\lim_{x \to a^-} f(x) = (a-1)^2 + 1 = f(a)$. Similarly, the right-hand limit is $\lim_{x \to a^+} f(x) = \lim_{x \to a^+}[(x-1)^2+1] = (a-1)^2+1 = f(a)$. See Fig. 1.6.

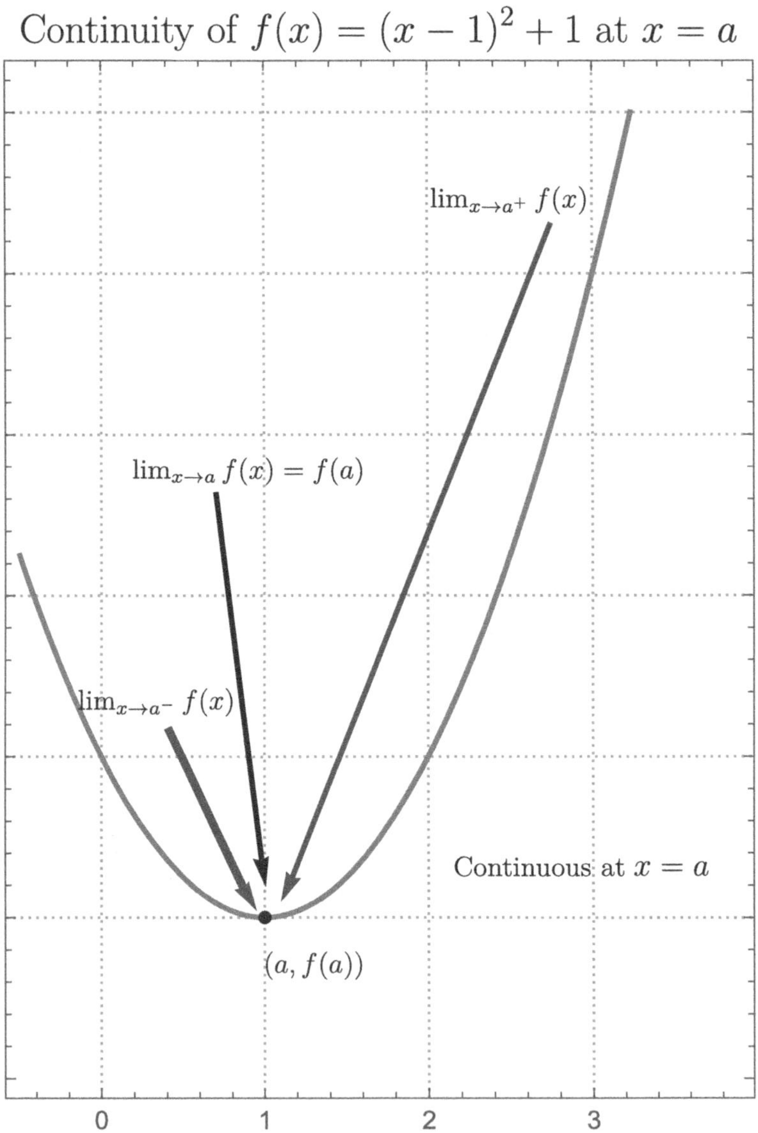

FIGURE 1.6
The function $f(x) = (x-1)^2 + 1$ is continuous at any point $x = a$ because the two-sided limit equals the value of the function at that point.

Because the left-hand limit equals the right-hand limit and both equal $f(a)$, the two-sided limit exists and $\lim_{x\to a} f(x) = f(a)$. Therefore, by the limit definition of continuity, f is continuous at $x = a$. This holds for every real number a, so f is continuous on all of $\mathbb{R}$.

Using the precise ϵ-δ definition and limits: $\lim_{x\to a} f(x) = f(a)$, which means for every $\epsilon > 0$, there exists $\delta > 0$ such that if $0 < |x - a| < \delta$, then $|f(x) - f(a)| < \epsilon$. Here, $|f(x) - f(a)| = |(x-1)^2 - (a-1)^2| = |x-a| \cdot |x+a-2|$.

Since $|x + a - 2|$ is bounded near a (for $|x - a| < 1$, it is less than $|a| + 3$), choosing $\delta = \min(1, \epsilon/(|a| + 3))$ ensures $|f(x) - f(a)| < |x - a| \cdot (|a| + 3) < \delta \cdot (|a| + 3) \leq \epsilon$. Thus, the limit condition holds, confirming continuity at $x = a$. Note: the repeated $||f(a) - f(a)||$ in your question is a typo and should be $|f(x) - f(a)|$.

Not all the continuous functions are so well behaved like polynomials, there are piecewise continuous which are also interesting for our proposes. A piecewise continuous function is a function that is built from several separate pieces, where each piece is continuous on its own interval, and at the points where the pieces meet, the function might jump up or down, but it always has a well-defined value from both the left and the right (no infinite blow-ups or wild oscillations). In other words, it's continuous inside each piece and doesn't go crazy at the connection points, even if the height suddenly changes there. An example for a piecewise continuous is shown in Fig. 1.7 (*left*).

But there is a more general function, which is piecewise but not continuous. A piecewise function is a function that is defined in different ways on different parts (or *pieces*) of its domain. Instead of having one single rule that works everywhere, it uses several rules, each applying only to a specific interval or set of input values — like a single function wearing different outfits depending on where you are on the number line. This types of functions are going to appear from time to time along the book, so it's good to remember them. An example of a piecewise function is displayed at *right* of Fig. 1.7.

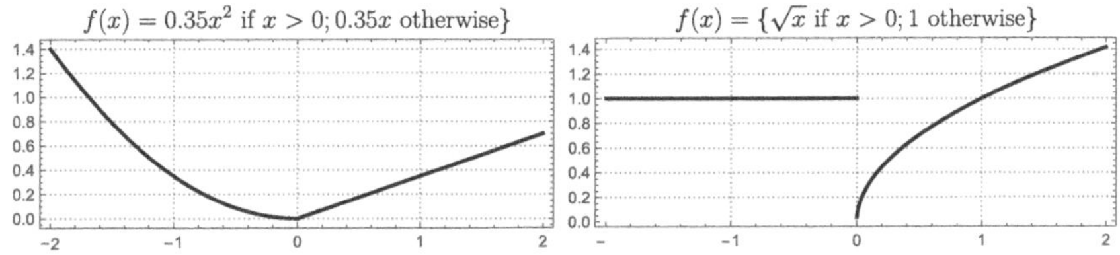

FIGURE 1.7
An example of a piecewise continuous function (*left*). An example of a piecewise function (*right*)

1.5 Vector calculus

The electric field is a vector function that depends on several variables and changes across these variables; this change can be understood and measured with vector calculus.

Vector calculus extends single-variable calculus to functions of several variables, particularly vector fields in three-dimensional Euclidean space ($\mathbb{R}^3$). It provides tools to describe and analyze phenomena involving motion, forces, and fields, such as fluid flow, electric fields (as we mentioned earlier), and gravitational forces. Central operations are the gradient for scalar fields, divergence, and curl for vector fields.

1.5.1 Gradient

The gradient of a scalar field $f(x, y)$ or $f(x, y, z)$, assumed differentiable, is a vector field that points in the direction of the steepest increase of f, with its magnitude representing the rate of change [3],

$$\nabla f(x, y) = \left[\frac{\partial f}{\partial x}, \frac{\partial f}{\partial y}\right], \tag{1.26}$$

or,

$$\nabla f(x, y, z) = \left[\frac{\partial f}{\partial x}, \frac{\partial f}{\partial y}, \frac{\partial f}{\partial z}\right], \tag{1.27}$$

where $\frac{\partial}{\partial x}$, $\frac{\partial}{\partial y}$, and $\frac{\partial}{\partial z}$ are the partial derivative respect to x, y, and z. Geometrically, the gradient is normal to the level surfaces $f(x, y, z) = c$.

Example: For $f(x, y, z) = x^2 + y^2 + z^2$, the gradient is

$$\nabla f = [2x, 2y, 2z]. \tag{1.28}$$

At $(1, 1, 1)$, $\nabla f = [2, 2, 2]$, pointing radially outward, consistent with the spherical symmetry of the function.

Example: For $f(x, y, z) = xyz$,

$$\nabla f = [yz, xz, xy]. \tag{1.29}$$

At $(1, 2, 3)$, $\nabla f = [6, 3, 2]$, indicating the direction of steepest increase of the function at that point.

We mentioned that the gradient is a vector field that points in the direction of the steepest increase of a function. Suppose the function $f(x, y)$ is plotted in the x-y plane and represented by a grayscale, where higher values of $f(x, y)$ appear brighter and lower values appear darker. With this in mind, if $f(x, y)$ increases along the y-direction, the gradient arrows will point toward increasing y (as shown in the *left* figure). If, instead, $f(x, y)$ increases along

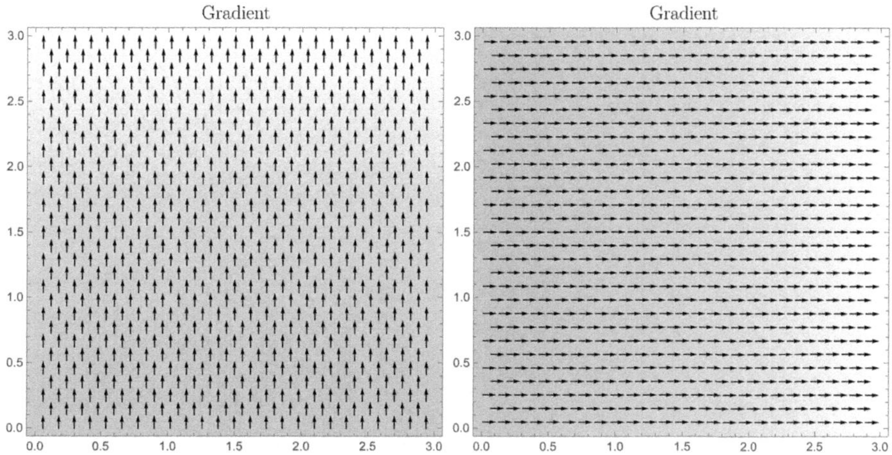

FIGURE 1.8
The gradient is a vector field pointing in the direction of steepest increase of $f(x, y)$. In a grayscale plot (brighter = higher values), arrows point toward brighter regions: upward along y if f rises with y (*left*), or rightward along x if f rises with x (*right*).

the x-direction, the arrows will point toward increasing x (as shown in the *right* figure).

With the representation presented in Fig. 1.8, it is clear that the gradient is a vector field pointing in the direction of steepest increase. For a more complicated function $f(x, y)$ you can imagine that the arrows are pointing where it gets brighter.

1.5.2 Divergence

The divergence of a vector field $\mathbf{F} = [P, Q]$ is given by [3],

$$\nabla \cdot \mathbf{F} = \frac{\partial P}{\partial x} + \frac{\partial Q}{\partial y}. \tag{1.30}$$

and the divergence of a vector field $\mathbf{F} = [P, Q, R]$

$$\nabla \cdot \mathbf{F} = \frac{\partial P}{\partial x} + \frac{\partial Q}{\partial y} + \frac{\partial R}{\partial z}. \tag{1.31}$$

In both cases, it measures the net *outflow* at a point. Positive divergence indicates a source and negative divergence indicates a sink. Fig. 1.9 shows a vector field that diverges represented by arrows (*left*) and as a stream (*right*).

Example For $\mathbf{F} = [x, y, z]$,

$$\nabla \cdot \mathbf{F} = 1 + 1 + 1 = 3. \tag{1.32}$$

This uniform divergence suggests the field expands outward, like a radial flow.

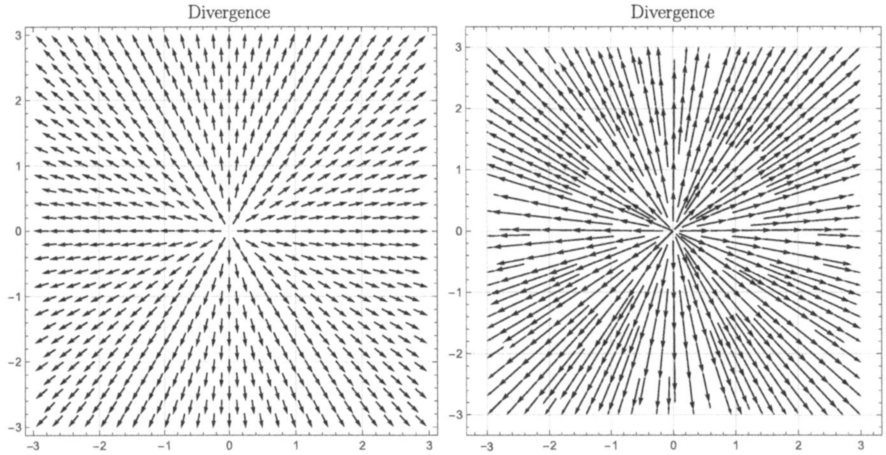

FIGURE 1.9
The divergence of a vector field measures the net *outflow* at a point. Positive divergence indicates a source, negative divergence indicates a sink. The figure shows a vector field that diverge represented by arrows (*left*) and as a stream (*right*).

Example For $\mathbf{F} = [x^2, y^2, z^2]$,

$$\nabla \cdot \mathbf{F} = 2x + 2y + 2z. \tag{1.33}$$

At $(1, 0, 0)$, $\nabla \cdot \mathbf{F} = 2$, indicating a local source.

1.5.3 Curl

The curl of a vector field $\mathbf{F} = [P, Q, R]$ measures its rotational tendency [3],

$$\nabla \times \mathbf{F} = \left[\frac{\partial R}{\partial y} - \frac{\partial Q}{\partial z}, \frac{\partial P}{\partial z} - \frac{\partial R}{\partial x}, \frac{\partial Q}{\partial x} - \frac{\partial P}{\partial y} \right]. \tag{1.34}$$

The curl is a vector whose direction aligns with the axis of rotation (via the right-hand rule just like the cross product).

Example: For $\mathbf{F} = [-y, x, 0]$,

$$\nabla \times \mathbf{F} = [0 - 0, 0 - 0, 1 - (-1)] = [0, 0, 2]. \tag{1.35}$$

This indicates rotation around the z-axis.

Example: For $\mathbf{F} = [yz, xz, xy]$,

$$\nabla \times \mathbf{F} = [x - x, y - y, z - z] = [0, 0, 0], \tag{1.36}$$

indicating an irrational field.

Fig. 1.10 shows a vector field with an axis of rotation at the center of the coordinates system. This rotation is represented by arrows (*left*) and as a stream (*right*) and it is clear it is clockwise.

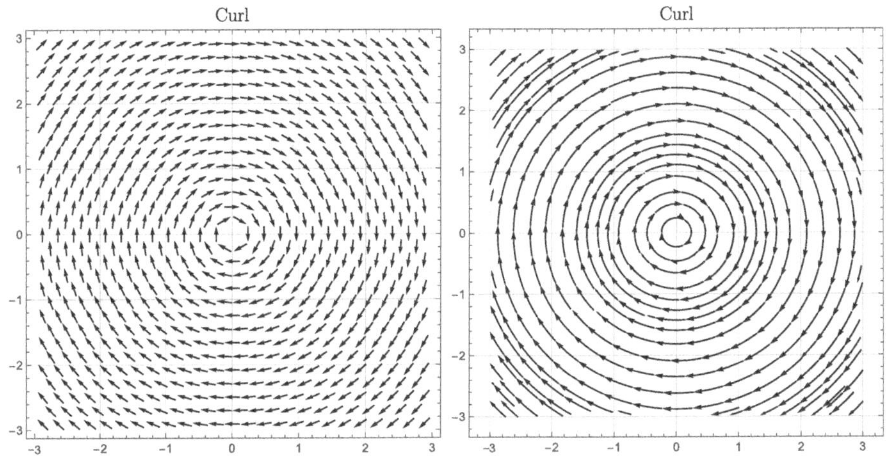

FIGURE 1.10
The figure shows a vector field with an axis of rotation at the center of the coordinates system. This is clockwise. rotations is represented by arrows (*left*) and as a stream (*right*).

1.6 Parametric representations

Another important mathematical tool will be to express the electric field **E** in terms of different independent variables, for example, from the independent variables (x, y) to the other two independent variables (u, v). The changes of variables can facilitate the computations and expressions related to **E**. This trick is called by mathematicians as parametric representation. **Parametric representations will be beneficial to represent the different integral equations and will be the key to generalizing the Richards-Wolf integrals to freeform wavefronts.** Therefore, in this section, we are going to give a refresher on parametric representations.

In multivariable calculus, **parametric representation** is a method to describe curves, surfaces, or higher-dimensional objects by expressing their coordinates as functions of one or more independent parameters [3]. This approach is particularly useful for modeling complex geometric objects, motion in space, or transformations that are difficult to express with a single equation. Below, I'll explain the concept, its applications, and key ideas, keeping the explanation clear and concise while incorporating relevant mathematical notation.

Parametric representation describes points in space using a set of parameters, typically denoted by variables like t, u, or v. Instead of relating coordinates directly (e.g., $y = f(x)$), each coordinate is expressed as a function of one or more parameters.

The two main objects that can be successfully parametrized are curves and surfaces. In our case, surfaces will be the most used in parts three, four, and five of this book. But we will first take a look at curves since these parameterizations are easier to digest. Then, we will continue with surfaces.

A **Curve** in $\mathbb{R}^2$ or $\mathbb{R}^3$ space can be represented parametrically by a vector-valued function:

$$\mathbf{r}(t) = [x(t), y(t)] \quad (\mathbb{R}^2) \quad \text{or} \quad \mathbf{r}(t) = [x(t), y(t), z(t)] \quad (\mathbb{R}^3), \tag{1.37}$$

where t is the parameter, often representing time or another independent variable, and $x(t), y(t), z(t)$ are functions defining the coordinates.

A **Surface** in $\mathbb{R}^2$ or $\mathbb{R}^3$ space is parameterized using two parameters, u and v:

$$\mathbf{r}(u, v) = [x(u, v), y(u, v)] \quad \text{or} \quad \mathbf{r}(u, v) = [x(u, v), y(u, v), z(u, v)], \tag{1.38}$$

where the parameters u and v vary over a region, mapping points to a surface, for example, a sphere or torus.

Parametric representations allow for curves and surfaces that are not functions in the Cartesian sense (e.g., circles or self-intersecting curves).

Example: a circle cannot be described by a single $y = f(x)$, but parametrically

$$\mathbf{r}(t) = [\cos t, \sin t], \quad t \in (0, 2\pi), \tag{1.39}$$

traces a unit circle. The plot is presented at *left*-side of Fig. 1.11.

Example: a spiral can be parametrically expressed as

$$\mathbf{r}(t) = [t \cos t, t \sin t], \quad t \in (0, 2\pi), \tag{1.40}$$

See *right*-side of the Fig. 1.11.

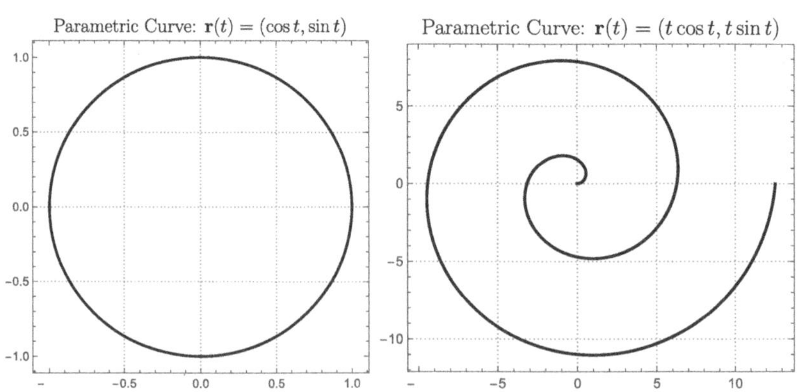

FIGURE 1.11
Parametric curves: A circle in the *left* and a spiral in the *right*.

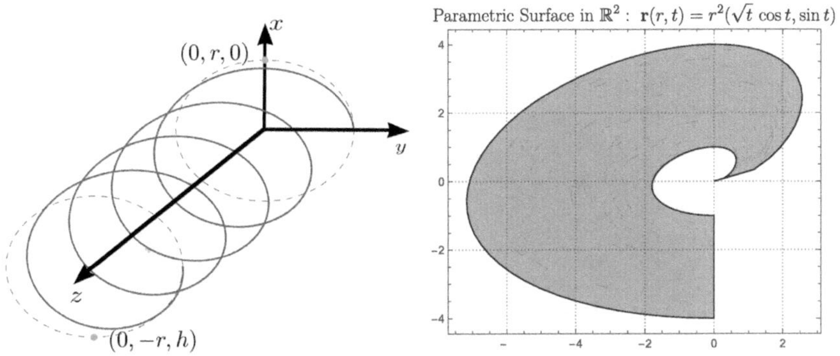

FIGURE 1.12
A helix (*left*) and a parametric surface *right.*

Example: A helix spiraling around the z-axis can be parameterized as

$$\mathbf{r}(t) = [\cos t, \sin t, t], \quad t \in \mathbb{R}. \tag{1.41}$$

See the *left*-side of Fig. 1.12.

Example: Surface in $\mathbb{R}^2$ is parameterized using two parameters, u and v. For the following surface, let $u \to r$ and $v \to t$.

$$\mathbf{r}(r,t) = r^2[\sqrt{t}\cos t, \sin t], \quad r \in (1,2),\ t \in (0.3\pi/2). \tag{1.42}$$

The area closed is a surface in $\mathbb{R}^2$. The plot is presented on *right* side of the Fig. 1.12.

Example: A sphere of radius R can be elegantly described using spherical coordinates, where just two angles specify every point on its surface: the polar angle θ (measured down from the positive z-axis) and the azimuthal angle ϕ (measured around the z-axis in the xy-plane). This natural parameterization sweeps out the entire sphere smoothly as ϕ runs from 0 to π and θ runs from 0 to 2π.

$$\mathbf{r}(\theta,\phi) = [R\sin\phi\cos\theta, R\sin\phi\sin\theta, R\cos\phi], \quad \theta \in (0,2\pi), \quad \phi \in (0,\pi). \tag{1.43}$$

Fig. 1.13 shows the parametric surface of a sphere on the xyz coordinate system.

Three fundamental concepts can be obtained more easily with parametric representations, which we will use throughout the book and find particularly useful in the last two parts. These concepts are the tangent vector of a surface, the normal vector of a surface, and the Jacobian matrix. All of them will be explained in separate subsections due to their relevance.

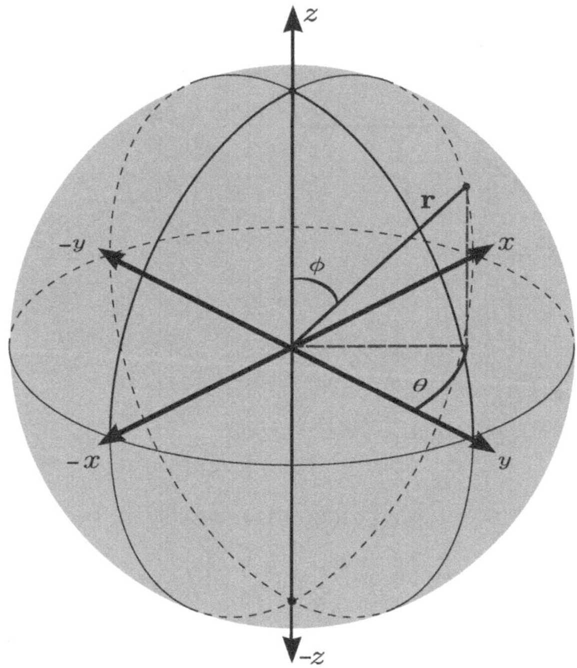

FIGURE 1.13
Parameterization of a spherical surface. A sphere of radius R is described in spherical coordinates using two angles: polar ϕ (from the positive z-axis) and azimuthal θ (around the z-axis in the xy-plane). It sweeps the surface as $\phi \in (0, \pi)$ and $\theta \in (0, 2\pi)$.

1.6.1 Tangent and normal vectors of parametric curve in $\mathbb{R}^2$

Consider a parametric curve in $\mathbb{R}^2$ given by

$$\mathbf{r}(t) = \big[x(t),\ y(t)\big]. \tag{1.44}$$

The tangent vector geometrically is the vector that lies along the curve at the point $\mathbf{r}(t_0)$ and indicates the instantaneous direction in which the curve is heading as the parameter increases. Then, it is the derivative of $\mathbf{r}(t)$,

$$\mathbf{r}'(t) = \big[x'(t),\ y'(t)\big]. \tag{1.45}$$

where $x'(t)$, and $y'(t)$ are the derivatives of $x(t)$, and $y(t)$ with respect to t, respectively. It points in the direction of motion and its length $\|\mathbf{r}'(t)\|$, it is sometimes called the *speed.*

The normal vector is any vector perpendicular to the tangent. A simple choice in $\mathbb{R}^2$ is obtained by rotating the tangent 90°,

$$\mathbf{N}(t) = \big[-y'(t),\ x'(t)\big] \quad \text{or} \quad \mathbf{N}(t) = \big[y'(t),\ -x'(t)\big]. \tag{1.46}$$

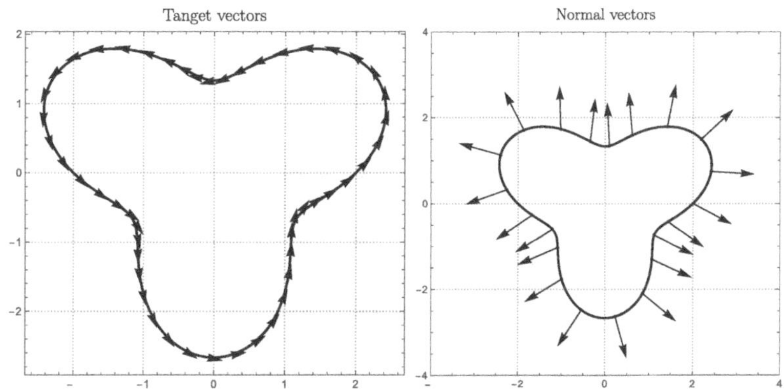

FIGURE 1.14
Tangent vectors $\mathbf{r}'(t)$ (left) and normal vectors (right) along the parametric curve; the vectors are orthogonal everywhere, as required.

The first points to the *left* of the direction of motion (with respect to the orientation), the second to the right.

Fig. 1.14 shows a parametric curve given by [.....]. On the *left*-side are the tangent vectors along the mentioned curve. On the *right*-side are plotted some normal vectors along the curve. Notice that on the plot, tangent vectors and normal vectors are orthogonal to each other, as expected by the definition.

1.6.2 Tangent vectors to a parametric surface in $\mathbb{R}^3$

Let $\mathbf{r}(u, v)$ be a parametric surface in $\mathbb{R}^3$, which is represented by a vector-valued function,

$$\mathbf{r}(u, v) = [x(u, v), y(u, v), z(u, v)], \tag{1.47}$$

where u and v are parameters varying over a domain in $\mathbb{R}^2$, and $x(u, v)$, $y(u, v)$, $z(u, v)$ are differentiable functions mapping (u, v) to points on the surface in $\mathbb{R}^3$ space.

The tangent plane at a point P on the surface is the plane that best approximates the surface locally. It is spanned by two tangent vectors, which are obtained by taking the partial derivatives of $\mathbf{r}(u, v)$ with respect to the parameters u and v. See Fig. 1.15, the derivatives are

$$\frac{\partial \mathbf{r}}{\partial u} = \left[\frac{\partial x}{\partial u}, \frac{\partial y}{\partial u}, \frac{\partial z}{\partial u}\right], \quad \frac{\partial \mathbf{r}}{\partial v} = \left[\frac{\partial x}{\partial v}, \frac{\partial y}{\partial v}, \frac{\partial z}{\partial v}\right]. \tag{1.48}$$

These vectors are tangent to the surface provided they are non-zero and linearly independent. The tangent plane at a point $P = \mathbf{r}(u_0, v_0)$ consists of all vectors of the form,

$$\mathbf{w} = a\frac{\partial \mathbf{r}}{\partial u}(u_0, v_0) + b\frac{\partial \mathbf{r}}{\partial v}(u_0, v_0), \tag{1.49}$$

where $a, b \in \mathbb{R}$.

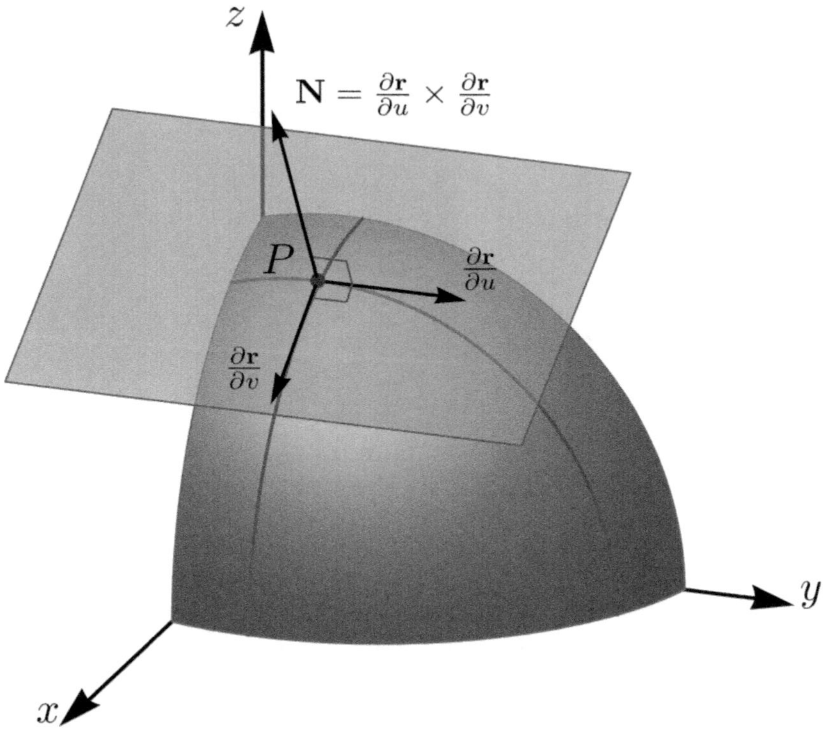

FIGURE 1.15
Diagram of the normal vector $\mathbf{N}$ of a parametric surface, $\mathbf{N}$ is orthogonal to the tangent plane spanned by the anger vectors $\frac{\partial \mathbf{r}}{\partial u}$ and $\frac{\partial \mathbf{r}}{\partial v}$.

To understand why $\frac{\partial \mathbf{r}}{\partial u}$ and $\frac{\partial \mathbf{r}}{\partial v}$ are tangent vectors, consider the parametric representation. Fixing $v = v_0$ and varying u traces a curve on the surface,

$$\mathbf{r}(u, v_0) = [x(u, v_0), y(u, v_0), z(u, v_0)]. \tag{1.50}$$

The tangent vector to this curve at $u = u_0$ is

$$\frac{d}{du}\mathbf{r}(u, v_0) = \left[\frac{\partial x}{\partial u}(u, v_0), \frac{\partial y}{\partial u}(u, v_0), \frac{\partial z}{\partial u}(u, v_0)\right] = \frac{\partial \mathbf{r}}{\partial u}(u_0, v_0). \tag{1.51}$$

This vector lies along the surface and points in the direction of the curve at $\mathbf{r}(u_0, v_0)$. Similarly, fixing $u = u_0$ and varying v produces a curve $\mathbf{r}(u_0, v)$, with tangent vector,

$$\frac{\partial \mathbf{r}}{\partial v}(u_0, v_0). \tag{1.52}$$

Since these curves lie on the surface, their tangent vectors are tangent to the surface. If $\frac{\partial \mathbf{r}}{\partial u}$ and $\frac{\partial \mathbf{r}}{\partial v}$ are linearly independent, they span the tangent plane. Just like the plane displayed in Fig. 1.15.

To rigorously prove that $\frac{\partial \mathbf{r}}{\partial u}$ and $\frac{\partial \mathbf{r}}{\partial v}$ are tangent vectors, we need to show that any curve on the surface passing through $\mathbf{r}(u_0, v_0)$ has a tangent vector that is a linear combination of these partial derivatives.

Consider a curve on the surface parameterized by a single parameter t, given by,

$$\mathbf{c}(t) = \mathbf{r}(u(t), v(t)), \tag{1.53}$$

where $u(t)$ and $v(t)$ are differentiable functions, and the curve passes through the point $\mathbf{r}(u_0, v_0)$ at $t = t_0$, i.e., $u(t_0) = u_0$, $v(t_0) = v_0$.

The tangent vector to the curve $\mathbf{c}(t)$ is its derivative with respect to t,

$$\frac{d\mathbf{c}}{dt} = \frac{d}{dt}\mathbf{r}(u(t), v(t)). \tag{1.54}$$

Using the chain rule for vector-valued functions:

$$\frac{d\mathbf{c}}{dt} = \frac{\partial \mathbf{r}}{\partial u}\frac{du}{dt} + \frac{\partial \mathbf{r}}{\partial v}\frac{dv}{dt}. \tag{1.55}$$

At $t = t_0$, this becomes,

$$\frac{d\mathbf{c}}{dt}(t_0) = \frac{\partial \mathbf{r}}{\partial u}(u_0, v_0)\frac{du}{dt}(t_0) + \frac{\partial \mathbf{r}}{\partial v}(u_0, v_0)\frac{dv}{dt}(t_0). \tag{1.56}$$

This shows that the tangent vector to any curve on the surface at $\mathbf{r}(u_0, v_0)$ is a linear combination of $\frac{\partial \mathbf{r}}{\partial u}(u_0, v_0)$ and $\frac{\partial \mathbf{r}}{\partial v}(u_0, v_0)$.

Since $\mathbf{c}(t)$ lies on the surface, its tangent vector $\frac{d\mathbf{c}}{dt}$ is tangent to the surface. The equation above demonstrates that $\frac{\partial \mathbf{r}}{\partial u}$ and $\frac{\partial \mathbf{r}}{\partial v}$ span the space of all possible tangent vectors to curves on the surface at that point. Thus, they are tangent vectors, and the tangent plane is the plane spanned by them (assuming linear independence).

For the tangent plane to be well-defined, $\frac{\partial \mathbf{r}}{\partial u}$ and $\frac{\partial \mathbf{r}}{\partial v}$ must be linearly independent, i.e., their cross product is non-zero:

$$\frac{\partial \mathbf{r}}{\partial u} \times \frac{\partial \mathbf{r}}{\partial v} \neq \mathbf{0}, \tag{1.57}$$

where $\mathbf{0} = [0, 0, 0]$. This cross product also gives the normal vector to the surface, which is perpendicular to the tangent plane.

Example: Consider a sphere of radius R, parameterized in spherical coordinates,

$$\mathbf{r}(\theta, \phi) = [R\sin\phi\cos\theta, R\sin\phi\sin\theta, R\cos\phi], \quad \theta \in (0, 2\pi), \quad \phi \in (0, \pi). \tag{1.58}$$

Compute the partial derivatives,

$$\frac{\partial \mathbf{r}}{\partial \theta} = [-R\sin\phi\sin\theta, R\sin\phi\cos\theta, 0], \tag{1.59}$$

$$\frac{\partial \mathbf{r}}{\partial \phi} = [R\cos\phi\cos\theta, R\cos\phi\sin\theta, -R\sin\phi]. \tag{1.60}$$

At a point, say (θ_0, ϕ_0), these vectors lie in the tangent plane. Their cross product,

$$\frac{\partial \mathbf{r}}{\partial \theta} \times \frac{\partial \mathbf{r}}{\partial \phi} = [R^2\sin^2\phi\cos\theta, R^2\sin^2\phi\sin\theta, R^2\sin\phi\cos\phi], \tag{1.61}$$

is normal to the surface (see next subsection) and non-zero for $0 < \phi < \pi$, confirming that $\frac{\partial \mathbf{r}}{\partial \theta}$ and $\frac{\partial \mathbf{r}}{\partial \phi}$ are tangent vectors spanning the tangent plane.

The partial derivatives $\frac{\partial \mathbf{r}}{\partial u}$ and $\frac{\partial \mathbf{r}}{\partial v}$ of a parametric surface $\mathbf{r}(u, v)$ are tangent vectors because they represent the tangent vectors to curves on the surface obtained by varying one parameter while fixing the other. The proof via the chain rule shows that any tangent vector to a curve on the surface is a linear combination of these partial derivatives, confirming that they span the tangent plane.

1.6.3 Normal vector of a parametric surface

From the last subsection we know that the tangent plane at a point $\mathbf{r}(u_0, v_0)$ on the surface is spanned by the partial derivatives,

$$\frac{\partial \mathbf{r}}{\partial u} = \left[\frac{\partial x}{\partial u}, \frac{\partial y}{\partial u}, \frac{\partial z}{\partial u}\right], \quad \frac{\partial \mathbf{r}}{\partial v} = \left[\frac{\partial x}{\partial v}, \frac{\partial y}{\partial v}, \frac{\partial z}{\partial v}\right], \tag{1.62}$$

evaluated at (u_0, v_0). These vectors are tangent to the surface, as they represent the directions of curves on the surface obtained by varying u or v while keeping the other parameter fixed.

The **normal vector** to the surface at $\mathbf{r}(u_0, v_0)$ is a vector perpendicular to the tangent plane. It is given by the cross product of the tangent vectors,

$$\mathbf{N} = \frac{\partial \mathbf{r}}{\partial u} \times \frac{\partial \mathbf{r}}{\partial v}. \tag{1.63}$$

This vector is normal to the surface, provided $\frac{\partial \mathbf{r}}{\partial u}$ and $\frac{\partial \mathbf{r}}{\partial v}$ are linearly independent (i.e., $\mathbf{N} \neq \mathbf{0}$).

The cross product $\frac{\partial \mathbf{r}}{\partial u} \times \frac{\partial \mathbf{r}}{\partial v}$ yields a vector perpendicular to both $\frac{\partial \mathbf{r}}{\partial u}$ and $\frac{\partial \mathbf{r}}{\partial v}$. Since these partial derivatives span the tangent plane (as shown in the tangent vector proof), their cross product is orthogonal to every vector in the tangent plane, making it a normal vector to the surface. The magnitude of the cross product, $\left\|\frac{\partial \mathbf{r}}{\partial u} \times \frac{\partial \mathbf{r}}{\partial v}\right\|$, also serves as the scaling factor in surface integrals, representing the area element.

To prove that $\mathbf{N} = \frac{\partial \mathbf{r}}{\partial u} \times \frac{\partial \mathbf{r}}{\partial v}$ is a normal vector to the surface at $\mathbf{r}(u_0, v_0)$, we must show that it is perpendicular to every vector in the tangent plane, i.e., perpendicular to both $\frac{\partial \mathbf{r}}{\partial u}$ and $\frac{\partial \mathbf{r}}{\partial v}$.

By definition of the cross product in $\mathbb{R}^3$, for any two vectors $\mathbf{a} = [a_1, a_2, a_3]$ and $\mathbf{b} = [b_1, b_2, b_3]$, the cross product $\mathbf{a} \times \mathbf{b}$ is orthogonal to both $\mathbf{a}$ and $\mathbf{b}$. This is verified by checking the dot product,

$$(\mathbf{a} \times \mathbf{b}) \cdot \mathbf{a} = 0, \quad (\mathbf{a} \times \mathbf{b}) \cdot \mathbf{b} = 0. \tag{1.64}$$

Applying this to our tangent vectors, let,

$$\mathbf{a} = \frac{\partial \mathbf{r}}{\partial u} = \left[\frac{\partial x}{\partial u}, \frac{\partial y}{\partial u}, \frac{\partial z}{\partial u}\right], \quad \mathbf{b} = \frac{\partial \mathbf{r}}{\partial v} = \left[\frac{\partial x}{\partial v}, \frac{\partial y}{\partial v}, \frac{\partial z}{\partial v}\right]. \tag{1.65}$$

The cross product is

$$\begin{aligned} \mathbf{N} = \frac{\partial \mathbf{r}}{\partial u} \times \frac{\partial \mathbf{r}}{\partial v} &= \begin{vmatrix} \hat{\mathbf{x}} & \hat{\mathbf{y}} & \hat{\mathbf{z}} \\ \frac{\partial x}{\partial u} & \frac{\partial y}{\partial u} & \frac{\partial z}{\partial u} \\ \frac{\partial x}{\partial v} & \frac{\partial y}{\partial v} & \frac{\partial z}{\partial v} \end{vmatrix} \\ &= \left[\frac{\partial y}{\partial u}\frac{\partial z}{\partial v} - \frac{\partial z}{\partial u}\frac{\partial y}{\partial v}, \frac{\partial z}{\partial u}\frac{\partial x}{\partial v} - \frac{\partial x}{\partial u}\frac{\partial z}{\partial v}, \frac{\partial x}{\partial u}\frac{\partial y}{\partial v} - \frac{\partial y}{\partial u}\frac{\partial x}{\partial v}\right]. \end{aligned} \tag{1.66}$$

To confirm $\mathbf{N}$ is normal, compute the dot product with $\frac{\partial \mathbf{r}}{\partial u}$,

$$\begin{aligned} \mathbf{N} \cdot \frac{\partial \mathbf{r}}{\partial u} = \left(\frac{\partial y}{\partial u}\frac{\partial z}{\partial v} - \frac{\partial z}{\partial u}\frac{\partial y}{\partial v}\right)\frac{\partial x}{\partial u} + \left(\frac{\partial z}{\partial u}\frac{\partial x}{\partial v} - \frac{\partial x}{\partial u}\frac{\partial z}{\partial v}\right)\frac{\partial y}{\partial u} \\ + \left(\frac{\partial x}{\partial u}\frac{\partial y}{\partial v} - \frac{\partial y}{\partial u}\frac{\partial x}{\partial v}\right)\frac{\partial z}{\partial u}. \end{aligned} \tag{1.67}$$

Expanding and simplifying,

$$\mathbf{N} \cdot \frac{\partial \mathbf{r}}{\partial u} = \frac{\partial x}{\partial u}\frac{\partial y}{\partial u}\frac{\partial z}{\partial v} - \frac{\partial x}{\partial u}\frac{\partial z}{\partial u}\frac{\partial y}{\partial v} + \frac{\partial y}{\partial u}\frac{\partial z}{\partial u}\frac{\partial x}{\partial v} - \frac{\partial y}{\partial u}\frac{\partial x}{\partial u}\frac{\partial z}{\partial v} + \frac{\partial z}{\partial u}\frac{\partial x}{\partial u}\frac{\partial y}{\partial v} - \frac{\partial z}{\partial u}\frac{\partial y}{\partial u}\frac{\partial x}{\partial v}. \tag{1.68}$$

Group terms,

$$\begin{aligned} \mathbf{N} \cdot \frac{\partial \mathbf{r}}{\partial u} = \left(\frac{\partial x}{\partial u}\frac{\partial y}{\partial u}\frac{\partial z}{\partial v} - \frac{\partial y}{\partial u}\frac{\partial x}{\partial u}\frac{\partial z}{\partial v}\right) + \left(\frac{\partial y}{\partial u}\frac{\partial z}{\partial u}\frac{\partial x}{\partial v} - \frac{\partial z}{\partial u}\frac{\partial y}{\partial u}\frac{\partial x}{\partial v}\right) \\ + \left(\frac{\partial z}{\partial u}\frac{\partial x}{\partial u}\frac{\partial y}{\partial v} - \frac{\partial x}{\partial u}\frac{\partial z}{\partial u}\frac{\partial y}{\partial v}\right). \end{aligned} \tag{1.69}$$

Each pair cancels (e.g., $\frac{\partial x}{\partial u}\frac{\partial y}{\partial u}\frac{\partial z}{\partial v} - \frac{\partial y}{\partial u}\frac{\partial x}{\partial u}\frac{\partial z}{\partial v} = 0$), so,

$$\mathbf{N} \cdot \frac{\partial \mathbf{r}}{\partial u} = 0. \tag{1.70}$$

Similarly, $\mathbf{N} \cdot \frac{\partial \mathbf{r}}{\partial v} = 0$, confirming that $\mathbf{N}$ is orthogonal to both tangent vectors.

Since $\frac{\partial \mathbf{r}}{\partial u}$ and $\frac{\partial \mathbf{r}}{\partial v}$ span the tangent plane (assuming linear independence), any vector in the tangent plane is a linear combination $a\frac{\partial \mathbf{r}}{\partial u} + b\frac{\partial \mathbf{r}}{\partial v}$. The dot product with $\mathbf{N}$ is

$$\mathbf{N} \cdot \left(a\frac{\partial \mathbf{r}}{\partial u} + b\frac{\partial \mathbf{r}}{\partial v} \right) = a(\mathbf{N} \cdot \frac{\partial \mathbf{r}}{\partial u}) + b(\mathbf{N} \cdot \frac{\partial \mathbf{r}}{\partial v}) = 0. \tag{1.71}$$

Thus, $\mathbf{N}$ is perpendicular to every vector in the tangent plane, proving that it is a normal vector.

For $\mathbf{N}$ to be a valid normal vector, $\frac{\partial \mathbf{r}}{\partial u}$ and $\frac{\partial \mathbf{r}}{\partial v}$ must be linearly independent, so:

$$\frac{\partial \mathbf{r}}{\partial u} \times \frac{\partial \mathbf{r}}{\partial v} \neq \mathbf{0}. \tag{1.72}$$

Points where $\mathbf{N} = \mathbf{0}$ are called *singular points*, where the tangent plane may not be well-defined.

Example: Consider a sphere of radius R, parameterized by:

$$\mathbf{r}(\theta, \phi) = [R\sin\phi\cos\theta, R\sin\phi\sin\theta, R\cos\phi], \quad \theta \in (0, 2\pi), \quad \phi \in (0, \pi). \tag{1.73}$$

The partial derivatives are

$$\frac{\partial \mathbf{r}}{\partial \theta} = [-R\sin\phi\sin\theta, R\sin\phi\cos\theta, 0], \tag{1.74}$$

$$\frac{\partial \mathbf{r}}{\partial \phi} = [R\cos\phi\cos\theta, R\cos\phi\sin\theta, -R\sin\phi]. \tag{1.75}$$

Compute the cross product,

$$\mathbf{N} = \frac{\partial \mathbf{r}}{\partial \theta} \times \frac{\partial \mathbf{r}}{\partial \phi} = \begin{vmatrix} \hat{\mathbf{x}} & \hat{\mathbf{y}} & \hat{\mathbf{z}} \\ -R\sin\phi\sin\theta & R\sin\phi\cos\theta & 0 \\ R\cos\phi\cos\theta & R\cos\phi\sin\theta & -R\sin\phi \end{vmatrix}. \tag{1.76}$$

Evaluate the determinant,

$$\mathbf{N} = [R^2\sin^2\phi\cos\theta, R^2\sin^2\phi\sin\theta, R^2\sin\phi\cos\phi]. \tag{1.77}$$

This vector is parallel to the position vector,

$$\mathbf{r}(\theta, \phi) = [R\sin\phi\cos\theta, R\sin\phi\sin\theta, R\cos\phi], \tag{1.78}$$

confirming it is normal (radial) to the sphere's surface for $0 < \phi < \pi$.

The cross product $\frac{\partial \mathbf{r}}{\partial u} \times \frac{\partial \mathbf{r}}{\partial v}$ is a normal vector to the parametric surface $\mathbf{r}(u, v)$ because it is perpendicular to the tangent vectors $\frac{\partial \mathbf{r}}{\partial u}$ and $\frac{\partial \mathbf{r}}{\partial v}$, which span the tangent plane. The proof relies on the properties of the cross product and the fact that the tangent plane contains all tangent vectors to curves on the surface.

1.6.4 Normal vector of an explicit surface

In the last section, we study the normals and tangential vectors of parametric surfaces. But it is also very important for our purposes to study these vectors in explicit surfaces.

An explicit surface in $\mathbb{R}^3$ is given by an equation of the form:

$$z = f(x, y), \tag{1.79}$$

where $f : D \subset \mathbb{R}^2 \to \mathbb{R}$ is a continuously differentiable function on an open domain D.

It is important to highlight that explicit surfaces can be also parametrized. Therefore, all the results valid from parametric surface are valid for the explicit surface. But the explicit surface is going to be used over and over along the book, so it is worth it to take special attention to it. Such a surface can be parametrized as

$$\mathbf{r}(x, y) = \big[x, y, f(x, y)\big], \quad (x, y) \in D. \tag{1.80}$$

This is also called the Monge patch. The unit normal vector to the surface at a point $[x, y, f(x, y)]$ is perpendicular to the tangent plane. Let's derive the standard formula and prove it rigorously using vector calculus.

Let S be the surface $z = f(x, y)$ with $f \in C^1(D)$. At a point $P = (x_0, y_0, z_0)$ where $z_0 = f(x_0, y_0)$, a normal vector to S if

$$\mathbf{N}(x_0, y_0) = \left[-\frac{\partial f}{\partial x}, -\frac{\partial f}{\partial y}, 1\right]\Bigg|_{(x_0, y_0)} \tag{1.81}$$

The unit normal is

$$\mathbf{n} = \frac{\mathbf{N}}{||\mathbf{N}||} = \frac{[-f_x, -f_y, 1]}{\sqrt{f_x^2 + f_y^2 + 1}}, \tag{1.82}$$

where $f_x = \frac{\partial f}{\partial x}\big|_{(x_0, y_0)}$, $f_y = \frac{\partial f}{\partial y}\big|_{(x_0, y_0)}$ are the partial derivatives with respect to x and y.

We prove this by constructing the tangent plane using partial derivatives and showing that $\mathbf{N}$ is perpendicular to all tangent vectors.

Consider the parametrization,

$$\mathbf{r}(u, v) = \big[u,\ v,\ f(u, v)\big], \quad (u, v) \in D. \tag{1.83}$$

Let $P = (x_0, y_0, z_0)$ correspond to parameters $[u_0, v_0) = (x_0, y_0]$.

The partial derivatives as we have seen before they are the tangent vectors. They are expressed as,

$$\mathbf{r}_u = \frac{\partial \mathbf{r}}{\partial u} = \left[1,\ 0,\ \frac{\partial f}{\partial u}\right], \qquad \mathbf{r}_v = \frac{\partial \mathbf{r}}{\partial v} = \left[0,\ 1,\ \frac{\partial f}{\partial v}\right]. \tag{1.84}$$

At the point of interest we have (u_0, v_0),

$$\mathbf{r}_u\Big|_{(x_0,y_0)} = [1,\ 0,\ f_x], \quad \mathbf{r}_v\Big|_{(x_0,y_0)} = [0,\ 1,\ f_y]. \tag{1.85}$$

These two vectors span the tangent plane to the surface at P.

Now we have also seen that the normal vector can be computed using the cross product between the tangent vectors,

$$\mathbf{N} = \mathbf{r}_u \times \mathbf{r}_v. \tag{1.86}$$

Computing,

$$\begin{aligned} \mathbf{r}_u \times \mathbf{r}_v = \begin{vmatrix} \hat{\mathbf{x}} & \hat{\mathbf{y}} & \hat{\mathbf{z}} \\ 1 & 0 & f_x \\ 0 & 1 & f_y \end{vmatrix} &= \hat{\mathbf{x}}(0 \cdot f_y - f_x \cdot 1) - \hat{\mathbf{y}}(1 \cdot f_y - f_x \cdot 0) + \hat{\mathbf{z}}(1 \cdot 1 - 0 \cdot 0) \\ &= \hat{\mathbf{x}}(-f_x) - \hat{\mathbf{y}}(f_y) + \hat{\mathbf{z}}(1) = \left[-f_x,\ -f_y,\ 1\right]. \end{aligned} \tag{1.87}$$

The answer is as expected,

$$\mathbf{N}(x_0, y_0) = \left[-\frac{\partial f}{\partial x}, -\frac{\partial f}{\partial y}, 1\right]\Bigg|_{(x_0,y_0)}. \tag{1.88}$$

We can verify the perpendicularity between the normal vector and the tangential plane. Let $\mathbf{T}_1 = (1, 0, f_x)$, $\mathbf{T}_2 = (0, 1, f_y)$ be tangent vectors. Checking dot products

$$\mathbf{N} \cdot \mathbf{T}_1 = (-f_x)(1) + (-f_y)(0) + (1)(f_x) = -f_x + f_x = 0, \tag{1.89}$$

and

$$\mathbf{N} \cdot \mathbf{T}_2 = (-f_x)(0) + (-f_y)(1) + (1)(f_y) = -f_y + f_y = 0. \tag{1.90}$$

Hence, $\mathbf{N} \perp \mathbf{T}_1$ and $\mathbf{N} \perp \mathbf{T}_2$, so $\mathbf{N}$ is normal to the tangent plane.

The surface $z = f(x, y)$ defines a graph over the xy-plane. The vector $\mathbf{N} = [-f_x, -f_y, 1]$ has three components. The x component is $-f_x$. It is the opposite of the slope in x-direction. The y component is given by $-f_y$, which in fact is opposite to the slope in the y direction. Finally the z component is $+1$. This means that it is an upward bias. This makes sense: the normal should tilt away from steep slopes and point upward relative to the xy-plane.

Example: Consider the plane $z = ax + by + c$

$$f_x = a, \quad f_y = b \quad \Rightarrow \quad \mathbf{N} = [-a, -b, 1]. \tag{1.91}$$

The actual normal to the plane $ax + by - z + d = 0$ is $[a, b, -1]$, or opposite $[-a, -b, 1]$. Notice that this matches!

Example: Consider the following hemisphere,

$$z = \sqrt{R^2 - x^2 - y^2}, \tag{1.92}$$

Let's compute its derivatives

$$f_x = \frac{-x}{\sqrt{R^2 - x^2 - y^2}}, \quad f_y = \frac{-y}{\sqrt{R^2 - x^2 - y^2}}. \tag{1.93}$$

Therefore,

$$\mathbf{N} = \left[\frac{x}{z}, \frac{y}{z}, 1\right] = \left[\frac{x}{z}, \frac{y}{z}, \frac{z}{z}\right] = \frac{[x, y, z]}{z}. \tag{1.94}$$

Since $z > 0$, $\mathbf{N} = \frac{[x,y,z]}{||[x,y,z]||}$ — outward unit normal.

In general, the unit normal vector in general is more used that normal vector. Remember a unit vector is the vector divided by its magnitude, in the case of the normal vector we have,

$$\mathbf{n} = \frac{[-f_x, -f_y, 1]}{\sqrt{f_x^2 + f_y^2 + 1}} \tag{1.95}$$

If the last expression is multiplied by minus one, the direction of the normal vector changes. Depending on the surface and the coordinate system, this may mean whether the vector is pointing outwards or inward.

1.7 Jacobian matrix

In this section, we direct our attention to the Jacobian, a mathematical tool that plays a critical role in the transformation of variables within the context of multiple integrals [3]. The Jacobian is of particular importance due to its ubiquitous application in the analytical models of scalar and vector diffraction theories, where double integrals frequently emerge as essential tools for modeling diffraction patterns [6, 7]. These theories, which underpin the study of wave propagation and diffraction, rely heavily on the accurate manipulation of such integrals to describe field distributions and their interactions across multidimensional domains [8]. By facilitating the change of variables in these integrals, the Jacobian ensures the preservation of the integral's value under coordinate transformations, thereby enabling precise computations in diverse coordinate systems tailored to the geometry of the problem at hand. This discussion provides a rigorous foundation for understanding the Jacobian's role,

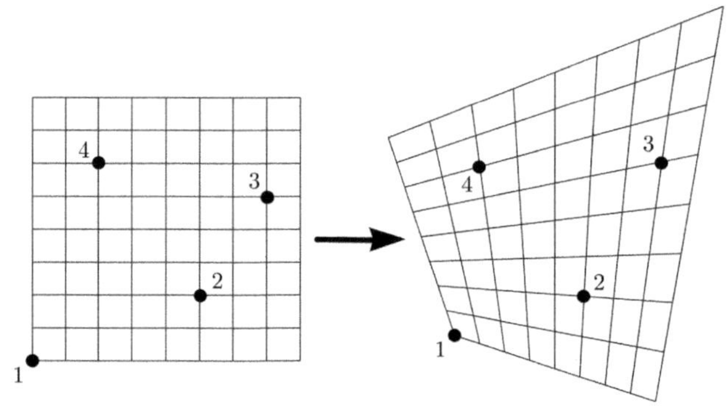

FIGURE 1.16
Transformation of a grid. The grid is distorted after the transformation as show how the label points 1, 2, 3, 4 have changed their positions.

equipping readers with the necessary mathematical tools to navigate the intricate double integrals prevalent in the subsequent exploration of scalar and vector diffraction theories.

So let's assume an transformation from one set of variables. In Fig. 1.16, we have a grid with several points, let's label them as 1, 2, 3, 4. After the transformation, the point are displased in different positions as shown in *right* side of Fig. 1.16. It is the same but it has been distorted by the transformation of coordinates. The Jacobean will ensure the preservation of the integral, in this case the grid under the transformation.

For a transformation from one set of variables to another, say from (u, v) to (x, y), where $x = x(u, v)$ and $y = y(u, v)$, the **Jacobian matrix** is the matrix of partial derivatives,

$$\mathbf{Jcb} = \begin{bmatrix} \frac{\partial x}{\partial u} & \frac{\partial x}{\partial v} \\ \frac{\partial y}{\partial u} & \frac{\partial y}{\partial v} \end{bmatrix}. \tag{1.96}$$

The **Jacobian determinant** (often just called the Jacobian) is

$$\det(\mathbf{Jcb}) = \frac{\partial x}{\partial u}\frac{\partial y}{\partial v} - \frac{\partial x}{\partial v}\frac{\partial y}{\partial u} \tag{1.97}$$

We are interested in preparing readers to effectively address the prevalent double integrals in the subsequent study of scalar and vector diffraction theories. Therefore, for a double integral, if we transform the region of integration from

(x, y)-coordinates to (u, v)-coordinates, the integral becomes,

$$\iint_R f(x, y)\ \mathrm{d}x\ \mathrm{d}y = \iint_S f(x(u, v), y(u, v))\,|\det(\mathbf{Jcb})|\ \mathrm{d}u\ \mathrm{d}v \tag{1.98}$$

Here, R is the region in the (x, y)-plane, S is the corresponding region in the (u, v)-plane, and $|\det(\mathbf{Jcb})|$ accounts for the scaling of the area element under the transformation.

To prove the change of variables formula for a double integral, we need to show why the area element $\mathrm{d}x\ \mathrm{d}y$ transforms to $|\det(\mathbf{Jcb})|\ \mathrm{d}u\ \mathrm{d}v$. The proof generalizes to higher dimensions, but we'll focus on the $\mathbb{R}^2$ case for clarity.

Consider a small rectangle in the (u, v)-plane with sides Δu and Δv. Under the transformation $x = x(u, v)$, $y = y(u, v)$, this rectangle maps to a region in the (x, y)-plane. For small Δu and Δv, the image is approximately a parallelogram, and we need to compute its area.

The vertices of the rectangle at (u_0, v_0), $(u_0 + \Delta u, v_0)$, $(u_0, v_0 + \Delta v)$, and $(u_0 + \Delta u, v_0 + \Delta v)$ map to points in the (x, y)-plane. The sides of the rectangle along Δu and Δv correspond to vectors. Along Δu we have,

$$\begin{aligned}\mathbf{r}_u &= [x(u_0 + \Delta u, v_0) - x(u_0, v_0), y(u_0 + \Delta u, v_0) - y(u_0, v_0)] \\ &\approx \left[\frac{\partial x}{\partial u}\Delta u, \frac{\partial y}{\partial u}\Delta u\right],\end{aligned} \tag{1.99}$$

and along Δv, similarly, we have,

$$\mathbf{r}_v \approx \left[\frac{\partial x}{\partial v}\Delta v, \frac{\partial y}{\partial v}\Delta v\right] \tag{1.100}$$

The area of the parallelogram in the (x, y)-plane spanned by $\mathbf{r}_u$ and $\mathbf{r}_v$ is given by the magnitude of their cross product. In $\mathbb{R}^2$, the cross product's magnitude corresponds to the determinant

$$\text{Area} \approx \left|\det\begin{bmatrix}\frac{\partial x}{\partial u}\Delta u & \frac{\partial x}{\partial v}\Delta v \\ \frac{\partial y}{\partial u}\Delta u & \frac{\partial y}{\partial v}\Delta v\end{bmatrix}\right| = \left|\frac{\partial x}{\partial u}\frac{\partial y}{\partial v} - \frac{\partial x}{\partial v}\frac{\partial y}{\partial u}\right| \Delta u \Delta v = |\det(\mathbf{Jcb})|\, \Delta u \Delta v \tag{1.101}$$

Thus, the area element $\mathrm{d}x\ \mathrm{d}y$ in the (x, y)-plane is approximately $|\det(\mathbf{Jcb})|\ \mathrm{d}u\ \mathrm{d}v$ in the (u, v)-plane.

For an integral over a region R in the (x, y)-plane, we partition the corresponding region S in the (u, v)-plane into small rectangles of area $\Delta u \Delta v$. The integral $\iint_R f(x, y)\ \mathrm{d}x\ \mathrm{d}y$ is approximated by a Riemann sum,

$$\sum f(x(u_i, v_j), y(u_i, v_j)) \cdot (\text{Area of mapped rectangle}). \tag{1.102}$$

Each small area in the (x, y)-plane is $|\det(\mathbf{Jcb})|\ \Delta u \Delta v$. As the partition becomes finer ($\Delta u, \Delta v \to 0$), the Riemann sum converges to

$$\iint_S f(x(u,v), y(u,v))\, |\det(\mathbf{Jcb})|\ du\ dv\,. \tag{1.103}$$

The absolute value $|\det(\mathbf{Jcb})|$ ensures the area is positive, accounting for orientation changes in the transformation (e.g., if the transformation reverses orientation, the determinant may be negative, but area is always positive).

In order to apply the Jacobian we need to assume that the transformation $(u, v) \to (x, y)$ is differentiable and one-to-one (invertible) on the region of interest. Also, we need to consider that the function f is continuous, and the region R is well-behaved (e.g., bounded with a piecewise smooth boundary).

Example: Consider polar coordinates: $x = r\cos\theta$, $y = r\sin\theta$. The Jacobian matrix is

$$\mathbf{Jcb} = \begin{bmatrix} \frac{\partial x}{\partial r} & \frac{\partial x}{\partial \theta} \\ \frac{\partial y}{\partial r} & \frac{\partial y}{\partial \theta} \end{bmatrix} = \begin{bmatrix} \cos\theta & -r\sin\theta \\ \sin\theta & r\cos\theta \end{bmatrix}. \tag{1.104}$$

The determinant is

$$\det(\mathbf{Jcb}) = \cos\theta \cdot r\cos\theta - (-r\sin\theta)\cdot\sin\theta = r\cos^2\theta + r\sin^2\theta = r \tag{1.105}$$

Thus, $\mathrm{d}x\ \mathrm{d}y = |\det(\mathbf{Jcb})|\ dr\, d\theta = r\, dr\, d\theta$, which is the familiar area element in polar coordinates.

The Jacobian matrix and its determinant provide the scaling factor for transforming integrals under a change of variables. The proof relies on the geometric interpretation of the Jacobian as the factor that adjusts the area (or volume) element, ensuring the integral's value remains consistent across coordinate systems.

1.8 Imaginary numbers

The main quantity we are going to study in this electric field $\mathbf{E}$, which is a vector as we mentioned already, and we have also given some useful references on vector theory. $\mathbf{E}$ is also a complex quantity, which means it has a real part that belongs to the real numbers and an imaginary part. So, it is also valid to give a quick refresh on imaginary numbers.

Imaginary numbers, rooted in the concept of $\sqrt{-1}$, extend the real number system to solve equations that have no solutions in the real numbers alone. Denoted by i, where $\mathrm{i}^2 = -1$, imaginary numbers form the basis of complex numbers [3]. A complex number is written as $z = a + b\mathrm{i}$, where a is the real part, b is the imaginary part, and i is the imaginary unit. The set of complex numbers, $\mathbb{C}$, includes both real numbers (when $b = 0$) and purely imaginary numbers (when $a = 0$).

1.8.1 Basic operations of complex numbers

The arithmetic of complex numbers follows these rules:

- **Addition**: $(a + b\mathrm{i}) + (c + d\mathrm{i}) = (a + c) + (b + d)\mathrm{i}$
- **Subtraction**: $(a + b\mathrm{i}) - (c + d\mathrm{i}) = (a - c) + (b - d)\mathrm{i}$
- **Multiplication**: $(a + b\mathrm{i})(c + d\mathrm{i}) = (ac - bd) + (ad + bc)\mathrm{i}$
- **Conjugate**: The conjugate of $z = a + b\mathrm{i}$ is $\overline{z} = a - b\mathrm{i}$
- **Modulus**: The modulus of $z = a + b\mathrm{i}$ is $|z| = \sqrt{a^2 + b^2}$

Example: Let $z_1 = 3 + 2\mathrm{i}$ and $z_2 = 1 - 4\mathrm{i}$. Compute the addition

$$z_1 + z_2 = (3 + 1) + (2 - 4)\mathrm{i} = 4 - 2\mathrm{i} \tag{1.106}$$

multiplication,

$$z_1 \cdot z_2 = (3+2\mathrm{i})(1-4\mathrm{i}) = 3-12\mathrm{i}+2\mathrm{i}-8\mathrm{i}^2 = 3-12\mathrm{i}+2\mathrm{i}+8 = 11-10\mathrm{i} \tag{1.107}$$

Modulus of z_1 is

$$|z_1| = \sqrt{3^2 + 2^2} = \sqrt{13}. \tag{1.108}$$

Notice that complex numbers can be seen as vectors in a very direct and natural way. Since, each complex number consists of a real part and an imaginary part, and when you place the real part along the horizontal axis and the imaginary part along the vertical axis in a plane (the complex plane), the complex number corresponds exactly to a point in that plane. From the origin, you can draw an arrow that goes to that point, and that arrow is the vector representation of the complex number.

Adding two complex numbers then works exactly like placing vectors tip-to-tail and finding the resultant vector. Multiplying a complex number by a positive real number simply stretches or shrinks that vector while keeping the same direction. Multiplying by the imaginary unit i rotates the vector ninety degrees counterclockwise. The length of the vector (its modulus) is the distance from the origin to the point, and the angle the vector makes with the positive real axis is the argument of the complex number. See Fig. 1.17.

In this way, every complex number is not merely a point in the plane, but fully behaves as a vector living in two-dimensional space, and the usual rules of complex arithmetic match perfectly with the geometric rules of vector addition, scaling, and rotation. This geometric interpretation is one of the reasons complex numbers are so intuitive and powerful in mathematics and physics.

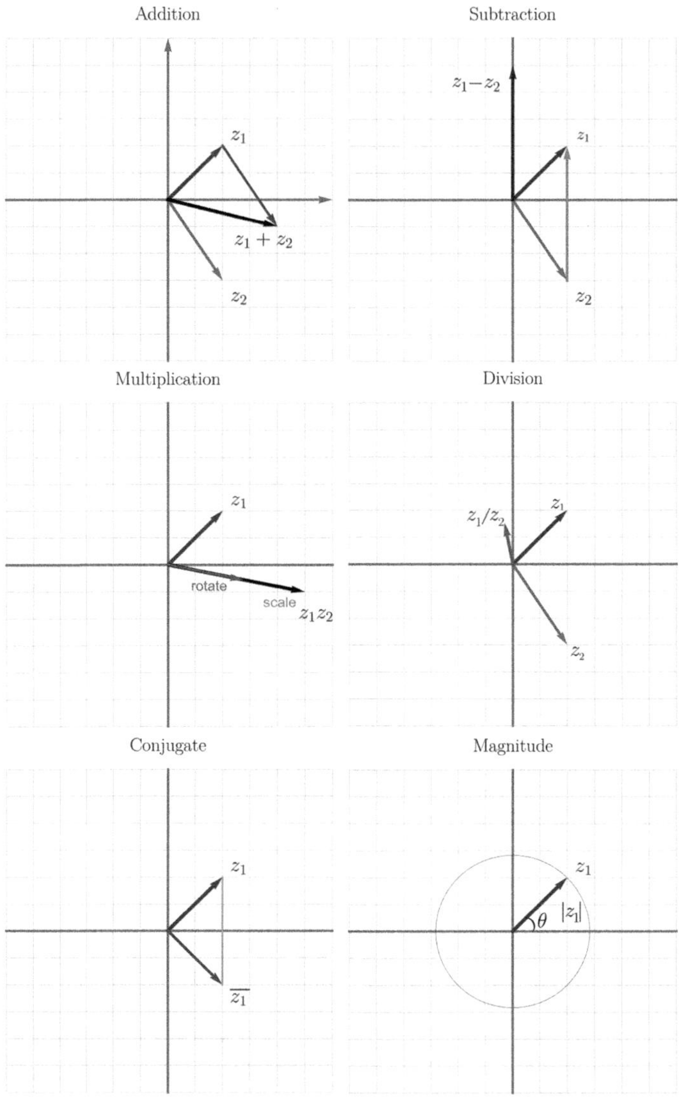

FIGURE 1.17
Vectorial representation of basic operations of complex numbers. For this example is $z_1 = 2 + 2\mathrm{i}$ and $z_2 = 2 - 3\mathrm{i}$. The horizontal axis is the real line, and the vertical axis is the imaginary line.

1.9 The Fourier transform

The Fourier transform is arguably the most widely recognized and utilized integral transform across the sciences and engineering disciplines. Originally introduced by the French mathematician Joseph Fourier in his seminal work on the propagation of heat in solid bodies [6], the Fourier transform establishes a fundamental link between the time or spatial domain and the frequency domain. Fourier transform is extensively applied in diverse areas, such as signal processing [9], quantum mechanics [10], acoustics, etc. Particularly in optics, the Fourier transform is widely used in diffraction analysis [11], imaging systems [12], microscopy [13, 14], spatial filtering [8], wavefront sensing and reconstruction, fiber optics, etc.

I want to be perfectly clear on this point, and I will repeat it: **the most foundational idea behind the Fourier transform is that it establishes a deep and fundamental connection between the time (or spatial) domain and the frequency domain.** Imagine you have a signal. If this signal can be expressed as a sum of simple harmonic waves (sine and cosine waves of different frequencies and amplitudes), the Fourier transform lets you move to the frequency domain. There, each peak in the spectrum tells you how much weight (amplitude) that particular frequency contributes when reconstructing the original signal back in the time domain. In other words, the Fourier transform reveals exactly which frequencies are present and how strongly they participate in building the original waveform.

Consider the signal $f(x)$ shown in Fig. 1.18. This signal is formed by adding three harmonic waves of different frequencies, each contributing with its own

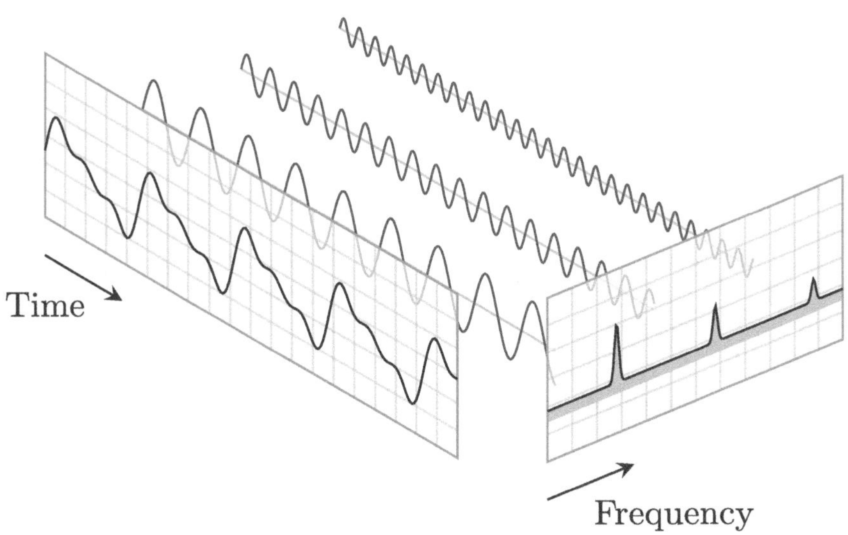

FIGURE 1.18
A signal $f(x)$ in time domain, is constructed by the sum of three different harmonic waves of different frequencies with its own amplitude and phase. The Fourier transform of the signal shows the strength (and phase) of each of these frequency components in the frequency domain.

amplitude and phase to the final shape of $f(x)$. When we plot the strength (and phase) of each of these frequency components in the frequency domain, we obtain the Fourier transform of the signal. The information contained in the original function $f(x)$ remains exactly the same; only the representation changes, it is now expressed in the *language* of frequencies rather than in the original time or spatial domain. The Fourier transform does not alter or distort the signal; it simply reveals how much each pure harmonic frequency participates in building the original waveform.

Nearly all diffraction integrals are inherently connected to the two-dimensional Fourier transform rather than its one-dimensional counterpart. For this reason, from now on we will focus exclusively on the two-dimensional Fourier transform.

The two-dimensional Fourier transform maps a function $f(x, y)$, defined in the spatial domain x, and y, to its corresponding representation in the spatial frequency domain k_x and, k_y. The standard definitions of the two-dimensional Fourier transform and its inverse are given by [15],

$$F(k_x, k_y) = \iint_{-\infty}^{\infty} f(x, y) e^{-\mathrm{i}(k_x x + k_y y)} \, \mathrm{d}x \, \mathrm{d}y \,, \tag{1.109}$$

where k_x and k_y are spatial frequencies, and $F(k_x, k_y)$ is the frequency domain representation.

The inverse transform is

$$f(x, y) = \frac{1}{(2\pi)^2} \iint_{-\infty}^{\infty} F(k_x, k_y) e^{\mathrm{i}(k_x x + k_y y)} \, \mathrm{d}k_x \, \mathrm{d}k_y \,, \tag{1.110}$$

where the pairs of x, y and k_x, k_y are assumed to range over the entire real line $(-\infty, +\infty)$. The Fourier transform of the function $f(x, y)$ is well-defined only when the integral converges for all values of k_x, and k_y. The factor $\frac{1}{(2\pi)^2}$ ensures normalization, though conventions may vary under the definition of the Fourier transform, please see Ref.[3] for more details about the mentioned factor. We will continue with the properties of the Fourier transform without given a deeper explanation on this topic.

1.9.1 Properties of the Fourier transform

The two-dimensional Fourier transform extends the concept of the one-dimensional Fourier transform to two-dimensional signals, such as images or spatial data, allowing us to analyze their frequency components in both dimensions. By transforming a function from the spatial domain to the frequency domain, it reveals critical insights into the structure and patterns of the signal. Below, we outline the definition of the two-dimensional Fourier transform and rigorously prove its key properties, including linearity, scaling, translation, rotation, and separability, with mathematical derivations to ensure clarity and precision.

1.9.1.1 Linearity

Property: For functions $f(x, y)$ and $g(x, y)$ with transforms $F(k_x, k_y)$ and $G(k_x, k_y)$, and constants a, b,

$$\mathcal{F}\{af(x,y) + bg(x,y)\} = aF(k_x, k_y) + bG(k_x, k_y). \tag{1.111}$$

Proof:

$$\mathcal{F}\{af(x,y) + bg(x,y)\} = \iint_{-\infty}^{\infty} [af(x,y) + bg(x,y)] e^{-\mathrm{i}(k_x x + k_y y)} \, \mathrm{d}x \, \mathrm{d}y \,, \tag{1.112}$$

$$= a \iint_{-\infty}^{\infty} f(x,y) e^{-\mathrm{i}(k_x x + k_y y)} \, \mathrm{d}x \, \mathrm{d}y \,,$$

$$+ b \iint_{-\infty}^{\infty} g(x,y) e^{-\mathrm{i}(k_x x + k_y y)} \, \mathrm{d}x \, \mathrm{d}y \,, \tag{1.113}$$

$$= aF(k_x, k_y) + bG(k_x, k_y). \tag{1.114}$$

The linearity of the integral ensures the property holds.

1.9.1.2 Scaling

Property: Scaling the spatial coordinates affects the frequency domain. If $f(x, y)$ has Fourier transform $F(u, v)$, then for scaling factors $a, b \neq 0$, the transform of $f(ax, by)$ is

$$\mathcal{F}\{f(ax, by)\} = \frac{1}{|ab|} F\left(\frac{k_x}{a}, \frac{k_y}{b}\right). \tag{1.115}$$

Proof: Substitute $u = ax$, $v = by$, so $x = \frac{u}{a}$, $y = \frac{v}{b}$, $\mathrm{d}x \, \mathrm{d}y = \frac{\mathrm{d}u \mathrm{d}v}{|ab|}$:

$$\mathcal{F}\{f(ax, by)\} = \iint_{-\infty}^{\infty} f(ax, by) e^{-\mathrm{i}(k_x x + k_y y)} \, \mathrm{d}x \, \mathrm{d}y \,, \tag{1.116}$$

$$= \iint_{-\infty}^{\infty} f(u, v) e^{-\mathrm{i}\left(k_x \frac{u}{a} + k_y \frac{v}{b}\right)} \frac{\mathrm{d}u \, \mathrm{d}v}{|ab|}, \tag{1.117}$$

$$= \frac{1}{|ab|} \iint_{-\infty}^{\infty} f(u, v) e^{-\mathrm{i}\left(\frac{k_x}{a} u + \frac{k_y}{b} v\right)} \, \mathrm{d}u \, \mathrm{d}v \,, \tag{1.118}$$

$$= \frac{1}{|ab|} F\left(\frac{k_x}{a}, \frac{k_y}{b}\right). \tag{1.119}$$

1.9.1.3 Shift

Property: Translating the function in the spatial domain introduces a phase shift in the frequency domain. So, for a shifted function $f(x - x_0, y - y_0)$, we have,

$$\mathcal{F}\{f(x - x_0, y - y_0)\} = e^{-\mathrm{i}(k_x x_0 + k_y y_0)} F(k_x, k_y). \tag{1.120}$$

Proof: Substitute $u = x - x_0$, $v = y - y_0$, so $\mathrm{d}x\,\mathrm{d}y = \mathrm{d}u\,\mathrm{d}v$,

$$\mathcal{F}\{f(x - x_0, y - y_0)\} = \iint_{-\infty}^{\infty} f(u, v)e^{-\mathrm{i}[k_x(u+x_0)+k_y(v+y_0)]}\,\mathrm{d}u\,\mathrm{d}v\,, \tag{1.121}$$

$$= e^{-\mathrm{i}(k_x x_0+k_y y_0)} \iint_{-\infty}^{\infty} f(u, v)e^{-\mathrm{i}(k_x u+k_y v)}, \mathrm{d}u\,\mathrm{d}v\,, \tag{1.122}$$

$$= e^{-\mathrm{i}(k_x x_0+k_y y_0)} F(k_x, k_y). \tag{1.123}$$

1.9.1.4 Convolution

Property: The convolution theorem for the two-dimensional Fourier transform states that the Fourier transform of the convolution of two functions equals the product of their individual Fourier transforms. In simpler terms, if you convolve two signals (like blurring an image with a filter), their Fourier transform is the pointwise multiplication of the transforms of each signal. This makes frequency-domain operations like filtering much faster and easier. The Fourier transform of the convolution $f * g$, defined as

$$(f * g)(x, y) = \iint_{-\infty}^{\infty} f(x', y')g(x - x', y - y')\,\mathrm{d}x'\,\mathrm{d}y' \tag{1.124}$$

is

$$\mathcal{F}\{f * g\} = F(k_x, k_y)G(k_x, k_y). \tag{1.125}$$

Proof:

$$\mathcal{F}\{f * g\} = \iint_{-\infty}^{\infty} \left[\iint_{-\infty}^{\infty} f(x', y')g(x - x', y - y')\,\mathrm{d}x'\,\mathrm{d}y'\right] e^{-\mathrm{i}(k_x x+k_y y)}\,\mathrm{d}x\,\mathrm{d}y\,, \tag{1.126}$$

$$= \iint_{-\infty}^{\infty} f(x', y') \left[\iint_{-\infty}^{\infty} g(x - x', y - y')e^{-\mathrm{i}(k_x x+k_y y)}\,\mathrm{d}x\,\mathrm{d}y\right] \mathrm{d}x'\,\mathrm{d}y'\,. \tag{1.127}$$

Substitute $u = x - x'$, $v = y - y'$:

$$\iint_{-\infty}^{\infty} g(u, v)e^{-\mathrm{i}[k_x(u+x')+k_y(v+y')]}\,\mathrm{d}u\,\mathrm{d}v = e^{-\mathrm{i}(k_x x'+k_y y')}G(k_x, k_y). \tag{1.128}$$

Thus,

$$\mathcal{F}\{f * g\} = \iint_{-\infty}^{\infty} f(x', y')e^{-\mathrm{i}(k_x x'+k_y y')}G(k_x, k_y)\,\mathrm{d}x'\,\mathrm{d}y' \tag{1.129}$$

$$= F(k_x, k_y)G(k_x, k_y). \tag{1.130}$$

1.9.1.5 Parseval's theorem

Property: Parseval's theorem, in the context of the two-dimensional Fourier transform, states that the total energy of a signal in the spatial domain is equal to the total energy in its frequency domain, establishing a conservation of energy principle,

$$\iint_{-\infty}^{\infty} |f(x,y)|^2 \, \mathrm{d}x \, \mathrm{d}y = \frac{1}{(2\pi)^2} \iint_{-\infty}^{\infty} |F(k_x,k_y)|^2 \, \mathrm{d}k_x \, \mathrm{d}k_y \,. \tag{1.131}$$

Proof: Compute the energy in the spatial domain,

$$\iint_{-\infty}^{\infty} |f(x,y)|^2 \, \mathrm{d}x \, \mathrm{d}y = \iint_{-\infty}^{\infty} f(x,y) f^*(x,y) \, \mathrm{d}x \, \mathrm{d}y \,. \tag{1.132}$$

Express $f^*(x,y)$ using the inverse transform,

$$f^*(x,y) = \frac{1}{(2\pi)^2} \iint_{-\infty}^{\infty} F^*(k'_x,k'_y) e^{-\mathrm{i}(k'_x x + k'_y y)} \, \mathrm{d}k'_x \, \mathrm{d}k'_y \,. \tag{1.133}$$

Thus,

$$\begin{aligned}
\iint_{-\infty}^{\infty} |f(x,y)|^2 \, \mathrm{d}x \, \mathrm{d}y &= \iint_{-\infty}^{\infty} f(x,y) f^*(x,y) \, \mathrm{d}x \, \mathrm{d}y \,, \\
&= \iint_{-\infty}^{\infty} f(x,y) \left[\frac{1}{(2\pi)^2} \iint_{-\infty}^{\infty} F^*(k'_x,k'_y) e^{-\mathrm{i}(k'_x x + k'_y y)} \, \mathrm{d}k'_x \, \mathrm{d}k'_y \right] \mathrm{d}x \, \mathrm{d}y \,, \\
&= \frac{1}{(2\pi)^2} \iint_{-\infty}^{\infty} F^*(k'_x,k'_y) \left[\iint_{-\infty}^{\infty} f(x,y) e^{-\mathrm{i}(k'_x x + k'_y y)} \, \mathrm{d}x \, \mathrm{d}y \right] \mathrm{d}k'_x \, \mathrm{d}k'_y \,, \\
&= \frac{1}{(2\pi)^2} \iint_{-\infty}^{\infty} F^*(k'_x,k'_y) F(k'_x,k'_y) \, \mathrm{d}k'_x \, \mathrm{d}k'_y \,, \\
&= \frac{1}{(2\pi)^2} \iint_{-\infty}^{\infty} |F(k_x,k_y)|^2 \, \mathrm{d}k_x \, \mathrm{d}k_y \,.
\end{aligned} \tag{1.134}$$

Parseval's theorem implies that the sum of the squared magnitudes of the signal (e.g., pixel intensities in an image) equals the sum of the squared magnitudes of its frequency components. So, in other words the theorem highlights the Fourier transform's ability to preserve the signal's energy across domains, which is crucial for tasks such as signal filtering and compression, for example.

The two-dimensional Fourier transform has several limitations that impact its potential applications. It assumes stationarity, meaning the function is defined over an infinite domain or is periodic, which may not hold for real-world signals like images. Additionally, numerical implementation, such as the Fast Fourier transform (FFT), introduces aliasing errors when dealing with finite or discretized data. Furthermore, the transform assumes that the function $f(x,y)$ is square-integrable, ensuring the transform is well-defined, which may not be true for all signals encountered in practice.

1.9.1.6 Classic examples

Let's explore more about the two-dimensional Fourier transform with some classic examples. Here, we compute the two-dimensional Fourier transform for a rectangular aperture, circular aperture, and Gaussian function, with numerical evaluations.

Example: The rectangular Aperture. Consider a rectangular function $f(x,y) = u_0$ for $|x| \leq \frac{a}{2}$, $|y| \leq \frac{b}{2}$, and zero elsewhere,

$$f(x,y) = u_0 \mathrm{rect}\left(\frac{x}{a}\right) \mathrm{rect}\left(\frac{y}{b}\right). \tag{1.135}$$

The Fourier transform of the last expression is

$$F(k_x,k_y) = u_0 \int_{-\frac{a}{2}}^{\frac{a}{2}} e^{-\mathrm{i}k_x x}\, \mathrm{d}x \int_{-\frac{b}{2}}^{\frac{b}{2}} e^{-\mathrm{i}k_y y}\, \mathrm{d}y\,. \tag{1.136}$$

Evaluating

$$\int_{-\frac{a}{2}}^{\frac{a}{2}} e^{-\mathrm{i}k_x x}\, \mathrm{d}x = \frac{e^{-ik_x\frac{a}{2}} - e^{\mathrm{i}k_x\frac{a}{2}}}{-\mathrm{i}k_x} = a\frac{\sin\left(\frac{k_x a}{2}\right)}{\frac{k_x a}{2}}. \tag{1.137}$$

Similarly for y,

$$F(k_x,k_y) = u_0 ab \frac{\sin\left(\frac{k_x a}{2}\right)}{\frac{k_x a}{2}} \frac{\sin\left(\frac{k_y b}{2}\right)}{\frac{k_y b}{2}}. \tag{1.138}$$

Therefore, the magnitude squared is given by

$$|F(k_x,k_y)|^2 \propto \left(\frac{\sin\left(\frac{k_x a}{2}\right)}{\frac{k_x a}{2}}\right)^2 \left(\frac{\sin\left(\frac{k_y b}{2}\right)}{\frac{k_y b}{2}}\right)^2. \tag{1.139}$$

This example makes it clear that the two-dimensional Fourier transform is computed separately along the x and y directions (separable property). We can therefore take a vertical slice of the original function $f(x,y)$ that depends only on x (fixing y), and compare it directly with the corresponding slice of the transform magnitude $|F(k_x,k_y)|$ that depends only on k_x (fixing k_y). These one-dimensional slices reveal the exact correspondence between a spatial profile and its frequency-domain representation. This one-dimensional cross-section and its frequency-domain counterpart are plotted side by side in Fig. 1.19.

Example: The Gaussian Function. Consider $f(x,y) = u_0 e^{-\frac{x^2+y^2}{w^2}}$. The Fourier transform is

$$F(k_x,k_y) = u_0 \int_{-\infty}^{\infty} e^{-\frac{x^2}{w^2}} e^{-\mathrm{i}k_x x}\, \mathrm{d}x \int_{-\infty}^{\infty} e^{-\frac{y^2}{w^2}} e^{-\mathrm{i}k_y y}\, \mathrm{d}y\,. \tag{1.140}$$

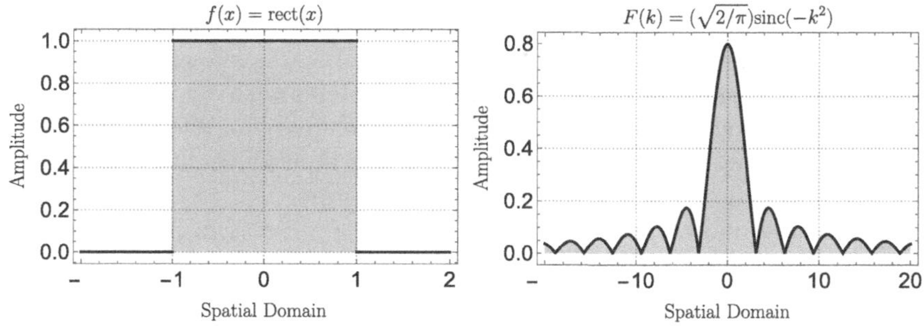

FIGURE 1.19
Vertical slice through the separable 2D test function $f(x, y)$ at fixed y (left) and the corresponding slice through the magnitude of its 2D Fourier transform $|F(k_x, k_y)|$ at the same fixed k_y (right). The plots demonstrate the direct one-dimensional Fourier correspondence along the x-direction.

For a two-dimensional Gaussian, $\int_{-\infty}^{\infty} e^{-\frac{x^2}{w^2}} e^{-\mathrm{i}k_x x}\, \mathrm{d}x = w\sqrt{\pi} e^{-\frac{k_x^2 w^2}{4}}$. Thus,

$$F(k_x, k_y) = u_0 w^2 \pi e^{-\frac{(k_x^2 + k_y^2) w^2}{4}}. \tag{1.141}$$

Squaring the magnitude,

$$|F(k_x, k_y)|^2 \propto e^{-\frac{(k_x^2 + k_y^2) w^2}{2}}. \tag{1.142}$$

We had seen that the two-dimensional Fourier transform decomposes a function of two variables into its spatial frequency components. It inherits fundamental properties—linearity, scaling, shift theorems, the convolution theorem, and Parseval's theorem—that make it an extraordinarily powerful tool in a wide range of applications. Simple yet revealing examples, such as the rectangular aperture (which yields sinc patterns) and the Gaussian function (which transforms to another Gaussian), clearly demonstrate its computational behavior. Although the transform formally requires certain technical conditions (e.g., square-integrability), these rarely limit its practical use. It remains a cornerstone of signal and image processing, optics, and many other fields.

We now turn to a particularly important special case that is so intimately linked to diffraction phenomena that it deserves separate and detailed attention: the circular aperture with a radially symmetric function $f(r)$.

1.10 Hankel transform

The next concept we are going to explore is the Hankel transform. The Hankel transform, named after Hermann Hankel, is an integral transform that expresses a radially symmetric function $f(r)$, defined for $r \geq 0$, as a weighted sum of Bessel functions of the first kind $\mathrm{J}_\nu(kr)$ [5]. Also known as the Fourier-Bessel transform, it is particularly useful in problems with cylindrical or spherical symmetry, such as those in optics, electromagnetism, and signal processing. We are going to use it a lot in computing image formation in spherical symmetry optical systems in the following chapters.

It is closely related to the two-dimensional Fourier transform for radially symmetric functions. In this section, we derive the Hankel transform and prove its key properties: linearity, scaling, Parseval's theorem, and the derivative property. Finally, we provide classic examples with numerical evaluations. The reader is invited to consult the following references for a deeper analysis of the Hankel transform, which are beyond the scope of this mathematical toolkit.

1.10.1 Definition of the Hankel transform

The Hankel transform of order ν for a function $f(r)$, defined for $r \geq 0$, is

$$F_\nu(k) = \mathcal{H}_\nu\{f(r)\} = \int_0^\infty f(r)\mathrm{J}_\nu(kr)r \, \mathrm{d}r \,. \tag{1.143}$$

where $k \geq 0$, $\mathrm{J}_\nu(kr)$ is the Bessel function of the first kind of order ν, and $F_\nu(k)$ is the transformed function. The inverse transform, for $\nu > -1/2$, is

$$f(r) = \mathcal{H}_\nu^{-1}\{F_\nu(k)\} = \int_0^\infty F_\nu(k)\mathrm{J}_\nu(kr)k \, \mathrm{d}k \,. \tag{1.144}$$

The transform is self-inverse under appropriate conditions (e.g., $f(r)$ is piecewise continuous, of bounded variation,

$$\int_0^\infty |f(r)|\sqrt{r} \, \mathrm{d}r < \infty. \tag{1.145}$$

The last expression ensures that the integration does not blow-up at infinity. A very similar condition we have for the Fourier transform in equation 1.131. It is because the Hankel transform is a special case of the two-dimensional Fourier transform. Lets explore this connection between them.

1.10.2 Derivation from the two-dimensional Fourier transform

The Hankel transform arises naturally from the two-dimensional Fourier transform for radially symmetric functions. Consider a function $f(x, y) = f(r)$, where $r = \sqrt{x^2 + y^2}$, independent of angle θ. The two-dimensional Fourier transform is

$$F(k_x, k_y) = \frac{1}{2\pi} \iint_{-\infty}^{\infty} f(r)e^{-\mathrm{i}(k_x x + k_y y)} \, \mathrm{d}x \, \mathrm{d}y \,. \tag{1.146}$$

In polar coordinates, $x = r\cos\theta$, $y = r\sin\theta$, $k_x = s\cos\phi$, $k_y = s\sin\phi$, $\mathrm{d}x\,\mathrm{d}y = r\,\mathrm{d}r\,\mathrm{d}\theta$,

$$F(s) = \frac{1}{2\pi}\int_0^{\infty}\int_0^{2\pi} f(r)e^{-\mathrm{i}sr\cos(\theta-\phi)}r\;\mathrm{d}\theta\,\mathrm{d}r\,. \tag{1.147}$$

The angular integral gives

$$\int_0^{2\pi} e^{-\mathrm{i}sr\cos(\theta-\phi)}\;\mathrm{d}\theta = 2\pi\mathrm{J}_0(sr), \tag{1.148}$$

where $\mathrm{J}_0(sr)$ is a Bessel functions of the first kind of order zero. Thus,

$$F(s) = \int_0^{\infty} f(r)\mathrm{J}_0(sr)r\;\mathrm{d}r \tag{1.149}$$

This is the Hankel transform of order $\nu = 0$. For general ν, the transform arises in higher-dimensional Fourier transforms with spherical harmonics, particularly for $\nu = n/2 - 1$ in n-dimensions.

The **Bessel functions of the first kind**, denoted $\mathrm{J}_\nu(x)$, are a family of special functions that arise as solutions to **Bessel's differential equation** [16],

$$x^2\frac{\mathrm{d}^2y}{\mathrm{d}x^2} + x\frac{\mathrm{d}y}{\mathrm{d}x} + (x^2 - \nu^2)y = 0,$$

where ν is the *order* of the function (real or integer). These functions are fundamental in physics, particularly in problems involving cylindrical symmetry.

The series definition of a Bessel function of the first kind is

$$\mathrm{J}_n(x) = \sum_{k=0}^{\infty}\frac{(-1)^k}{k!\,\Gamma(\nu+k+1)}\left(\frac{x}{2}\right)^{2k+\nu}, \quad \text{for } \mathrm{Re}(\nu) > -1.$$

For integer n, the gamma function $\Gamma(\nu+k+1)$ turns to $\Gamma(\nu+k+1) = (\nu+k)!$ In the context of the Hankel transform, the Bessel function is referred to as the kernel because it forms the fundamental building block of the acts as the weight function, or the kernel, because it dictates how the function $f(r)$ is *modified* during the transformation. In the next section, we are going to show some of the main properties of the Hankel transform.

For the moment let's focus more on the Bessel function. Bessel functions of first kind for orders zero, one, and two, are plotted in Fig. 1.20.

Bessel functions of the first kind with the same order but different zeros are orthogonal over the interval from 0 to a fixed radius a because they naturally arise as eigenfunctions of a Sturm-Liouville problem on that disk with a boundary condition that forces the function to vanish at $r = a$. Bessel's differential equation, where the Bessel function comes, is a special case of Sturm-Liouville problem. The Sturm-Liouville problem is explained in Ref. [17]. The reader

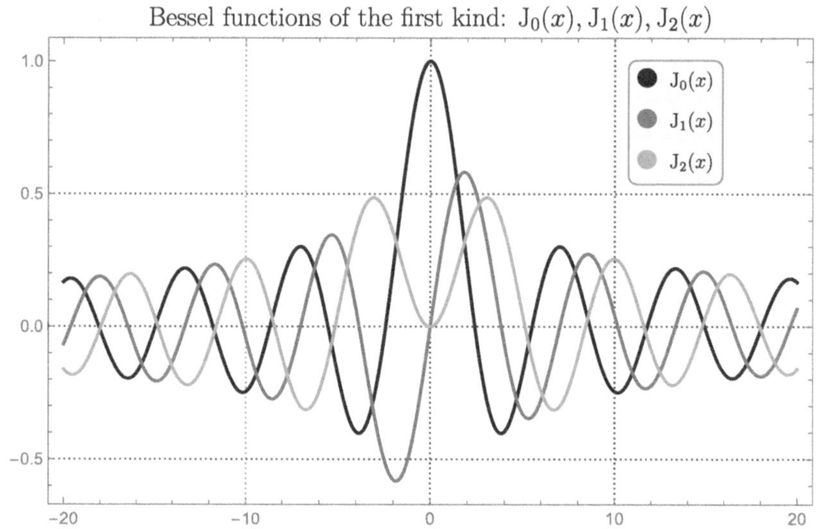

FIGURE 1.20
Plot of Bessel functions of first kind for orders zero, one, and two.

is gently invited to read the mentioned reference for more information of the Sturm-Liouville problem.

In that self-adjoint differential equation problem, each different zero corresponds to a different eigenvalue, and Sturm-Liouville theory guarantees that eigenfunctions belonging to different eigenvalues are orthogonal with respect to the weight function r (the natural weight in radial coordinates) on the interval $[0, a]$ [16].

Physically, this orthogonality means that circular waves of different radial frequencies (different numbers of rings) do not interfere with each other when you expand a radial pattern on a disk; mathematically, it is the direct consequence of the functions solving the same regular singular differential equation under the same boundary conditions. Fig. 1.21 illustrates the orthogonally between $J_0(x)$ and $J_1(x)$. It is clear from the figure that they are below the curve of multiplication between $J_0(x)$ and $J_1(x)$, $J_0(x)J_1(x)$ is zero. Therefore, $J_0(x)$ and $J_1(x)$ are orthogonal. Unfortunately the rigorous proof is not a part of the scope of this text, but the reader is encouraged to read the following reference [17] for a deep discussion on this topic.

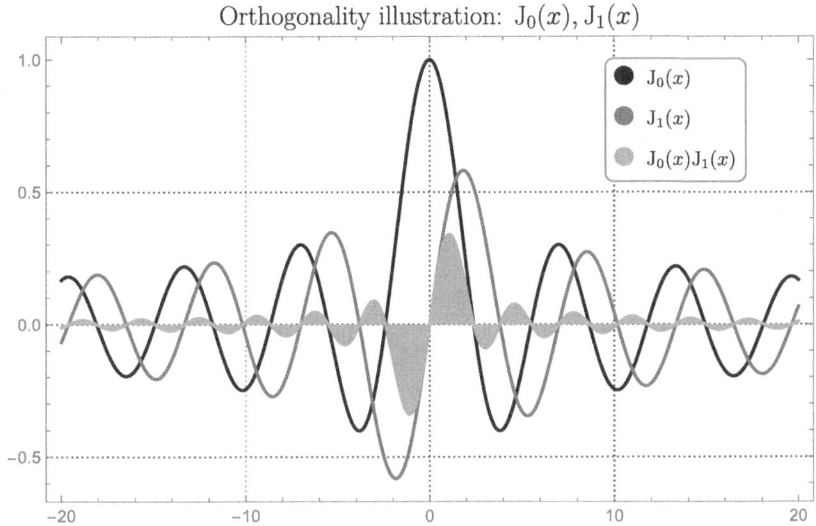

FIGURE 1.21
The orthogonally between $J_0(x)$ and $J_1(x)$ is clear since they are below the curve of $J_0(x)J_1(x)$ is zero.

1.11 Properties of the Hankel transform

We are now going to prove key properties of the Hankel transform, focusing on order ν.

1.11.1 Linearity

The linearity property of the Hankel transform states that the transform of a linear combination of functions is the same linear combination of their individual transforms.

Property: For functions $f(r)$, $g(r)$, and constants a, b:

$$\mathcal{H}_\nu\{af(r) + bg(r)\} = a\mathcal{H}_\nu\{f(r)\} + b\mathcal{H}_\nu\{g(r)\}. \tag{1.150}$$

Proof:

$$\mathcal{H}_\nu\{af(r) + bg(r)\} = \int_0^\infty [af(r) + bg(r)]\mathrm{J}_\nu(kr)r \,\mathrm{d}r \tag{1.151}$$

$$= a\int_0^\infty f(r)\mathrm{J}_\nu(kr)r \,\mathrm{d}r + b\int_0^\infty g(r)\mathrm{J}_\nu(kr)r \,\mathrm{d}r \tag{1.152}$$

$$= aF_\nu(k) + bG_\nu(k). \tag{1.153}$$

The linearity is expected since the Hankel transform at the end is an integral and, overall all integrals are linear.

1.11.2 Scaling

The scaling property of the Hankel transform states that scaling the argument of a function by a factor scales the transform by the inverse of that factor raised to square.

Property: For function $f(ar)$, where $a > 0$, we have then,

$$\mathcal{H}_\nu\{f(ar)\} = \frac{1}{a^2} F_\nu\left(\frac{k}{a}\right). \tag{1.154}$$

In order to probe the last expression, we need some substitutions.

Proof: Let's substitute the variable $u = ar$, so r becomes $r = \frac{u}{a}$, and its derivative is $\mathrm{d}r = \frac{\mathrm{d}u}{a}$,

$$\mathcal{H}_\nu\{f(ar)\} = \int_0^\infty f(ar)\mathrm{J}_\nu(kr)r \;\mathrm{d}r\,, \tag{1.155}$$

$$= \int_0^\infty f(u)\mathrm{J}_\nu\left(k\frac{u}{a}\right)\frac{u}{a}\frac{\mathrm{d}u}{a}, \tag{1.156}$$

$$= \frac{1}{a^2}\int_0^\infty f(u)\mathrm{J}_\nu\left(\frac{k}{a}u\right) u \;\mathrm{d}u\,, \tag{1.157}$$

$$= \frac{1}{a^2} F_\nu\left(\frac{k}{a}\right). \tag{1.158}$$

As expected scaling the argument of a function by a factor scales the transform.

1.11.3 Parseval's theorem

Parseval's theorem for the Hankel transform states that the total energy of a function in the spatial domain is equal to the total energy of its Hankel transform in the frequency domain, with an appropriate scaling factor, r. Let's prove it.

Property: For functions $f(r)$, $g(r)$ with transforms $F_\nu(k)$, $G_\nu(k)$,

$$\int_0^\infty f(r)g(r)r \;\mathrm{d}r = \int_0^\infty F_\nu(k)G_\nu(k)k \;\mathrm{d}k\,, \tag{1.159}$$

where a special case is when $f(r)$=$g(r)$,

$$\int_0^\infty |f(r)|^2 r \;\mathrm{d}r = \int_0^\infty |F_\nu(k)|^2 k \;\mathrm{d}k\,. \tag{1.160}$$

Proof: The proof begins with using the inverse transform over Equation 1.143,

$$f(r) = \int_0^\infty F_\nu(k) \mathrm{J}_\nu(kr) k \, \mathrm{d}k \,, \tag{1.161}$$

and the reordering the integrals,

$$\int_0^\infty f(r)g(r)r \, \mathrm{d}r = \int_0^\infty \left[\int_0^\infty F_\nu(k) \mathrm{J}_\nu(kr) k \, \mathrm{d}k \right] g(r) r \, \mathrm{d}r \,, \tag{1.162}$$

$$= \int_0^\infty F_\nu(k) k \left[\int_0^\infty g(r) \mathrm{J}_\nu(kr) r \, \mathrm{d}r \right] \mathrm{d}k \,, \tag{1.163}$$

$$= \int_0^\infty F_\nu(k) G_\nu(k) k \, \mathrm{d}k \,. \tag{1.164}$$

The last expression expresses that the total energy is preserved under the Hankel transformation. From last expression, it is easy to see that if $f(r){=}g(r)$ ends with Eq. 1.160.

1.11.4 Derivative property

The Bessel operator property of the Hankel transform states that applying the Bessel differential operator to a function in the spatial domain corresponds to multiplying its Hankel transform by a factor involving the order of the Bessel function in the frequency domain.

Property: The Bessel differential operator is given by

$$B_\nu(f) = r^2 \frac{\mathrm{d}^2 f}{\mathrm{d}r^2} + r \frac{\mathrm{d}f}{\mathrm{d}r} + (r^2 - \nu^2) f, \tag{1.165}$$

where $\frac{\mathrm{d}}{\mathrm{d}r}$ is the derivative with respect to the radial variable r, and $\frac{\mathrm{d}^2}{\mathrm{d}r^2}$ is the second derivative with respect to the radial variable r.

The Hankel transform of $B_\nu(f)$ is

$$\mathcal{H}_\nu\{B_\nu(f)\} = -k^2 F_\nu(k). \tag{1.166}$$

Proof: Since $\mathrm{J}_\nu(kr)$ satisfies the Bessel operator

$$r^2 \frac{\mathrm{d}^2 \mathrm{J}_\nu(kr)}{\mathrm{d}r^2} + r \frac{\mathrm{d}\mathrm{J}_\nu(kr)}{\mathrm{d}r} + (k^2 r^2 - \nu^2) \mathrm{J}_\nu(kr) = 0, \tag{1.167}$$

we have $B_\nu(\mathrm{J}_\nu(kr)) = k^2 r^2 \mathrm{J}_\nu(kr)$. So let's consider the Hankel transform of $B_\nu(f)$,

$$\mathcal{H}_\nu\{B_\nu(f)\} = \int_0^\infty \left[r^2 \frac{\mathrm{d}^2 f}{\mathrm{d}r^2} + r \frac{\mathrm{d}f}{\mathrm{d}r} + (r^2 - \nu^2) f \right] \mathrm{J}_\nu(kr) r \, \mathrm{d}r \,. \tag{1.168}$$

Use integration by parts, assuming $f(r) \to 0$, $\frac{\mathrm{d}f}{\mathrm{d}r} \to 0$ as $r \to \infty$, and $f(r)$ is well-behaved at $r = 0$. Split the integral,

$$I_1 = \int_0^\infty r^2 \frac{\mathrm{d}^2 f}{\mathrm{d}r^2} \mathrm{J}_\nu(kr) r \ \mathrm{d}r = \int_0^\infty \frac{\mathrm{d}^2 f}{\mathrm{d}r^2} \mathrm{J}_\nu(kr) r^3 \ \mathrm{d}r\,, \tag{1.169}$$

$$I_2 = \int_0^\infty r \frac{\mathrm{d}f}{\mathrm{d}r} \mathrm{J}_\nu(kr) r \ \mathrm{d}r = \int_0^\infty \frac{\mathrm{d}f}{\mathrm{d}r} \mathrm{J}_\nu(kr) r^2 \ \mathrm{d}r\,, \tag{1.170}$$

$$I_3 = \int_0^\infty (r^2 - \nu^2) f(r) \mathrm{J}_\nu(kr) r \ \mathrm{d}r\,. \tag{1.171}$$

For I_1, integrate by parts twice,

$$I_1 = \left[r^3 \frac{\mathrm{d}f}{\mathrm{d}r} \mathrm{J}_\nu(kr) \right]_0^\infty - \int_0^\infty \frac{\mathrm{d}f}{\mathrm{d}r} \frac{\mathrm{d}}{\mathrm{d}r} [r^3 \mathrm{J}_\nu(kr)] \ \mathrm{d}r\,, \tag{1.172}$$

$$= - \int_0^\infty \frac{\mathrm{d}f}{\mathrm{d}r} (3r^2 \mathrm{J}_\nu(kr) + r^3 k \mathrm{J}'_\nu(kr)) \ \mathrm{d}r\,. \tag{1.173}$$

Integrate the second term again,

$$\int_0^\infty \frac{\mathrm{d}f}{\mathrm{d}r} r^3 k \mathrm{J}'_\nu(kr) \ \mathrm{d}r = \left[r^3 f \mathrm{J}'_\nu(kr) \right]_0^\infty - \int_0^\infty f \frac{\mathrm{d}}{\mathrm{d}r} [r^3 \mathrm{J}'_\nu(kr)] \ \mathrm{d}r\,, \tag{1.174}$$

$$= - \int_0^\infty f (3r^2 \mathrm{J}'_\nu(kr) + r^3 k \mathrm{J}''_\nu(kr)) \ \mathrm{d}r\,. \tag{1.175}$$

Combine terms, using $\mathrm{J}''_\nu(kr) = \frac{1}{r} \mathrm{J}'_\nu(kr) - k^2 \mathrm{J}_\nu(kr) + \frac{\nu^2}{r^2} \mathrm{J}_\nu(kr)$. After simplification, the result is

$$\mathcal{H}_\nu\{B_\nu(f)\} = -k^2 \int_0^\infty f(r) \mathrm{J}_\nu(kr) r \ \mathrm{d}r = -k^2 F_\nu(k). \tag{1.176}$$

This last property is useful for solving differential equations. The Hankel transform has notable limitations that restrict its applicability. It assumes radial symmetry, meaning the function $f(r)$ depends solely on the radial distance r, which limits its use to radially symmetric systems. Additionally, it requires the function to be integrable in a specific sense, satisfying $\int_0^\infty |f(r)| \sqrt{r} \, dr < \infty$, which may not hold for all signals.

Example: Disk Function. Let's consider $f(r) = u_0$ for $r \leq a$, and 0 otherwise, thus,

$$F_0(k) = \int_0^a u_0 \mathrm{J}_0(kr) r \ \mathrm{d}r\,. \tag{1.177}$$

Let $u = kr$, so $r = \frac{u}{k}$, $\mathrm{d}r = \frac{\mathrm{d}u}{k}$,

$$F_0(k) = u_0 \int_0^{ka} \mathrm{J}_0(u) \frac{u}{k} \frac{\mathrm{d}u}{k} = \frac{u_0}{k^2} \int_0^{ka} \mathrm{J}_0(u) u \ \mathrm{d}u\,. \tag{1.178}$$

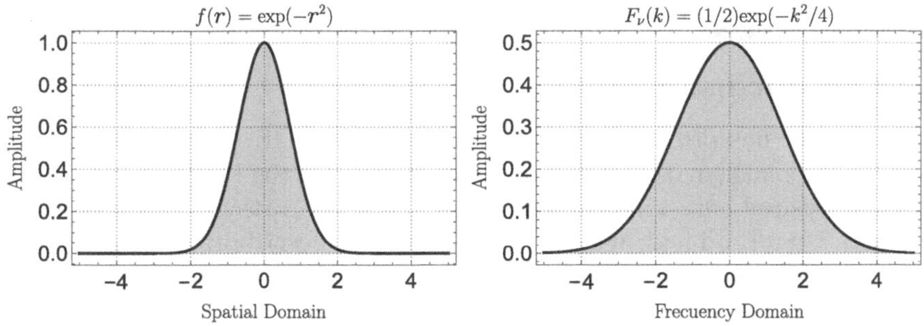

FIGURE 1.22
The function $f(r)$ is plotted on the *left*, and its Hankel transform is plotted on the *right*.

Using $\int u\mathrm{J}_0(u)\,du = u\mathrm{J}_1(u)$ [16],

$$F_0(k) = \frac{u_0}{k^2}\left[u\mathrm{J}_1(u)\right]_0^{ka} = \frac{u_0 ka}{k^2}\mathrm{J}_1(ka) = \frac{u_0 a}{k}\mathrm{J}_1(ka). \tag{1.179}$$

Example: Gaussian Function. Consider $f(r) = e^{-r^2/w^2}$. The Hankel transform of order $\nu = 0$ is

$$F_0(k) = \int_0^\infty e^{-r^2/w^2}\mathrm{J}_0(kr)r\;dr\,. \tag{1.180}$$

Using the identity $\int_0^\infty e^{-\alpha r^2}\mathrm{J}_0(kr)r\;dr = \frac{1}{2\alpha}e^{-\frac{k^2}{4\alpha}}$, with $\alpha = \frac{1}{w^2}$ [16],

$$F_0(k) = \frac{w^2}{2}e^{-\frac{k^2 w^2}{4}}. \tag{1.181}$$

The transform is a Gaussian, decaying rapidly for large k. Let be $w = 1$, then $f(r)$ turns to $f(r) = e^{-r^2}$. The function $f(r)$ is plotted in the *left*-side and its Hankel transform is plotted in the *right*-side of Fig. 1.22.

Example: Exponential Decay. Consider $f(r) = e^{-ar}$, $a > 0$. The Hankel transform of order $\nu = 0$ is

$$F_0(k) = \int_0^\infty e^{-ar}\mathrm{J}_0(kr)r\;dr\,. \tag{1.182}$$

Using the identity $\int_0^\infty e^{-ar}\mathrm{J}_0(kr)r\;dr = \frac{a}{(a^2+k^2)^{3/2}}$ from Ref. [16],

$$F_0(k) = \frac{a}{(a^2+k^2)^{3/2}}. \tag{1.183}$$

The transform decays as k increases.

1.12 Conclusions

This chapter has provided a self-contained mathematical toolkit that underpins all the mathematical tools needed throughout the book.

We revisited core vector algebra in $\mathbb{R}^3$, paying particular attention to the dot and cross products, which are essential for describing orthogonality between propagation and polarization directions. Rodrigues' rotation formula was derived geometrically. This formula will be used many times in the later chapters of the book to show how different polarizations can be obtained from a freeform wavefront.

We studied parametric representations of curves and surfaces, together with tangent planes, normal vectors, and the Jacobian determinant. These concepts appear repeatedly in diffraction integrals, as we will see in Chapter 7. We also reviewed complex numbers and reinforced their natural vector interpretation in the phase–amplitude plane.

A quick refresher on the two-dimensional Fourier and Hankel transforms was presented. These transforms are the central analytical tools for propagating fields from the pupil to the observation plane, especially in systems with radial symmetry (this will be explored in more detail in Chapters 3 and 4). For now, we can conclude this chapter with the following conceptual problems.

Problems

Problem 1: Vectors, dot & cross product
Let $\mathbf{u} = [2, -1, 3]$ and $\mathbf{v} = [4, 5, -2]$.

(a) Compute $\mathbf{u} \cdot \mathbf{v}$ and the angle θ between $\mathbf{u}$ and $\mathbf{v}$ (give the answer in degrees, rounded to one decimal place).

(b) Compute $\mathbf{u} \times \mathbf{v}$.

(c) Verify that $(\mathbf{u} \times \mathbf{v}) \cdot \mathbf{u} = 0$ and $(\mathbf{u} \times \mathbf{v}) \cdot \mathbf{v} = 0$.

(d) Find a unit vector that is perpendicular to both $\mathbf{u}$ and $\mathbf{v}$.

Problem 2: Rodrigues' rotation formula
Consider the vector $\mathbf{a} = [3, 0, 4]$. Rotate $\mathbf{a}$ by $\alpha = 120° = \frac{2\pi}{3}$ radians around the unit axis

$$\mathbf{k} = \left[\frac{1}{\sqrt{3}}, \frac{1}{\sqrt{3}}, \frac{1}{\sqrt{3}}\right].$$

(a) Use Rodrigues' rotation formula to compute the rotated vector $\hat{\mathbf{a}}$.

(b) Check that $\|\hat{\mathbf{a}}\| = \|\mathbf{a}\|$.

(c) Compute the dot product $\mathbf{k} \cdot \hat{\mathbf{a}}$ and compare it with $\mathbf{k} \cdot \mathbf{a}$. Explain the result geometrically.

Problem 3: Parametric curve, tangent, & normal
A parametric curve in the plane is given by

$$x(t) = t^2 - 4t, \qquad y(t) = t^3 - 6t, \qquad t \in \mathbb{R}.$$

(a) Find the tangent vector $\mathbf{r}'(t) = [x'(t), y'(t)]$.

(b) At the point corresponding to $t = 3$, compute the unit tangent vector and one possible normal vector.

Problem 4: Parametric surface, tangent plane, & normal
A surface is parametrized by

$$\mathbf{r}(u, v) = [u \cos v,\ u \sin v,\ u^2], \qquad u \geq 0,\ v \in [0, 2\pi).$$

(a) Compute the partial derivatives $\mathbf{r}_u$ and $\mathbf{r}_v$.

(b) Find the normal vector $\mathbf{N} = \mathbf{r}_u \times \mathbf{r}_v$ at the point where $u = 2$, $v = \pi/4$.

(c) Write the equation of the tangent plane at that point.

(d) What kind of surface is this?

Problem 5: Normal vector to an explicit surface
Consider the explicit surface $z = f(x, y) = \sqrt{9 - x^2 - y^2}$ (upper hemisphere of radius 3).

(a) Compute the gradient-based normal vector $\mathbf{N} = [-f_x, -f_y, 1]$ at the point $[1, 1, \sqrt{7}]$.

(b) Normalize it to obtain the unit normal $\mathbf{n}$ pointing upwards.

(c) Show that this normal is parallel to the position vector $[x, y, z]$ at every point on the hemisphere (except possibly at the equator). Explain why geometrically.

Problem 6: Problem 6 Jacobian & change of variables
Consider the coordinate transformation

$$x = u^2 - v^2, \qquad y = 2uv.$$

(a) Compute the Jacobian matrix and its determinant $\det \mathbf{Jcb}$.

(b) Evaluate $|\det \mathbf{Jcb}|$ at the point $(u, v) = [2, 1]$.

(c) What is the geometric meaning of the absolute value of the Jacobian in this context?

(d) If you integrate a function $g(x, y)$ over a region in the xy-plane, how does the area element change under this transformation?

Problem 7: Complex numbers & geometric interpretation
Let $z_1 = 2 + 3\mathrm{i}$ and $z_2 = 1 - 4\mathrm{i}$.

(a) Compute $z_1 + z_2$, $z_1 - z_2$, $z_1 z_2$, z_1/z_2, and $|z_1 z_2|$.

(b) Find the polar form (modulus and argument in radians) of z_1 and z_2.

(c) Compute $z_1 z_2$ using polar form and verify that the result matches part (a).

(d) Interpret the multiplication $z_1 z_2$ geometrically (rotation + scaling).

Problem 8: Fourier & Hankel transforms

(a) Explain in 3–4 sentences why the Hankel transform of order 0 can be regarded as a special case of the 2D Fourier transform.

(b) Write the mathematical relation that connects the two transforms for a radially symmetric function $f(r)$.

(c) You are told that the 2D Fourier transform of $\mathrm{rect}(x/a)$ rect (y/b) is proportional to $\mathrm{sinc}(k_x a/2)\,\mathrm{sinc}(k_y b/2)$. What is the most important qualitative difference you expect between the Fourier transform of a square aperture and the Hankel transform (order 0) of a circular disk of the same area?

(d) It is known that

$$\mathcal{H}_0\{\mathrm{rect}(r/a)\} = \frac{a}{k}\mathrm{J}_1(ka).$$

At which value of k does the first zero of this transform occur?

2

Geometrical optics toolkit

This chapter presents the foundational toolkit of geometrical optics, centered on reflection and refraction as governed by their respective laws. Fermat's principle of least time principle unifies them asserting that light follows the path minimizing optical path length. Then text introduces the refractive index, paraxial approximations, vectorial formulations of Snell's law, and stigmatic imaging with conic mirrors (elliptic, parabolic, hyperbolic). Essential design concepts include cardinal points, pupils, stops, focal length, f-number, numerical aperture, field of view, and various ray types (chief, marginal, meridional, sagittal, skew, paraxial). The chapter also covers monochromatic and chromatic aberrations, spot diagrams, and Zernike polynomials for wavefront and image analysis in optical instruments.

2.1 Introduction

Geometrical optics, a cornerstone of optical science, elucidates the manipulation of light's trajectory and ray propagation through two fundamental mechanisms: reflection and refraction. These principles underpin the design and analysis of optical systems, relying on the foundational axioms of Euclidean geometry to model the behavior of light [18, 19, 20]. This exposition aims to provide a comprehensive and formal exploration of these mechanisms, emphasizing their theoretical underpinnings and practical significance within the domain of geometrical optics.

Reflection involves the redirection of incident light rays upon encountering a polished surface, such as a mirror, where the angle of incidence equals the angle of reflection, as governed by the law of reflection [7], which we are going to study later in this chapter. On the other hand, we have refraction. Refraction, conversely, describes the bending of light as it passes from one medium to another with a different propagation speed, a process dictated by

DOI: 10.1201/9781003726494-3

the refractive index of the respective media. The refractive index, a dimensionless quantity, is defined as the ratio of the speed of light in a vacuum to that in the medium, serving as a pivotal parameter in geometrical optics [1]. This index quantifies the degree to which light is decelerated and redirected upon entering a medium, such as glass or water, thereby altering its path in accordance with Snell's law. The interplay of refractive indices across interfaces enables the functionality of lenses, prisms, and other optical components essential to focusing, dispersing, or redirecting light in sophisticated optical assemblies.

In the framework of geometrical optics, the dimensions of optical elements are presumed to significantly exceed the wavelength of the interacting light, rendering diffraction effects negligible [7]. Consequently, light is modeled as a collection of rays—lines perpendicular to the wavefront—that trace the path of propagation through sequential ray tracing. This sequential approach defines the ordered progression of rays through an optical system, providing a robust method for predicting image formation and system performance [21]. By abstracting light as rays, geometrical optics facilitates the design and optimization of complex optical systems, from microscopes to imaging devices, with a focus on precision and efficiency.

The refractive index of a medium, a dimensionless quantity, quantifies the speed at which light traverses said medium, a parameter central to geometrical optics,

$$n \equiv \frac{c}{v}, \tag{2.1}$$

in this framework, the variable c represents the speed of light in a vacuum, whereas v denotes the phase velocity of light within a given medium. A medium is classified as homogeneous when its refractive index remains constant throughout its volume. In such homogeneous media, light rays follow linear trajectories [22].

Refractive surfaces comprise optical elements characterized by a refractive index that differs from that of their surrounding medium. Such elements include lenses and prisms, each serving distinct functions in optical systems. Lenses are engineered to either converge or diverge light rays, thereby modifying their trajectories to form focused or dispersed beams. Prisms, in contrast, are designed to manipulate light through refraction, altering its path without inducing convergence or divergence [23].

Fundamentally, geometrical optics centers on the study of mirrors, which operate as reflective surfaces, and lenses and prisms, which function as refractive bodies. Despite the apparent simplicity of these components, the theoretical framework of geometrical optics is complex. The mathematical underpinnings often involve nonlinear equations, which can be intricate and cumbersome to solve. To address these challenges, optical practitioners frequently employ approximations to simplify their analyses. These approximations are derived from fundamental definitions involving trigonometric functions, such as $\sin\theta$ and $\cos\theta$, which facilitate the modeling of light behavior in optical systems,

$$\sin\theta \equiv \sum_{m=0}^{\infty} \frac{(-1)^m}{(2m+1)!}\theta^{2m+1} = \theta - \frac{\theta^3}{3!} + \frac{\theta^5}{5!} - \frac{\theta^7}{7!} + \cdots \tag{2.2}$$

and,

$$\cos\theta \equiv \sum_{m=0}^{\infty} \frac{(-1)^m}{(2m)!}\theta^{2m} = 1 - \frac{\theta^2}{2!} + \frac{\theta^4}{4!} - \frac{\theta^6}{6!} + \cdots \tag{2.3}$$

The paraxial approximation, widely regarded as a cornerstone in paraxial optics, is employed when the angles of incidence are sufficiently small [24]. This approach simplifies calculations, making it a practical choice for modeling light ray behavior with ease and precision,

$$\sin\theta \approx \theta, \tag{2.4}$$

and,

$$\cos\theta \approx 1. \tag{2.5}$$

Approximations in optics are generally categorized according to the number of terms incorporated in the computation. For example, the third-order approximation of $\sin\theta$ is designated when $m = 3$, signifying that terms up to the third order are included, keeping the math precise yet manageable [24].

$$\sin\theta \approx \theta - \frac{\theta^3}{3!}. \tag{2.6}$$

Even though the paraxial approximation skips out on a ton of the juicy details of an optical system, it's still super handy for figuring out the basic vibe of a system [25, 26].

2.1.1 The principle of least action

To elucidate Fermat's principle, it is instructive to first consider mechanics, a core discipline within physics that examines the behavior of objects in motion or at rest. By determining an object's initial position, velocity, and the forces acting upon it, mechanics provides a comprehensive framework for predicting the system's dynamics [27]. This approach is deeply rooted in the contributions of Sir Isaac Newton, the eminent English polymath, who formulated the laws governing the motion of bodies using the rigorous tools of vector calculus [28, 29]. While vector calculus enables precise predictions of a system's evolution, deriving the equations of motion can be a bit of a slog in complex scenarios, often resulting in nonlinear systems that are tricky to solve. To navigate these challenges, some physicists have adopted an alternative approach grounded in the law of conservation of energy, which states that energy is neither created nor destroyed but merely transformed. In 1662, Pierre de Fermat, the French mathematician and lawyer, laid the groundwork for the principle of minimum action [30]. Fermat proposed that light rays traveling between two points follow the path that minimizes travel time. Known as the principle of least action, this concept suggests that nature, in its chill efficiency, opts for the path of least energy expenditure. In the realm of optics, this idea manifests as Fermat's principle, which formally asserts that light rays traverse paths that minimize the time required to travel between two points, keeping things sleek and straightforward [31].

The optical path length (OPL) between two distinct fixed points, traversed by light, corresponds to the global minimum. It is calculated by multiplying the physical length of the path by the refractive index of the medium, defining the optical length.

Let's dive into the concept of the global minimum, which is just a fancy way of saying the smallest value in a whole set. Picture this: you've got a bunch of possible optical paths light could take between two points, each with its own OPL. According to Fermat's principle, the path that actually gets used is the one with the shortest optical length—pretty straightforward, right? It's like light's picking the quickest route to vibe from one spot to another. This book will keep circling back to Fermat's principle, since it's the bedrock for designing stigmatic systems that keep things sharp and clear. In a more formal sense, Fermat's principle can be expressed as an integral over time, capturing the duration it takes for light to travel from point x_1 to point x_2. Let t_1 and t_2 represent the moments when light is at points x_1 and x_2, respectively, making the math nice and tidy for tracking light's journey [31].

$$\int_{t_1}^{t_2} \mathrm{d}t = \frac{1}{c} \int_{t_1}^{t_2} \frac{c}{v} \frac{\mathrm{d}s}{\mathrm{d}t} \, \mathrm{d}t = \frac{1}{c} \int_{x_1}^{x_2} n \, \mathrm{d}s \,. \tag{2.7}$$

Here, c is the speed of light zooming through a vacuum, $\mathrm{d}s$ is just a tiny step along the light ray's path, $v = \mathrm{d}s \,/\, \mathrm{d}t$ is how fast light's moving through a medium, and $n = c/v$ is the refractive index, basically telling us how much the medium slows light down. With that in mind, the OPL for light going from point x_1 to point x_2 can be written out as,

$$\mathrm{OPL} = \int_{x_1}^{x_2} n \, \mathrm{d}s \,. \tag{2.8}$$

Notice that the optical path is a geometrical quantity, since,

$$\mathrm{OPL} = c \int_{t_1}^{t_2} \mathrm{d}t\,. \tag{2.9}$$

Let's take a look at a set that includes all possible travel times for light moving between points x_1 and x_2, alongside another set covering all the OPL for that same journey. According to Fermat's principle, the shortest travel time for light going from x_1 to x_2 aligns perfectly with the smallest OPL in that set, keeping things elegantly simple [31].

2.1.2 Reflection

Fermat's principle totally nails how light behaves when it bounces off a surface. Check out Fig. 2.1—it shows a light ray hitting a surface and getting reflected. Picture this: the ray starts at point (x_1, y_1), zips in a straight line to the spot where it bounces off the surface, and then heads to point (x_2, y_2). With that in mind, we can jot down the time it takes for light to travel from (x_1, y_1) to

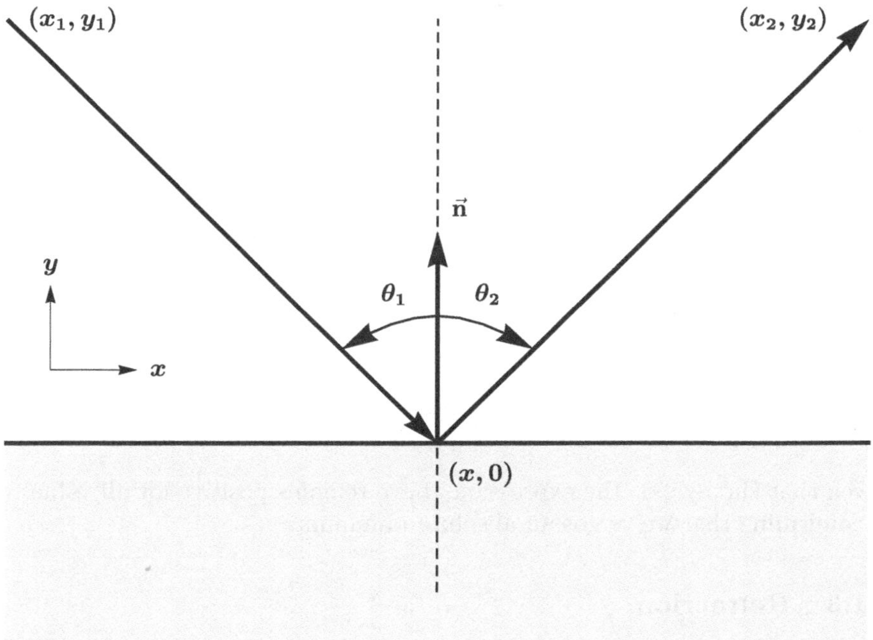

FIGURE 2.1
Law of reflection, the incident angle θ_1 is equal to the reflected angle θ_2.

(x_2, y_2) with this slick expression,

$$T(x) = \frac{1}{c}\left[\sqrt{(x-x_1)^2 + y_1^2} + \sqrt{(x_2-x)^2 + y_2^2}\right], \tag{2.10}$$

x is where the light hits the plane on the horizontal axis. From the ray equation, we're sure that $T(x)$ is spot-on for spots where the refractive index stays the same, meaning light just cruises in straight lines. Now, we're still figuring out how the light bounces. The point $(x, 0)$ is super key here—it's where the reflection happens. The value of x basically sets the slope of those straight-line paths. Thanks to Fermat's principle, we know light picks the quickest route possible. So, to nail this down, we take the derivative with respect to x, set it to zero, and boom, we're on track to find the shortest path.

$$\frac{\partial T(x)}{\partial x} = 0, \tag{2.11}$$

derivating,

$$\frac{x-x_1}{\sqrt{(x-x_1)^2 + y_1^2}} = \frac{x_2-x}{\sqrt{(x_2-x)^2 + y_2^2}}. \tag{2.12}$$

From Fig. 2.1, we can see that the last equation can be re-formulated as,

$$\theta_1 = \theta_2. \tag{2.13}$$

As depicted in Fig. 2.1, which gives a clean and simple sketch of light bouncing off a flat mirror, the incident angle θ_1 matches the reflected angle θ_2, so we've got $\theta_1 = \theta_2$. To confirm this path, we can compute the second derivative and check that it's a true minimum,

$$c\frac{\partial^2 T}{\partial x^2} = (1-\sin^2\theta_2)\left[\frac{1}{\sqrt{(x-x_1)^2 + y_1^2}} + \frac{1}{\sqrt{(x_2-x)^2 + y_2^2}}\right]. \tag{2.14}$$

Given that $\sin^2\theta_2 \leq 1$, the expression above remains positive for all values of x, confirming that we've got an absolute minimum.

2.1.3 Refraction

Fig. 2.2 illustrates a diagram depicting refraction at a planar interface separating two media with distinct refractive indices, offering a straightforward view of this optical phenomenon.

Snell's law, while seemingly simple, can be a bit of a beast to work with mathematically. Its complexity often pushes optical scientists toward paraxial

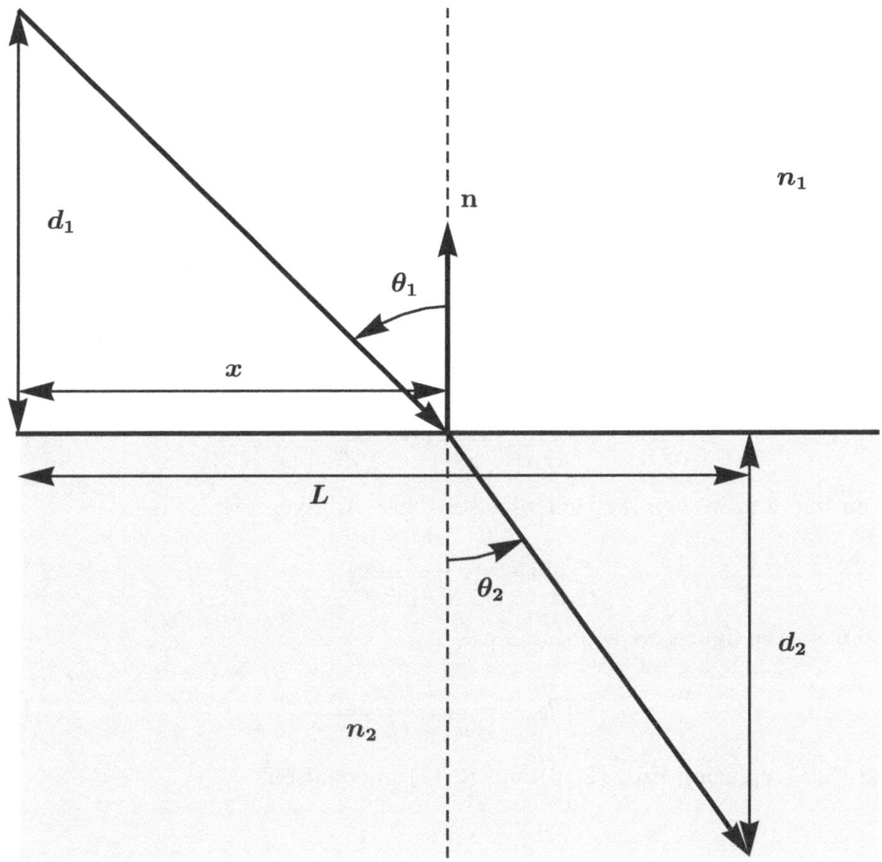

FIGURE 2.2
Refraction takes place on a flat interface between two media with different refractive indexes n_1 and n_2. The incident vector is $\mathbf{s}_1$, the refracted vector is $\mathbf{s}_2$, and the normal vector of the surface is $\mathbf{n}$.

approximations for ease. However, as outlined in this chapter's introduction, we've chosen to steer clear of paraxial approximations entirely, keeping our analysis sharp and unfiltered. Refraction occurs when light passes through a region where the refractive index varies. The simplest scenario involves an interface between two uniform media, one with a refractive index n_1 and the other with n_2. In such cases, Snell's Law, also known as the Law of Refraction, governs the reflection of the light ray as it crosses the boundary.

As shown in Fig. 2.2, we present a streamlined diagram of refraction occurring at a flat interface between two media with constant refractive indices.

From this figure, we can express the trajectory of the light ray as follows, keeping the math precise yet approachable [31].

$$R(x) = n_1\sqrt{d_1^2 + x^2} + n_2\sqrt{d_2^2 + (L - x)^2}, \tag{2.15}$$

where d_1 denotes the vertical height of the light ray's initial position, x represents the horizontal distance from the initial position to the origin of the coordinate system, L signifies the horizontal distance spanning from the initial to the final position, and d_2 indicates the vertical height of the light ray's final position. To determine the path that minimizes the travel time of the light ray, we take the derivative of Eq. (2.15) with respect to x, ensuring a precise yet relaxed approach to finding the optimal trajectory.

$$\frac{dR(x)}{dx} = n_1\frac{x}{\sqrt{d_1^2 + x^2}} - n_2\frac{(L - x)}{\sqrt{d_2^2 + (L - x)^2}}. \tag{2.16}$$

From Fig. 2.2, we can see that precisely $\sin\theta_1$ is given by,

$$\sin\theta_1 = \frac{x}{\sqrt{d_1^2 + x^2}}, \tag{2.17}$$

also from the figure, we see that $\sin\theta_2$ is

$$\sin\theta_2 = \frac{L - x}{\sqrt{d_2^2 + (L - x)^2}}, \tag{2.18}$$

therefore, replacing Eqs. (2.16) and (2.17) in equation,

$$n_1 \sin\theta_1 = n_2 \sin\theta_2. \tag{2.19}$$

In a homogeneous medium, light rays propagate along linear paths. When light transitions from one homogeneous medium to another with a different refractive index, the resulting deviation is described by the aforementioned equation, formally known as Snell's law, expressed in terms of angular notation.

2.1.4 Vectorial form of the Snell's law

To streamline the application of Snell's law, we reformulate it in vector form, which is particularly practical for two-dimensional scenarios, especially given that many chapters address rotationally symmetric surfaces.

Consider an orthonormal basis $\hat{\mathbf{x}}, \hat{\mathbf{y}}, \hat{\mathbf{z}}$ for the Euclidean space $\mathbb{R}^3$. Let $\mathbf{s}_1 \in \mathbb{R}^2$ denote the unit vector of the incident ray, $\mathbf{s}_2 \in \mathbb{R}^2$ represent the unit vector of the refracted ray, and $\mathbf{n} \in \mathbb{R}^2$ signify the unit normal vector to the surface. The angle θ_1 is defined as the incident angle relative to $\mathbf{n}$, while θ_2 is

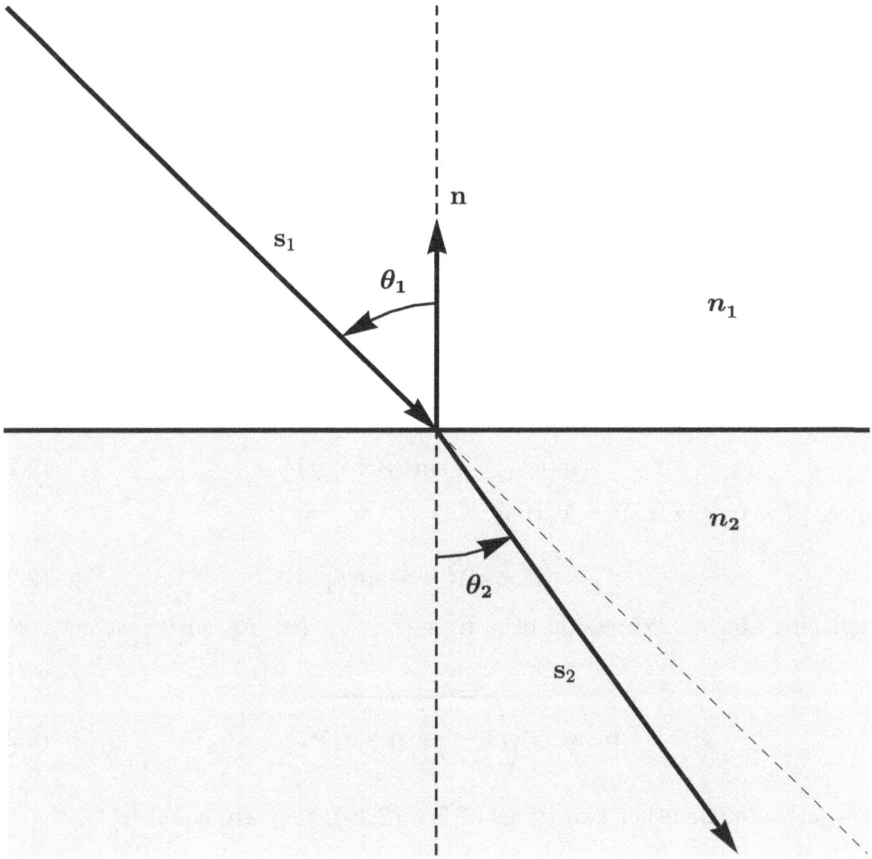

FIGURE 2.3
Diagram of the refraction of light of the vectorial Snell law.

the refracted angle relative to $-\mathbf{n}$. With these definitions, the refracted vector can be expressed as follows, keeping the formulation precise yet approachable

$$\mathbf{s}_2 = \frac{n_1}{n_2}\mathbf{s}_1 + \left(\frac{n_1}{n_2}|\mathbf{n}||\mathbf{s}_1|\cos\theta_1 - |\mathbf{n}||\mathbf{s}_2|\cos\theta_2\right)\mathbf{n}. \tag{2.20}$$

Looking Fig. 2.3, we can see the vertical component of the refract ray $\mathbf{s}_2$ is given by

$$\mathbf{n}\cdot\mathbf{s}_2 = |\mathbf{n}||\mathbf{s}_2|\cos\theta_2, \tag{2.21}$$

Since they are unit vectors $|\mathbf{n}| = |\mathbf{s}_2| \equiv 1$,

$$\mathbf{n}\cdot\mathbf{s}_2 = \cos\theta_2, \tag{2.22}$$

squaring the last expression,

$$(\mathbf{n} \cdot \mathbf{s}_2)^2 = \cos\theta_2^2 = 1 - \sin\theta_2^2 = 1 - \frac{n_1^2}{n_2^2}\sin\theta_1^2. \tag{2.23}$$

computing the cross-product,

$$\mathbf{n} \times \mathbf{s}_1 = |\mathbf{n}||\mathbf{s}_1|\sin\theta_1(\hat{\mathbf{x}} \times \hat{\mathbf{y}}), \tag{2.24}$$

also we have, $|\mathbf{n}| = |\mathbf{s}_1| = 1$.

$$\mathbf{n} \times \mathbf{s}_1 = \sin\theta_1(\hat{\mathbf{x}} \times \hat{\mathbf{y}}), \tag{2.25}$$

squaring,

$$(\mathbf{n} \times \mathbf{s}_1)^2 = \sin\theta_1^2(\hat{\mathbf{x}} \times \hat{\mathbf{y}})^2. \tag{2.26}$$

Since, $(\hat{\mathbf{x}} \times \hat{\mathbf{y}}) = \hat{\mathbf{z}}$, $(\hat{\mathbf{z}})^2 = 1$, thus,

$$(\mathbf{n} \times \mathbf{s}_1)^2 = -\sin\theta_1^2. \tag{2.27}$$

Replacing above's expression in $(-\mathbf{n} \cdot \mathbf{s}_2)^2 = 1 - (n_1^2/n_2^2)\sin\theta_1^2$, we get,

$$\mathbf{n} \cdot \mathbf{s}_2 = \sqrt{1 - \frac{n_1^2}{n_2^2}(\mathbf{n} \times \mathbf{s}_1)^2}. \tag{2.28}$$

Now let's see the other two terms of Eq. (2.20), they are equal to

$$\frac{n_1}{n_2}[\mathbf{s}_1 + |\mathbf{n}||\mathbf{s}_1|\cos\theta_1\mathbf{n}] = \frac{n_1}{n_2}[\mathbf{s}_1 - (\mathbf{n} \cdot \mathbf{s}_1)\mathbf{n}]. \tag{2.29}$$

Thus, the Snell law in vector form takes the form of,

$$\mathbf{s}_2 = \frac{n_1}{n_2}[\mathbf{s}_1 - (\mathbf{n} \cdot \mathbf{s}_1)\mathbf{n}] - \mathbf{n}\sqrt{1 - \frac{n_1^2}{n_2^2}(\mathbf{n} \times \mathbf{s}_1)^2} \quad \text{for } \mathbf{s}_2, \mathbf{s}_1, \mathbf{n} \in \mathbb{R}^3. \tag{2.30}$$

2.1.5 Stigmatism

There is a natural connection between Snell's law and Fermat's principle. Snell's law describes how light bends when passing between media, ensuring it follows the path of least OPL. Fermat's principle expresses essentially the same idea: light always chooses the path of least OPL. In both cases, the geometry of lenses and mirrors matters significantly, since their surfaces determine the orientation of the normal vector $\mathbf{n}$.

The central role of mirrors and lenses is to either converge or disperse light, depending on whether reflection or refraction is involved. Converging systems bring rays together at a focus, while dispersing systems spread rays as if they originated from a single point. If both the object and the image correspond to single points rather than extended regions, the system is referred to as stigmatic [23].

When we look at line segments that satisfy the conditions of conic sections, we notice they intersect at the corresponding foci. If these segments are interpreted as optical paths, it becomes clear why conic mirrors—those shaped according to conic curves—naturally concentrate light at their foci. This perspective is entirely consistent with Fermat's principle, since the paths can be understood as the ones light would take to minimize travel time.

For example, an elliptic mirror is stigmatic for a finite object and a finite image, as shown in Fig. 2.5. A parabolic mirror, on the other hand, reflects rays from an object at infinity to a single finite focus (see Fig. 2.4). Hyperbolic mirrors form a stigmatic pair when the object is real and the image is virtual (Fig. 2.6). In such cases, rays corresponding to virtual points are drawn as dashed lines. A virtual ray represents the extension of a physical ray in the forward direction of propagation. With hyperbolic mirrors, the law of reflection ensures that these extensions intersect at the virtual image point.

Whether an image or an object is considered real or virtual ultimately depends on the chosen global coordinate system. In the following section, we will examine how the definition of the coordinate system influences the classification of images and objects as real or virtual.

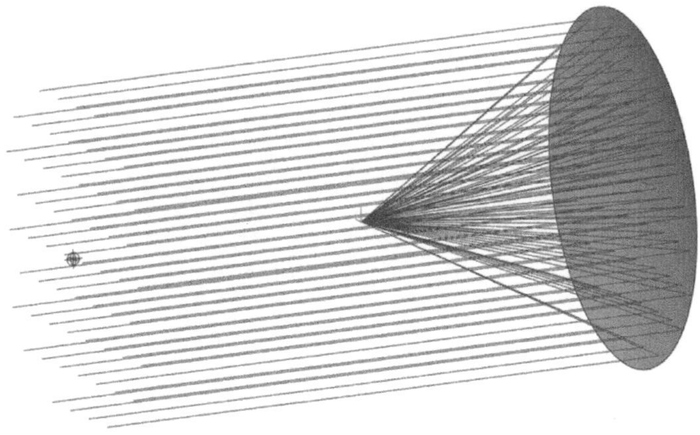

FIGURE 2.4
A parabolic mirror is stigmatic for an infinite object and a finite image.

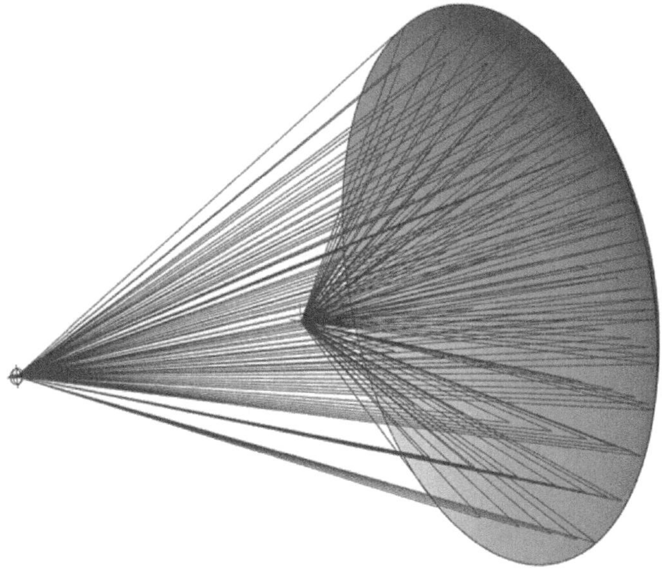

FIGURE 2.5
An elliptic mirror is stigmatic for a finite object and a finite image.

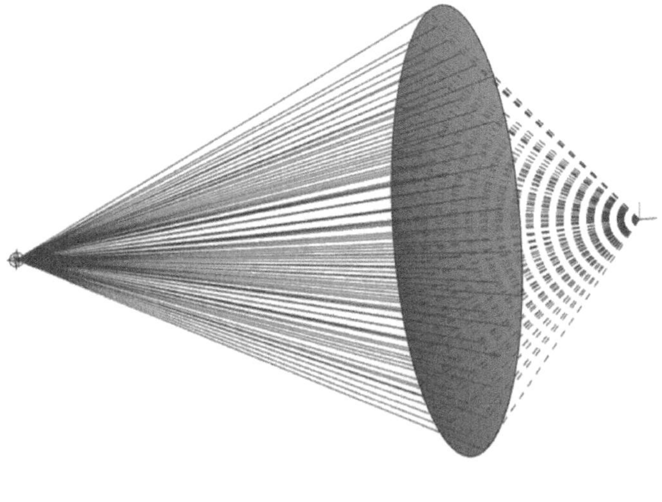

FIGURE 2.6
A hyperbolic mirror is stigmatic for a finite object and a finite image. But in this case, the image is virtual.

Elliptic, parabolic, and hyperbolic mirrors represent closed-form solutions that arise naturally from Fermat's principle. By closed-form, we mean that their shapes can be expressed in finite mathematical terms. A detailed derivation of these results, based on Fermat's principle, is provided in *Stigmatic Optics* [23]. Since that level of detail lies beyond the scope of this book, we present these results here primarily to establish the definitions of object and image points.

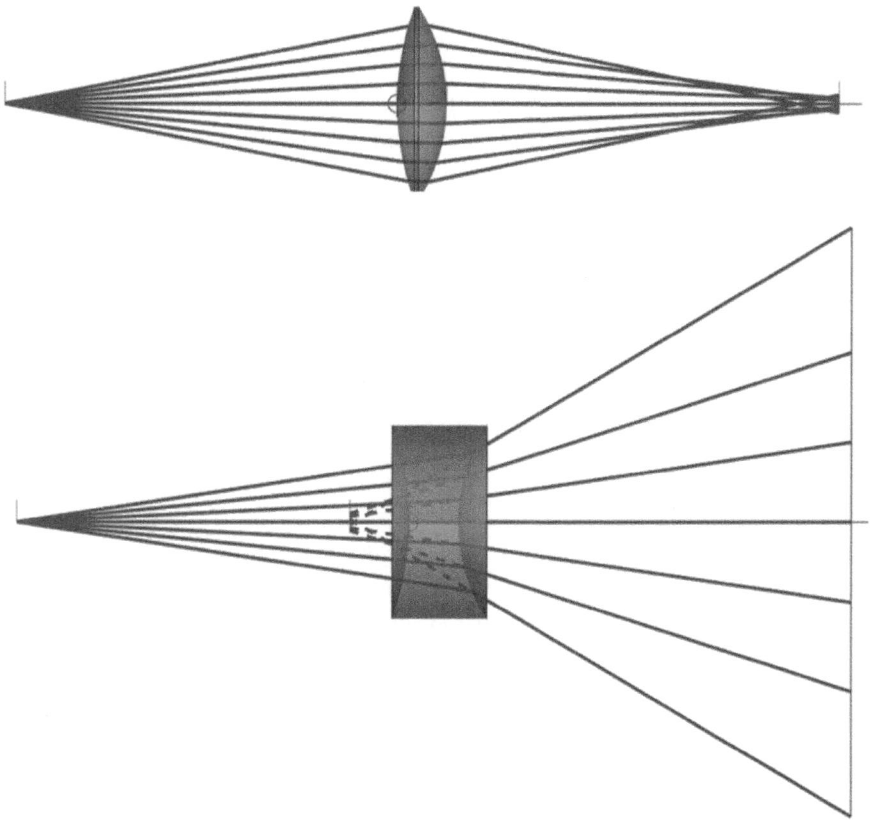

FIGURE 2.7
A positive lens converges light rays passing through it, effectively focusing them. Such lenses possess a positive focal length. Conversely, a negative lens diverges the light rays, resulting in dispersion. Negative lenses are characterized by a negative focal length. *Top:* a positive lens. *Bottom:* a negative lens.

2.2 Concepts in optical design

This section outlines essential concepts used by lens designers to understand and define optical systems. As these ideas are interconnected, revisiting this section may clarify their relationships, as a purely linear explanation can be challenging.

2.2.1 Notation and style convention

Most optical systems have some symmetry, which helps us understand their properties. In radially symmetric systems, the optical axis is an imaginary line through the center, key for devices like camera lenses or microscopes.

We use right-handed coordinate systems, with the z-axis along the optical axis pointing right, the x-axis toward the page, and the y-axis upward. Lenses often have spherical surfaces. A positive radius of curvature means the surface bulges right (toward $+z$), while a negative radius curves left (toward -z). This convention is standard in optical design software.

Lenses are either positive or negative. Positive lenses focus rays and have a positive focal length, while negative lenses diverge rays with a negative focal length. For negative lenses, rays diverge and don't cross the optical axis, but tracing them backward reveals a focal point. Images, whether real or virtual, are shown as regions, not points, since lenses aren't perfectly stigmatic.

2.2.2 Cardinal points

In paraxial geometrical optics, cardinal points are three pairs of points along the optical axis of a rotationally symmetric focal system: focal points, principal points, and nodal points, each with two corresponding points [25].

Focal points are where light converges in a stigmatic system, forming an infinitesimally small point. In non-stigmatic systems with aberrations, the focus becomes a region (that we are going to study later in this chapter [25]).

The principal point is where an incident ray's propagation intersects its corresponding emergent ray on the optical axis, connecting object and image space. Graphically, it's shown in Fig. 2.8 as the intersection of input/output rays, typically for marginal rays. The marginal ray is the ray at the margin of the bundle of rays, later, ray details are covered later in this chapter [25]. The principal point acts as the optical system's center, with the principal plane perpendicular to the optical axis, as depicted in Fig. 2.8 for a three-lens system.

An optical system has two principal points: one for the standard configuration (object space to image space) and one for the reversed setup, though the latter is rarely used.

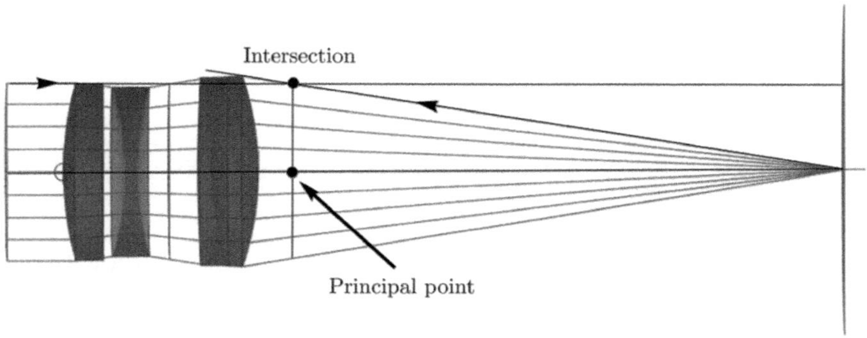

FIGURE 2.8
The principal point is where the incident ray's propagation intersects its corresponding emergent ray on the optical axis, marking the meeting point of ray propagation from object space to image space.

Nodal points have a unique property: a ray aimed at one nodal point appears to emerge from the other, maintaining the same angle relative to the optical axis. They relate to angles as principal planes do to transverse distances. When the medium is the same on both sides (e.g., vacuum), nodal and principal points align. Fig. 2.9 illustrates nodal points in a three-lens system, where a ray appears un-refracted, with identical entrance and exit angles.

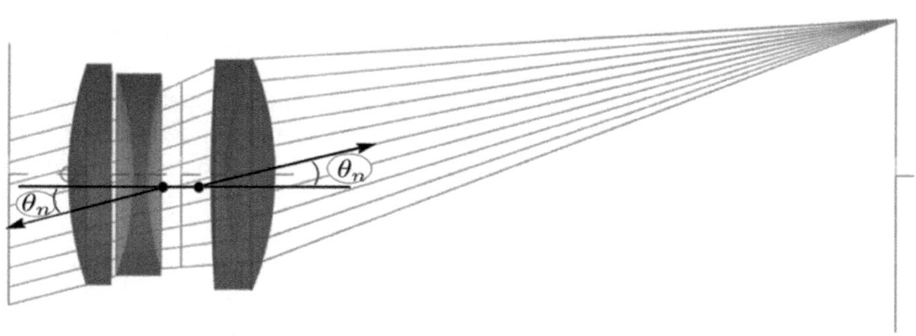

FIGURE 2.9
The front and rear nodal points of a lens have a unique trait: a ray directed at one nodal point is refracted to appear as if it originated from the other, preserving the same angle relative to the optical axis.

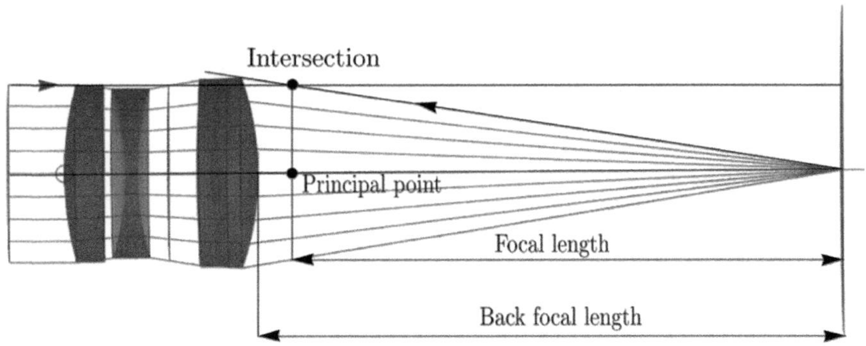

FIGURE 2.10
The effective focal length, commonly referred to as the focal length, is the distance from the rear principal point to the rear focal point of a lens. The back focal length is the distance from the lens's rear vertex to the rear focal point. The principal point lies on the optical axis, where the incident ray's propagation intersects its corresponding emergent ray.

2.2.3 Focal length and back focal length

Focal length, also called effective focal length f_{effl} or just f. From now on, the symbol f will be called the focus, and it will not be related to a mathematical function like in Chapter 1. The focal length is a key lens parameter, determining how strongly an optical system converges or diverges light, inversely related to its optical power [26]. Positive focal lengths indicate convergence, while negative ones signify divergence. Shorter focal lengths bend rays more sharply, focusing or diverging them faster.

The effective focal length is the distance from the rear principal point to the rear focal point, while the back focal length is from the lens's rear vertex to the rear focal point [26]. Fig. 2.10 illustrates these distances in a three-lens system, matching Fig. 2.8.

2.2.4 Aperture stop and field stop

An aperture is an opening that allows light to pass through an optical system. The aperture stop, a critical component, restricts the angular range of light received by the imaging system, as observed from the object plane's central point where the system is focused [25]. It may be a physical element, like a lens with a finite size, or an intentional optical aperture, such as an iris. In systems with multiple apertures, the aperture stop is the one imposing the strictest limit on the angular range, while other apertures, known as field stops, block rays from specific fields.

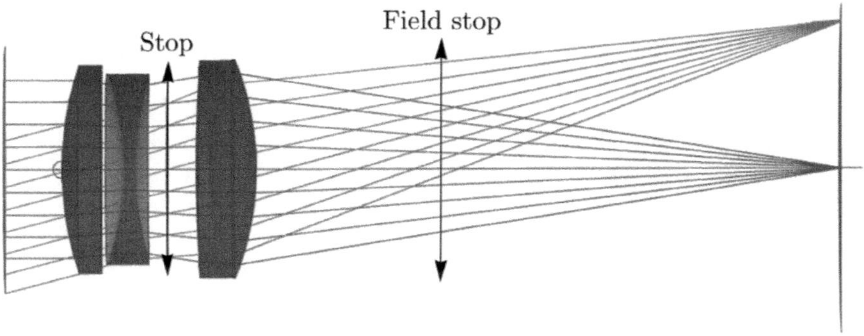

FIGURE 2.11
n a three-lens optical system, the aperture stop is positioned between the second and third elements as a standalone component, distinct from the lens surfaces. A field stop, located on the right side of the system, blocks rays from off-axis fields while allowing on-axis rays to pass unimpeded.

Fig. 2.11 illustrates the aperture stop in a three-lens system, positioned between the second and third elements as a standalone component, not a lens surface. The figure also shows a field stop on the right, affecting only off-axis fields while allowing on-axis rays to pass unblocked.

The number of field stops varies across optical systems, but each system requires at least one stop. This stop is vital for defining key optical properties, which we will explore in later sections of this chapter.

2.2.5 Entrance pupil

The entrance pupil of an optical system is formally defined as the image of the aperture stop as observed from the object space, giving the aperture stop its apparent form in that perspective [24]. This critical element shapes how light enters the system and influences its overall performance.

To determine the entrance pupil's position, one must trace the ray from the incident field with the largest angle that, after undergoing refraction or reflection, passes through the center of the aperture stop. This ray is then extended backward, prior to any refraction or reflection, until it intersects the optical axis, marking the entrance pupil's location. The entrance pupil's diameter is established by tracing the marginal rays of the on-axis field to the plane of the entrance pupil, which lies perpendicular to the optical axis.

Fig. 2.12 illustrates this in a system with three singlet lenses, featuring two fields: an on-axis field and an off-axis field, the latter having the largest incident angle. To maintain clarity, additional off-axis fields are omitted. From the off-axis field, we select the ray that, post-refraction or reflection, passes through the aperture stop's center. This ray is extended backward to the

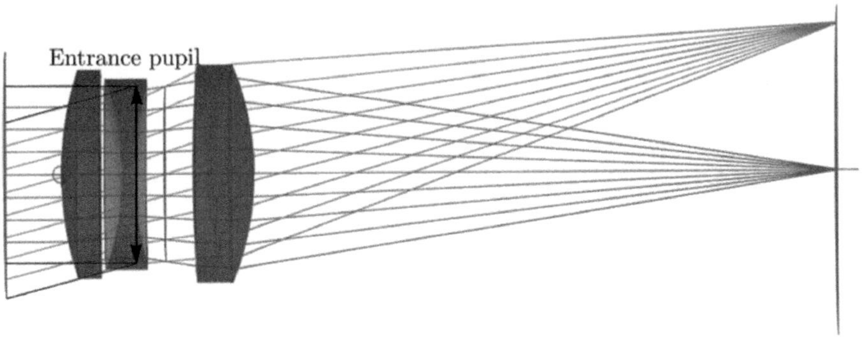

FIGURE 2.12
The entrance pupil of an optical system is defined as the image of the aperture stop in the object space. This means that from object space, the aperture stop has the appearance of the entrance pupil. From the off-axis field, we select the ray that passes (after suffering refraction/reflections) through the center of the aperture. Before this ray suffers refractions/reflections, we extend it until it reaches the optical axis.

optical axis to define the entrance pupil's position. The diameter is determined by the on-axis field's marginal rays intersecting the entrance pupil plane.

2.2.6 Exit pupil

The exit pupil of an optical system is formally defined as the image of the aperture stop as observed from the image space, effectively representing how the aperture stop appears from that perspective [25]. This key feature plays a significant role in determining how light exits the system, influencing the final image quality and system performance.

To locate the exit pupil, the process mirrors that of finding the entrance pupil but is applied in the image space. We begin by identifying the ray from the outcome field with the largest angle that passes through the center of the aperture stop. This ray is then traced forward in the image space until it intersects the optical axis, marking the position of the exit pupil. The diameter of the exit pupil is determined by the extent of the marginal rays from the on-axis outcome field, which define the boundaries of the exit pupil on a plane perpendicular to the optical axis.

Fig. 2.13 provides a clear example of this in a three-lens optical system. We first select the ray from the field with the largest angle that passes through the aperture stop's center. This ray is extended in the image space until it meets the optical axis, establishing the exit pupil's location. A plane perpendicular to the optical axis, referred to as the exit pupil plane, can be visualized at this intersection (though not depicted in the figure). The exit pupil's extent

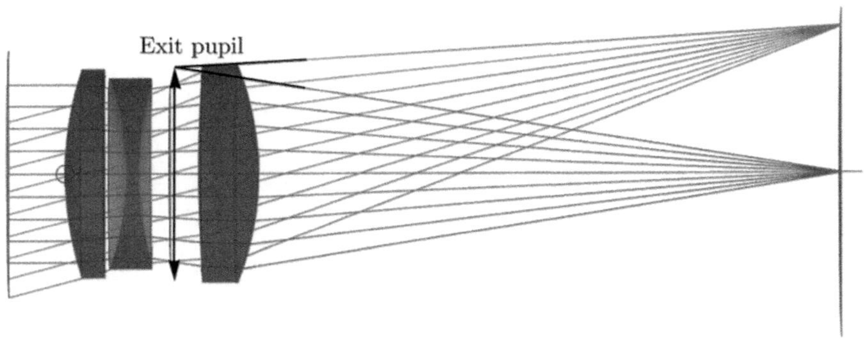

FIGURE 2.13
The exit pupil of an optical system is the image of the aperture stop as observed in the image space. To locate it, the ray from the outcome field with the largest angle, passing through the aperture stop's center, is traced until it intersects the optical axis. The exit pupil's diameter is determined by the extent of the marginal rays from the on-axis outcome field at this plane, perpendicular to the optical axis.

is defined by the marginal rays of the on-axis outcome field intersecting this plane, encapsulating the full scope of the exit pupil.

Chief rays pass through the center of the aperture, while marginal rays travel along its edges. Various types of rays, including their roles and properties, will be further explored in the following subsections.

2.2.7 Raytracing

Ray tracing is the cornerstone of optical design and analysis, rooted in geometric optics, which simplifies light propagation to straight rays, bypassing its wave nature. Though an approximation of wave optics, it remains highly accurate for many applications. By tracing select rays from an object through a lens, we can determine the image's position, size, and orientation [32, 33, 34].

We've referenced rays extensively without fully defining them, as rays and pupils are interdependent concepts. One could define rays first, but this would require mentioning stops and pupils prematurely. Instead, we now focus on rays, noting that their proper definition requires specifying the planes in which they lie.

2.2.8 Tangential and sagittal plane

In optical systems, rays are defined in two key planes that reveal system properties: the sagittal and tangential planes. With the optical axis aligned along the $+z$ direction, the tangential plane corresponds to the y-z plane, while the sagittal plane, perpendicular to it, aligns with the x-z plane [24].

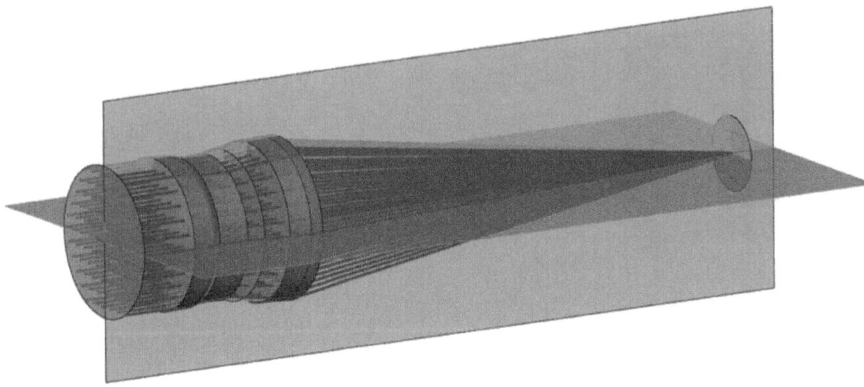

FIGURE 2.14
The tangential plane y/z is in light-gray and the sagittal plane is in dark-gray x/z.

Consider a three-lens optical system, as shown in Fig. 2.14, where the tangential plane is depicted in light-gray and the sagittal plane in dark-gray. These planes are relevant not only to this system but to all on-axis optical systems [24].

2.2.9 Meridional rays

The tangential plane, also known as the meridional plane, is critical in rotationally symmetric optical systems. The lens axis forms this plane and an object point offset from the axis, corresponding to the y-z plane in our coordinate system. When the object point lies on the axis, all rays are meridional [24].

Meridional rays are those whose trajectories remain within the meridional plane. Fig. 2.15 illustrates these rays in light-gray hue for a three-lens optical system. Here, we set the meridional rays to be less dense, with fewer than the sagittal rays.

2.2.10 Sagittal rays

In rotationally symmetric optical systems, the sagittal plane, defined as the x-z plane in our coordinate system, contains rays from an object point offset from the lens axis. When the object point is on-axis, all rays are meridional, but sagittal rays specifically follow trajectories within the sagittal plane [24].

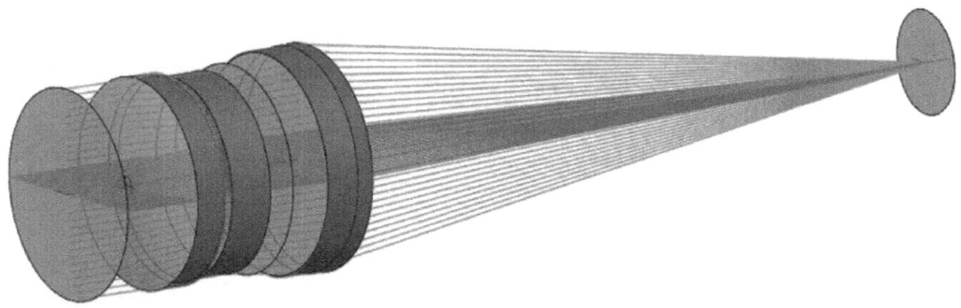

FIGURE 2.15
The meridional rays (light-gray) are those whose trajectories lie entirely in the meridional plane, i.e., the y/z plane. The sagittal rays (dark-gray) are those whose trajectories lie entirely in the sagittal plane, i.e., the x/z plane. Here, we set the meridional rays to be less dense, with fewer than the sagittal rays.

Fig. 2.15 illustrates the sagittal rays in a three-singlet optical system, depicted in dark-gray. In this configuration, the meridional rays are set to a lower density than the sagittal rays.

2.2.10.1 Chief ray

The so-called chief ray refers to the particular trajectory followed by a light ray that originates from a point located away from the optical axis and subsequently passes through the center of the aperture stop. This ray plays an important role in describing how off-axis object points are imaged, since it effectively defines the direction of the field relative to the system's stop. To illustrate this concept, Fig. 2.16 presents the chief ray (highlighted in black)

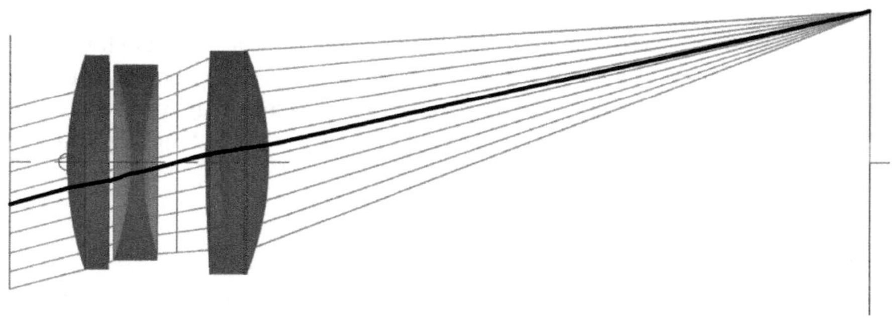

FIGURE 2.16
Chief ray (in black) of an off-axis field (in gray) of the optical system with three elements.

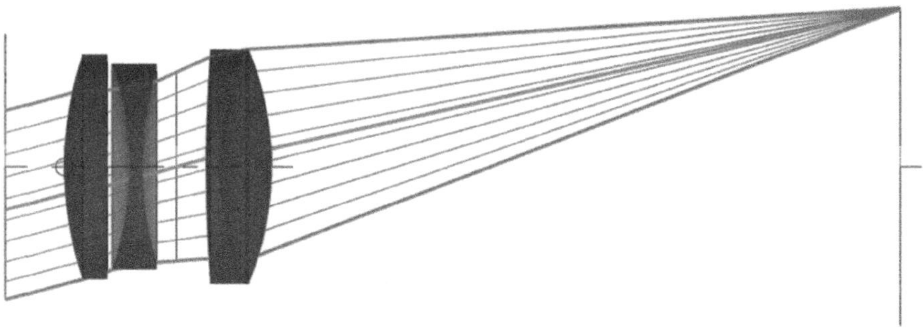

FIGURE 2.17
In thicker dark-gray, the marginal rays (edge rays) and the chief ray (center ray).

corresponding to an off-axis field within an optical system composed of three elements [25]

2.2.11 Marginal ray

The marginal ray of a given field is defined as the ray that lies along the outer boundary of the bundle of rays associated with that field. In other words, it traces the edge of the cone of rays extending from an object point as they pass through the optical system. This ray, together with the chief ray, provides a convenient way to characterize the geometry of image formation. Fig. 2.17 illustrates this idea, showing in thicker dark-gray both the marginal rays and the chief ray for a three-element optical system [25].

The combination of the marginal rays and the chief ray plays a fundamental role in determining several key optical properties of an imaging system. For instance, these rays are indispensable for establishing the effective focal length of the system, as well as for locating the positions of both the entrance pupil and the exit pupil. In this sense, it should now be clear why, in the preceding subsections, we relied on the marginal and chief rays to introduce and define these essential concepts [25].

2.2.12 Skew ray

A skew ray is defined as a ray whose trajectory neither runs parallel to nor intersects any of the principal planes of symmetry, namely the tangential (or meridional) plane and the sagittal plane [25]. Such rays, by their very nature, follow more general paths through the system, in contrast to the highly constrained behavior of meridional and sagittal rays. To better visualize this,

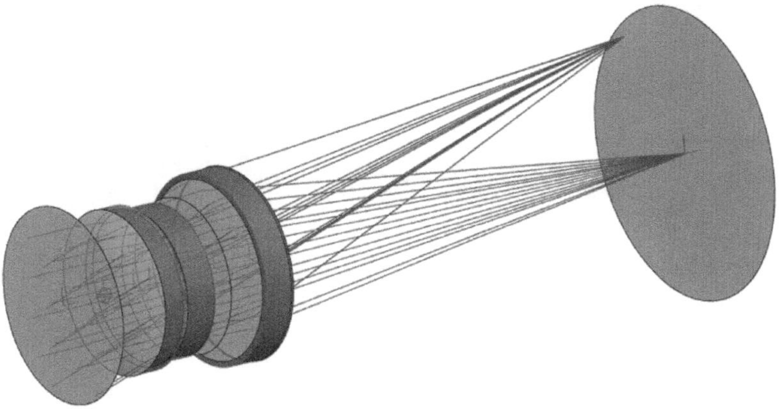

FIGURE 2.18
The rays of the off-axis field are classified as skew rays, as their trajectories do not lie within either the tangential (meridional) plane or the sagittal plane.

Fig. 2.18 illustrates an optical system consisting of three singlet lenses and two fields. The first field is an on-axis field, while the second field is an off-axis field. The rays associated with this off-axis field provide a clear example of skew rays, since their trajectories do not remain confined to either the tangential (meridional) plane or the sagittal plane.

2.2.13 Paraxial ray

A paraxial ray is defined as a ray that propagates at a very small angle relative to the optical axis and, as a result, remains in close proximity to it throughout the optical system. Because of this behavior, such rays can be accurately described using the paraxial, or small-angle, approximation, which greatly simplifies the mathematical treatment of imaging systems. It is important to note, however, that the rays discussed in the preceding sections may or may not fall into the category of paraxial rays, depending on their angle and distance from the axis. As a concrete example, the rays of the on-axis field in Fig. 2.18, being located very near the optical axis, can be regarded as paraxial rays [24].

2.2.14 Rays, object, and aperture space

A final remark that applies to all the rays and reference planes discussed so far is that they are always defined relative to a specific object point and aperture. In other words, the designation of a given plane as *tangential* or *sagittal* depends directly on the position of the object point under consideration. For example, if the object point lies in the x/z plane, then this plane is identified

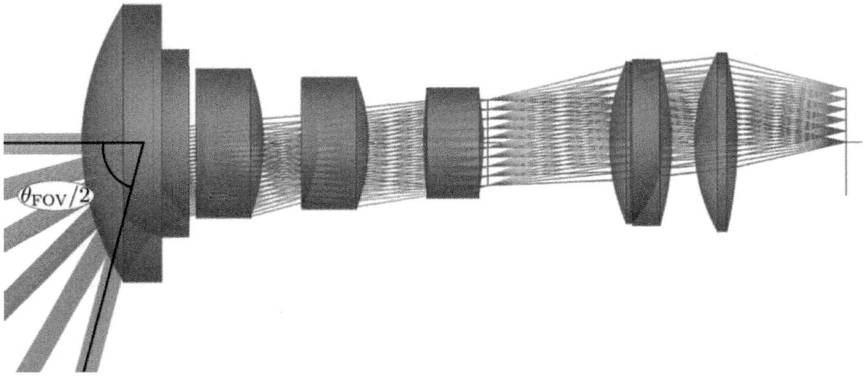

FIGURE 2.19
The half field of view, denoted by $\theta_{\rm FOV}/2$, is defined as the angle formed between the optical axis and the ray that enters the system with the largest incident angle.

as the tangential plane. Likewise, if the object point is located in the y/z plane, the y/z plane becomes the tangential plane instead. For clarity and consistency, it is generally advisable to select a coordinate system in which the chosen object point lies either in the x/z or the y/z plane.

2.2.15 Field of view, objects, and images sizes

The entrance and exit pupils of an optical system are defined by the rays that enter the system at the largest incident angles, as these rays effectively determine the system's limiting aperture. Closely associated with this concept is the field of view, which quantifies the extent of object space that can be projected onto the image plane. In other words, it identifies the portion of the object that the optical system is capable of capturing. Often, the half-field of view is used, representing half of the total angular span of the observable region. This concept is illustrated in Fig. 2.19, where the angle $\theta_{\rm FOV}/2$ denotes the angle between the optical axis and the ray with the largest incident angle [24].

In lens system design, multiple fields are considered, typically a finite set chosen to span the full range of incident angles within the system's field of view.

The image height can then be calculated using the following expression,

$$Im = f_{\rm efl} \tan(\theta_{\rm FOV}/2) \tag{2.31}$$

where Im denotes the image height, $f_{\rm efl}$ the effective focal length, and $\theta_{\rm FOV}/2$ the half field of view. This expression is derived from the paraxial approximation. Note that the actual image height may be influenced by distortion, an optical aberration that will be addressed later in this chapter.

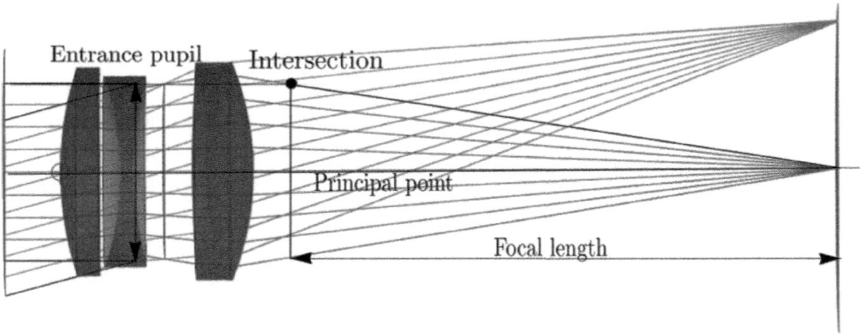

FIGURE 2.20
The f-number is the ratio of the effective focal length and entrance pupil diameter.

2.2.16 F-number and numerical aperture

The f-number, $f\#$, of a lens determines both the amount of light entering the system and the lens's ability to render contrast at a given resolution [24]. It is defined as the ratio of the effective focal length, f_{efl}, to the diameter of the entrance pupil, D

$$f\# = \frac{f_{\text{efl}}}{D} \tag{2.32}$$

In most lenses, the f-number is adjusted by rotating the iris ring, which opens or closes the diaphragm to modify the entrance pupil diameter. Fig. 2.20 illustrates the parameters used to calculate the f-number in an optical system.

The numerical aperture NA is mathematically defined as

$$\text{NA} = \frac{D}{2f_{\text{efl}}} = \frac{1}{2f\#} \tag{2.33}$$

The numerical aperture, denoted as NA, provides a quantitative measure of a lens's light-gathering ability. It can be understood as representing the ratio between the size of the entrance pupil and the lens's focal length. In essence, a larger entrance pupil or a shorter focal length results in a higher numerical aperture, indicating that the lens can capture more light and potentially achieve higher resolution in the image.[24].

2.3 Image analysis

Image formation is the process by which a three-dimensional object is projected onto a two-dimensional surface, creating a representation of the object that can be captured by a sensor or perceived by the human eye. In an ideal optical system, this process would concentrate the light from a single object point into a corresponding single point in the image, known as a stigmatic image point. For a given field, the rays emanating from the object should converge precisely at this point, and for multiple fields, each set of rays should converge at their respective image points. The collection of all these points forms the complete image of the object [25].

In real-world optical systems, however, imperfections prevent rays from converging perfectly at a single point. Instead, they focus over a small region, producing a blurred or distorted image. These deviations from the ideal image are referred to as optical aberrations. They are generally classified into two primary categories: monochromatic aberrations, which occur independently of wavelength, and chromatic aberrations, which result from wavelength-dependent variations in refractive index [33, 34, 32]. In this chapter, we will examine both types in detail, exploring their origins, effects, and methods of mitigation within optical systems.

2.4 Monochromatic aberrations

Monochromatic aberrations are a type of image imperfection that arise independently of the wavelength of light. These aberrations are intrinsically linked to the aperture size and field of an optical system: larger apertures or wider fields typically exacerbate the effect, since rays entering the system at higher angles experience greater nonlinearity according to Snell's law [22, 23]. In such cases, rays deviate significantly from the paraxial region, and the linear approximations commonly used in paraxial optics no longer hold.

There are two primary approaches for quantifying monochromatic aberrations, corresponding to the ray and wave perspectives. These approaches are fundamentally connected, as rays are always perpendicular to wavefronts, providing a bridge between geometrical and physical optics [7, 35, 36].

The first approach, based on ray optics, relies largely on geometrical methods. In this framework, rays are traced through the optical system, and the deviation of their actual intersection points from the ideal reference point is used to quantify aberrations. Within this perspective, monochromatic aberrations are commonly classified as either longitudinal or transverse [37].

Fig. 2.21 illustrates longitudinal aberration. In this figure, the black lines represent the reference rays, which indicate the ideal path that rays should follow, while the light-gray lines depict the actual rays. Longitudinal aberration is measured as the axial distance along the optical axis between the point where a ray intersects and the corresponding reference point. In special cases, such as when imaging a collimated beam at infinity, longitudinal aberration

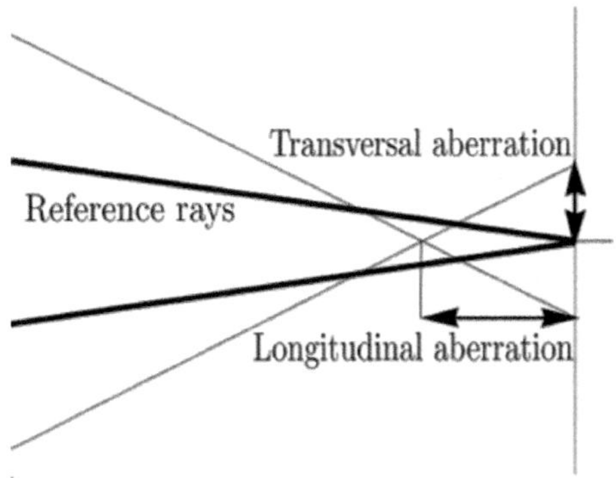

FIGURE 2.21
Longitudinal aberration and Transverse aberration.

can also be quantified by the angular deviation between a ray and the chief ray [38].

Transverse aberrations, on the other hand, are measured in the plane perpendicular to the optical axis. They correspond to the lateral displacement between the actual point where a ray intersects the image plane and the ideal reference point, as shown in Fig. 2.21 [39]. Both longitudinal and transverse measurements provide complementary information, together offering a comprehensive description of how monochromatic aberrations affect image formation.

The wave-optics perspective evaluates aberrations by comparing the actual wavefront emerging from the optical system with an ideal reference spherical wavefront, corresponding to a stigmatic system. If the rays at the exit pupil do not coincide with the normals to this reference wavefront, the system exhibits aberrations. Because rays are always perpendicular to wavefronts, these deviations can be quantified in terms of the OPL of each ray relative to the reference. In practice, both the ray and wave approaches are employed to analyze optical systems, providing complementary information. The system can then be characterized through a variety of measurements, graphical representations, and analytical tools widely available in the literature [40].

2.4.1 Defocus and tilt

Defocus occurs when the image plane is not located at the precise position where an optical system forms a sharp image—that is, when it is either closer to or farther from the point where the rays ideally converge [26, 39]. As a result, a single object point is imaged not as a sharp point but over a small region, causing a blurred or soft image. This issue can be corrected by adjusting the position of the image plane or the optical elements of the system to bring

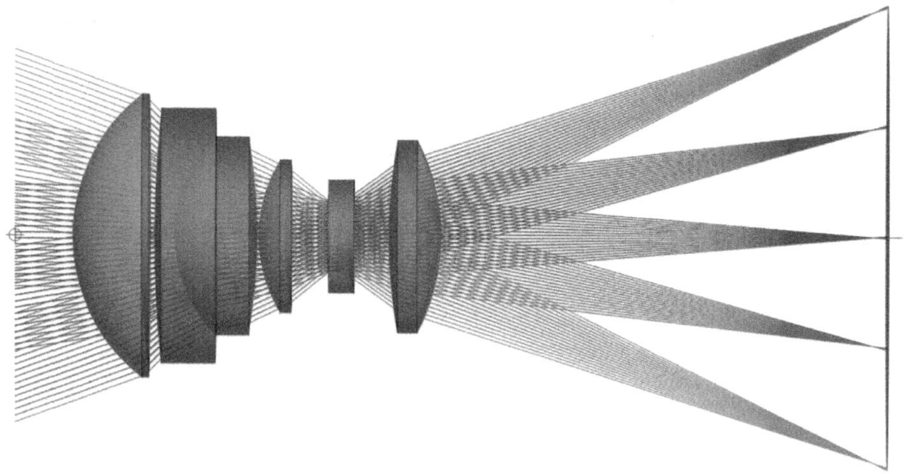

FIGURE 2.22
An optical system with defocus. The image is not where the fields converge it is a little bit away from it, where the fields start to diverge again.

it into proper alignment with the convergence point of the rays. Fig. 2.22 illustrates this situation: the image plane does not coincide with the region where the rays from the fields converge, demonstrating that the system is affected by defocus [40].

As illustrated in Fig. 2.22, the image plane in the system exhibits defocus. An optical system is said to suffer from defocus when the image plane is not positioned at the circle of least confusion. The circle of least confusion is defined as the smallest circle that encompasses all the rays emerging from the system. Its size and location can be determined by identifying the point along the optical axis where the caustic waist is minimal. A caustic represents the envelope formed by light rays that have been reflected or refracted by an optical system, or the projection of this envelope onto a surface. It is a curve or surface to which each ray is tangent, effectively outlining the region of concentrated light [41, 42].

Another common aberration is tilt, which occurs when the wavefront produced by the optical system is rotated in the tangential orientation. This rotation introduces an angular mismatch between the aberrated wavefront and the ideal reference spherical wavefront. Fig. 2.23 shows an optical system exhibiting tilt, where the system is misaligned relative to the image plane, resulting in a skewed image formation.

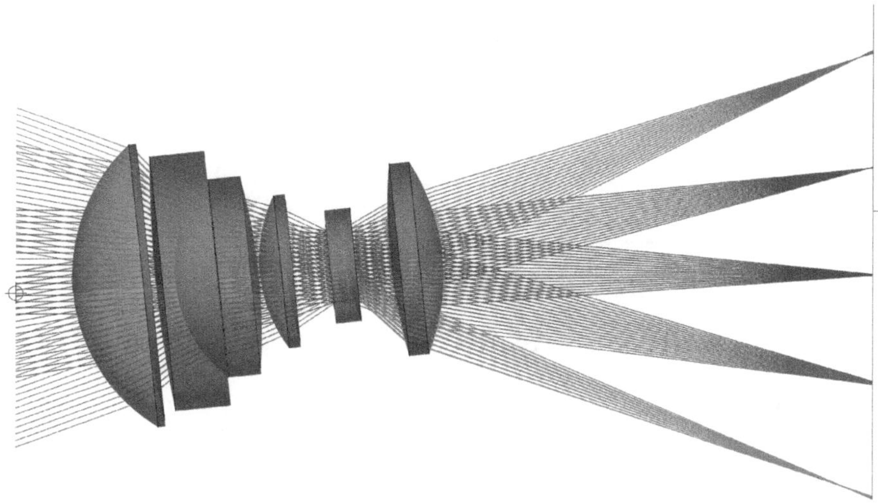

FIGURE 2.23
Optical system suffering from tilt aberration.

2.4.2 Spherical aberration

The radius of the pupil plays a crucial role in determining spherical aberration, which is the primary on-axis monochromatic aberration. This aberration arises because the optical power of a spherical surface varies with the angle at which rays strike it relative to the local surface normal, and this effect depends on the radial distance of the rays across the pupil. Consequently, rays passing through different regions of the pupil converge at different points along the optical axis, rather than a single focal point. In practical terms, larger pupil diameters exacerbate spherical aberration, which can reduce image sharpness and contrast. Therefore, careful control of the pupil size is essential in lens design to balance light-gathering capability with image quality [43].

Figures 2.24 and 2.25 show the ray tracing of a single lens, where the rays initially appear to converge at a single point. However, a closer look, as illustrated in Fig. 2.26, reveals that the angle at which each ray intersects the lens surface depends on its height within the aperture. In the central region, this variation is small, and the paraxial approximation remains valid, allowing rays passing through the center to converge closely along the optical axis.

As the rays move toward the edge of the lens, the angle between the rays and the surface normal increases substantially, causing the paraxial approximation to break down. According to Snell's law, these marginal rays bend more strongly than the paraxial rays, resulting in different focal points along the axis: the paraxial rays focus farther from the lens, while the marginal rays focus closer. This difference in convergence positions leads to a blurred image

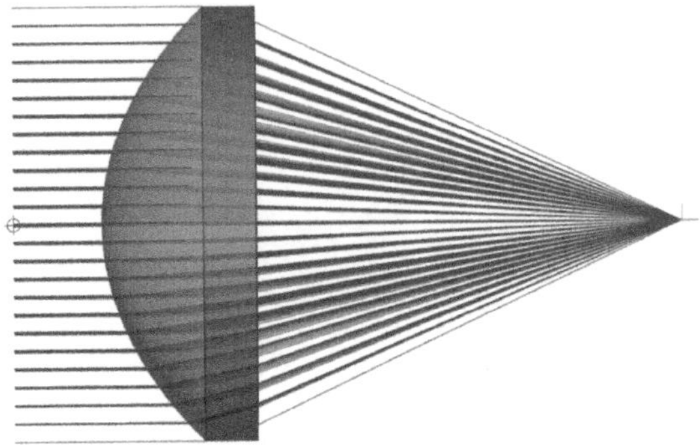

FIGURE 2.24
Raytracing of a singlet lens with a single on-axis field.

rather than a sharp point, which is the hallmark of spherical aberration. The effect is clearly depicted in Fig. 2.26, providing a visual explanation of why lens edges contribute more to aberration than the central region.

To introduce some familiar terminology, an optical system is described as 'fast' when it has a small f-number, meaning it gathers more light. However, fast systems are generally more susceptible to spherical aberration. Likewise, lenses with a smaller radius of curvature—implying a more strongly curved surface—also tend to exhibit greater spherical aberration. This occurs because rays passing through different regions of the aperture encounter increasingly different angles relative to the surface normal, which causes them to focus at different points along the optical axis. As a result, faster systems or highly

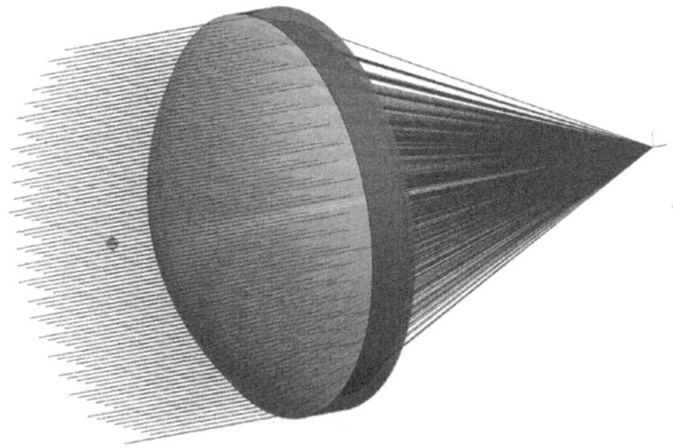

FIGURE 2.25
Side view of a singlet lens with a single on-axis field.

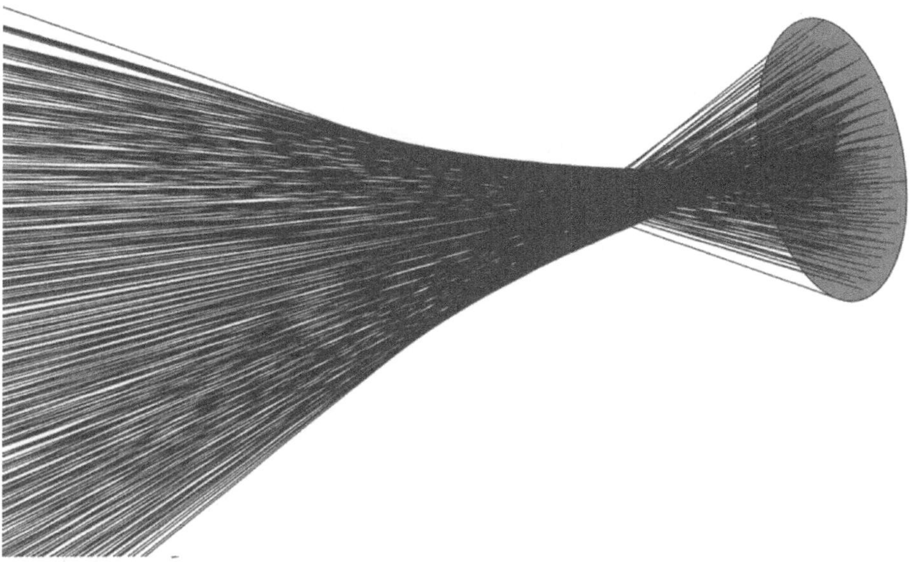

FIGURE 2.26
Side view of the zoom-up of the image plane. The rays do not converge in a single point due to the spherical aberration.

curved lenses often require careful design to balance brightness with image sharpness

Because rays from different parts of the aperture focus at different points along the optical axis, choosing the correct image plane is critical for achieving the sharpest possible image. The optimal plane is a compromise between the focal points of paraxial rays (near the center of the aperture) and marginal rays (near the edge), and is typically located where the cross-sectional area of the beam is smallest. This location is known as the circle of least confusion, as it represents the point where the image blur is minimized. Fig. 2.27 provides a graphical representation of this concept.

The collection of ray heights along the optical axis forms the caustic curve, which represents the envelope of the beam as it converges and diverges. In Fig. 2.27, this curve is shown in black, illustrating how rays from different regions of the aperture intersect the axis at varying distances, and why the circle of least confusion is chosen to optimize image sharpness [41, 42].

2.4.3 Coma

When an optical element is illuminated by an off-axis field, the symmetry of its surface is effectively broken. As a result, the angle at which rays intersect the surface varies across its annular zones. This causes rays from different regions of the aperture to focus at different lateral positions in the image plane,

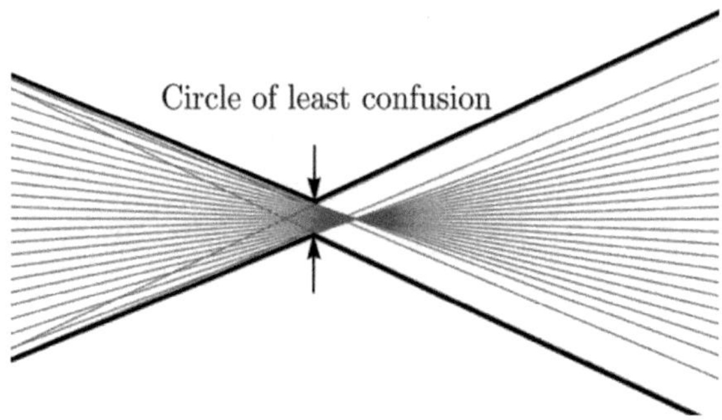

FIGURE 2.27
Circle of least confusion of an on-axis field.

creating a characteristic comet-like blur known as coma. Similar to spherical aberration, marginal rays focus differently than paraxial rays. However, unlike spherical aberration, coma depends not only on the aperture size but also on the angle of the incoming rays. Therefore, larger apertures or higher field angles produce more pronounced coma.

Visually, coma resembles a tiny comet: the image of a point source appears stretched with a bright head and a trailing tail pointing away from the optical axis. From an intuitive standpoint, coma can be seen as the off-axis counterpart to spherical aberration—while spherical aberration affects on-axis fields, coma emerges for off-axis fields. This analogy helps to relate the two phenomena, though it is more of a personal interpretation than a formal definition [26, 39].

Fig. 2.28 shows a lens diagram with multiple fields: the on-axis field and two off-axis fields. By focusing on the image plane and examining only the marginal off-axis field, the presence of coma aberration becomes apparent, as illustrated in Fig. 2.29.

2.4.4 Astigmatism

Off-axis rays are affected not only by the pupil's aperture zone but also by the plane in which they travel. When a ray bundle lies in the tangential plane, the rays strike the lens surface with a tilt relative to the sagittal plane. Because of this, tangential and sagittal rays follow different paths through the lens and come to focus at different points along the optical axis. This difference in focal positions causes the lens to exhibit varying optical power depending on the ray orientation—a phenomenon known as astigmatism [26, 39]. Fig. 2.30 illustrates this effect.

In the figure, rays in both the sagittal and tangential planes are traced to their respective focal points. These rays correspond to the off-axis field of the singlet lens shown at the bottom of Fig. 2.28. The diameter of the circle of

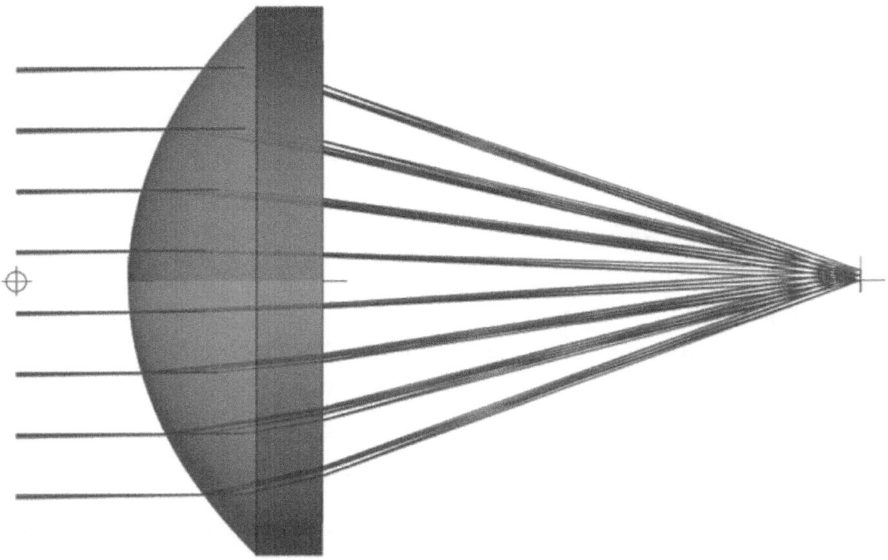

FIGURE 2.28
A singlet lens with one on-axis, and two off-axis fields.

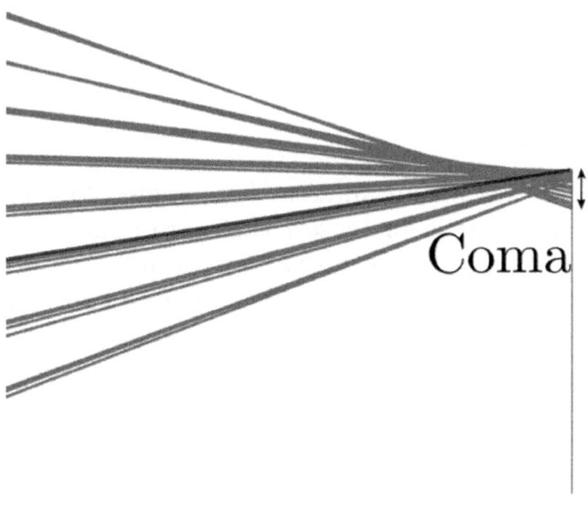

FIGURE 2.29
A zoom on the image plane of the singlet presented in Fig. 2.28. Here, we just show the ray of the marginal off-axis field. The pattern formed by the rays, which does not converge in a single point, is called coma.

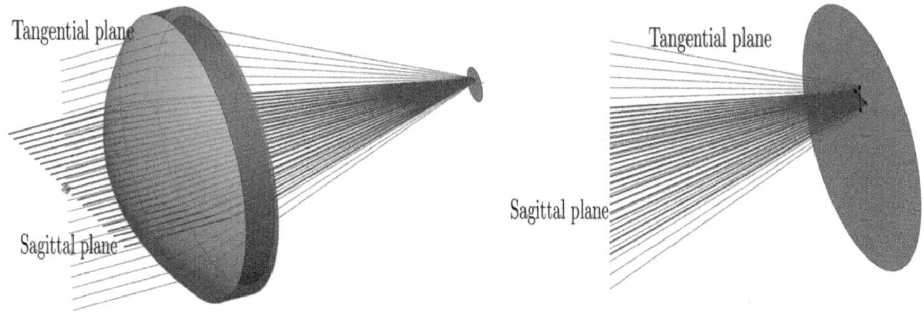

FIGURE 2.30
A singlet lens suffering from astigmatism.

least confusion differs between the two planes, which is a visual signature of astigmatism.

Sagittal rays travel a longer optical path within the lens than tangential rays, effectively making the lens 'appear thicker' in that plane. If the image plane is set at the focal point of the sagittal rays, the image appears as a vertical line: the sagittal rays have converged, while the tangential rays are either still converging or already diverging. This interplay between sagittal and tangential focusing creates the distinctive line-shaped blur characteristic of astigmatism, helping to visualize how the lens focuses light differently in orthogonal planes.

2.4.5 Field curvature

The image formed by rays from different field angles does not lie on a single flat plane. As the field angle increases, the corresponding focal points shift further from the nominal image plane, as shown in Fig. 2.31. This figure provides a zoomed-in view of the three fields traced through the singlet lens from Fig. 2.28. Notice that the circles of least confusion for the three fields do not align on the same plane, indicating that the image naturally forms on a curved surface rather than a flat one. This effect is known as field curvature [26, 39]. Unlike astigmatism, field curvature is independent of the azimuthal angle of the rays.

Field curvature is also referred to as Petzval curvature, describing the curvature of the surface on which the image from an object is sharply focused. Conceptually, one can imagine the ideal image surface as a slightly curved dome, rather than a flat screen, so that all field points are in focus simultaneously. Fig. 2.31 clearly illustrates this curvature in the focused image.

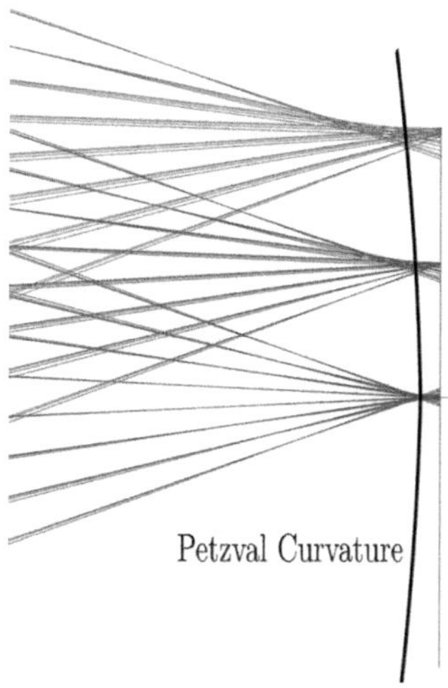

FIGURE 2.31
A zoom-up of the image plane of the singlet of Fig. 2.28. The points where the rays converge of the three fields are not aligned in a single plane. The surface that passes through all these converging points is the field curvature.

2.4.6 Distortion

Object points located at different distances from the optical axis can experience varying transverse magnifications. This means that while individual points may remain sharply focused, the overall geometric shape of the image is altered. Points farther from the paraxial region may still appear sharp, but they no longer align with their ideal image locations, creating a distortion effect.

Fig. 2.32 shows the distortion for the singlet lens from Fig. 2.28. The black rays indicate the ideal paths that would converge on perfect image points, while the colored rays from off-axis fields deviate from these points. The degree of this deviation, often expressed as a percentage, defines the amount of distortion. Conceptually, distortion is similar to what you see in wide-angle photographs, where straight lines appear to bend near the edges of the image [26, 39].

Mathematically speaking, we can measure the distortion with the following equation,

$$\text{Dist} = \frac{\text{AD} - \text{PD}}{\text{PD}} \times 100\% \tag{2.34}$$

where, D is the distortion, the actual distance is AD and the predicted distance is PD. The actual distance AD is where the actual rays lay in the image plane. The predicted distance PD is the predicted by the model. This one depends on the parading presented by the designer. There are several ways to express

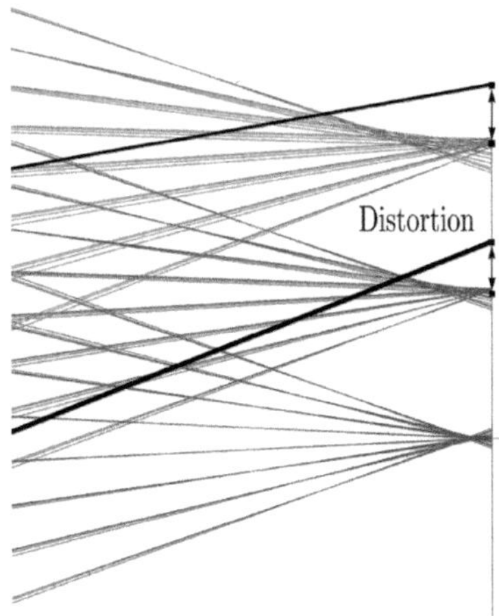

FIGURE 2.32
Distortion of the singlet of Fig. 2.28.

PD, but the two main used are given by

$$PD = f_{\text{efl}} \tan(\theta_{\text{FOV}}/2) \tag{2.35}$$

and

$$PD = f_{\text{efl}}(\theta_{\text{FOV}}/2). \tag{2.36}$$

Here, f_{efl} is the effective focal length of the system, and $\theta_{\text{FOV}}/2$ is the half field of view of the system. There is extensive discussion regarding which definition of PD should be used. In practice, however, lens designers typically use $PD = f_{\text{efl}} \tan(\theta_{\text{FOV}}/2)$ for systems with a half field of view around 45° or less, while $PD = f_{\text{efl}}(\theta_{\text{FOV}}/2)$ is generally applied for systems with a half field of view around 60° or more [40].

The shape of an image is influenced by how magnification changes across the field, resulting in distortion. There are two main types: pincushion distortion (positive distortion) and barrel distortion (negative distortion). Pincushion distortion occurs when magnification increases with distance from the optical axis, causing straight lines—such as the edges of a grid—to bow outward, stretching the corners of the image away from the center. Barrel distortion, on the other hand, occurs when magnification decreases toward the edges, making lines curve inward and giving the image a barrel-like appearance.

In practice, distortion is generally less harmful to image sharpness than other aberrations because it can often be corrected using post-processing techniques or calibration. In some cases, distortion is even desirable: for example, fish-eye lenses intentionally introduce strong barrel distortion to achieve a very wide field of view.

2.5 Chromatic aberrations

Chromatic aberrations arise when rays of different wavelengths fail to converge at the same focal point, leading to color-dependent blurring of the image. This effect originates from the wavelength dependence of the refractive index, as discussed in the previous chapter [26, 39].

Because the refractive index of a medium varies with wavelength, each color experiences a slightly different optical path through a lens. In contrast, mirrors, being purely reflective surfaces, are not affected by chromatic aberration. The refractive index is approximately inversely proportional to the square of the wavelength, meaning that two wavelengths that are equally spaced in the spectrum do not produce equally spaced image points in the focal plane. This differential focusing leads to color fringing and is a key consideration in the design of achromatic and apochromatic lens systems.

2.5.1 Longitudinal chromatic aberration

Longitudinal Chromatic Aberration occurs for on-axis rays, where different wavelengths come to focus at distinct positions along the optical axis. This arises because each wavelength experiences a different optical power, resulting in unique focal lengths. As the name suggests, this aberration affects only the longitudinal dimension: if the image plane is set at the focal point of one wavelength, rays of other wavelengths will be out of focus, either converging before or diverging beyond the image plane. Consequently, each wavelength has a different back focal length in the system.

Fig. 2.33 illustrates the effect of longitudinal chromatic aberration at the output of an optical system. The figure focuses on a zoomed-in view of the image plane rather than the entire optical system. In this example, three on-axis fields are shown, each represented in a different color corresponding to a different wavelength, highlighting how the focal positions vary across the spectrum.

It is important to note that the system's image may resemble defocus, but with variations depending on the wavelength. When the image plane is set at the focal point of a particular wavelength, that wavelength comes into sharp focus at the center of the image plane, while rays of other wavelengths remain defocused. This wavelength-dependent shift in focal positions is referred to as longitudinal chromatic aberration.

By exploiting differences in refractive indices and Abbe numbers among optical materials, it is possible to combine lenses to reduce dispersion and thereby correct longitudinal chromatic aberration. A common approach involves pairing a converging crown glass lens with a diverging flint glass lens to form a doublet. This configuration effectively compensates for chromatic aberration between two selected wavelengths at opposite ends of the spectrum.

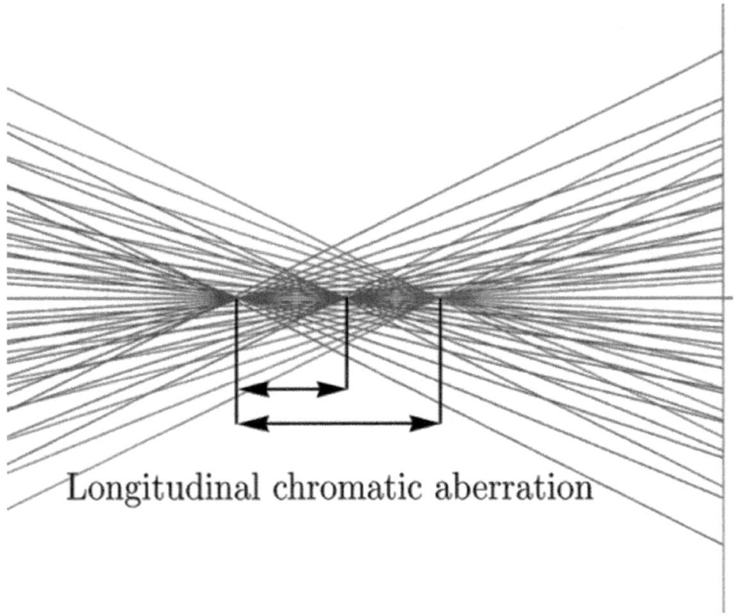

FIGURE 2.33
Longitudinal chromatic aberration at the output of an optical system is illustrated in the figure. The figure provides a zoomed-in view of the image plane rather than showing the entire optical system. In this example, three on-axis fields are displayed, each represented by a different color corresponding to a distinct wavelength. When the image plane is positioned at the focal point of one wavelength, that wavelength comes into sharp focus at the center of the image plane, while the other wavelengths appear defocused. This wavelength-dependent variation in focal positions is precisely what is referred to as longitudinal chromatic aberration.

However, correcting additional wavelengths requires the use of more complex lens combinations, as a single doublet can only correct chromatic aberration for two specific wavelengths.

2.5.2 Transverse chromatic aberration

The magnification of an optical system is also wavelength-dependent. For off-axis rays, the principal ray corresponding to each wavelength intersects the image plane at a different lateral position, resulting in a transverse displacement [26, 39]. As a consequence, the apparent image size varies across different wavelengths. This effect is known as transverse chromatic aberration, or lateral color.

Fig. 2.34 illustrates this phenomenon. The figure presents a zoomed-in view of the image plane for an off-axis field, with rays of different wavelengths

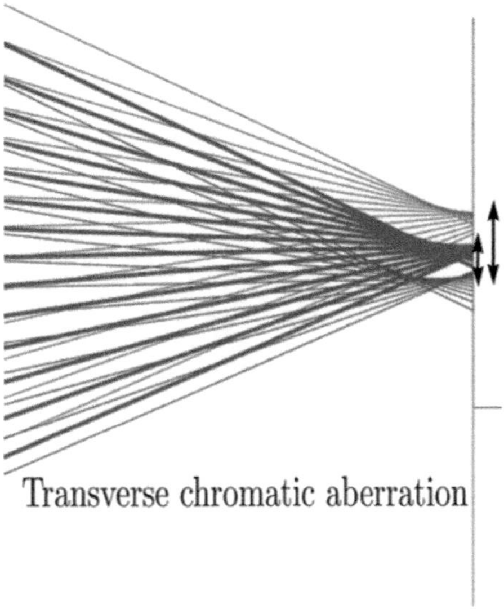

FIGURE 2.34
The magnification of an optical system also varies with wavelength. The principal ray of off-axis rays occupies distinct positions along the image plane for each wavelength, resulting in transverse displacement. Consequently, the image size varies across different wavelengths. This displacement is referred to as transverse chromatic aberration or lateral color.

shown in distinct colors. Each wavelength exhibits a different amount of lateral displacement, highlighting the wavelength-dependent variation in image position caused by angular deviations in the optical system.

2.5.3 Spot diagram

Arguably one of the most widely used tools for geometrically analyzing an optical system is the spot diagram. The spot diagram represents the front view of the image plane where rays intersect [26, 39]. It provides a convenient means to quickly visualize and identify potential aberrations within the system. Although it is not as precise as other analytical methods, the spot diagram serves as an effective initial diagnostic tool, as different types of aberrations produce characteristic patterns and shapes within the diagram.

The spot diagram provides a visualization of the two-dimensional distribution of rays on the image surface within a sequence. An example of such a diagram is shown in Fig. 2.35.

Both defocus and spherical aberration produce spot diagrams that exhibit concentric rings around a reference point. In the case of defocus, however, the energy distribution across these rings is relatively uniform. As seen in

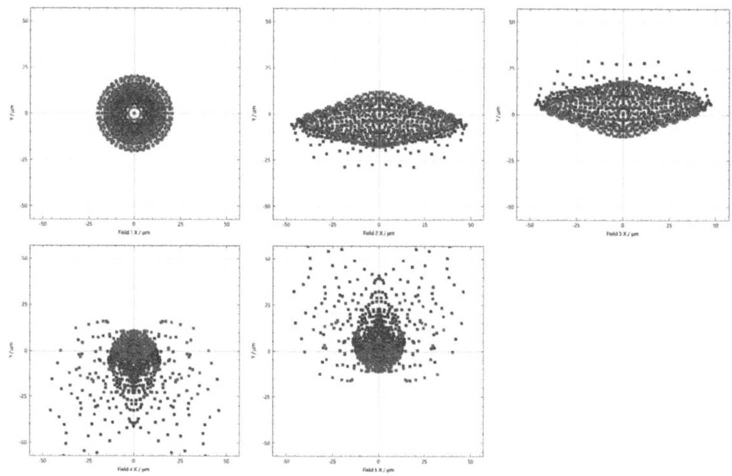

FIGURE 2.35
Spot diagram of the design of Fig. 2.22. The spot diagram shows a grind of spot diagrams, each diagram for each field. The color represents the wavelength. In this particular case.

the top-left of Fig. 2.35. It shows the effects of both defocus and spherical aberration, which is expected for the on-axis field.

For off-axis aberrations, the reference point may no longer coincide with the optical axis. In such cases, the centroid of the spot diagram can serve as the reference point. The coordinates of the centroid can be calculated using Eqs. (2.37) and (2.38).

$$\Delta x_c = \frac{1}{N}\sum_j \Delta x_j \tag{2.37}$$

and

$$\Delta y_c = \frac{1}{N}\sum_j \Delta y_j \tag{2.38}$$

In these equations, N denotes the total number of rays being traced, and j serves as the index for each individual ray. The Root-Mean-Square (RMS) radius, also referred to as the Gaussian moment, uses the centroid of the spot diagram as a reference point to quantify the deviation of rays in the image plane. The RMS radius is calculated according to Eq. (2.39).

$$\Delta r_{RMS} = \sqrt{\frac{1}{N}\sum_j [(\Delta x_c - \Delta x_j)^2 + (\Delta y_c - \Delta y_j)^2]} \tag{2.39}$$

Sagittal coma refers to the displacement between the reference point and the nearest point on the circle defined by the outermost annular zone of the pupil.

In contrast, tangential coma describes the displacement between the reference point and the farthest point on the same circle corresponding to that aperture zone.

Astigmatism, when present outside the sagittal or tangential planes, produces a characteristic ellipsoidal spot shape. Many optical systems exhibiting astigmatism display this elliptical pattern, oriented either horizontally or vertically. Fig. 2.35 shows fields affected by astigmatism in the top-center and top-right diagrams. Additionally, the bottom diagrams in the same figure demonstrate significant coma aberration, which is evident from the comet-like shape of the corresponding spot diagrams.

2.6 Zernike polynomials

The Zernike polynomials constitute a complete set of continuous functions that are orthogonal over a unit circle. The term complete indicates that these polynomials can represent any possible wavefront at the exit pupil of an optical system. The term orthogonal signifies that each term in the polynomial expansion is independent of the others; in other words, each term encodes unique information that is not shared with any other term. In practice, this orthogonality and completeness make Zernike polynomials particularly well-suited for representing wavefronts at the exit pupil [44, 45, 46].

The even symmetric Zernike polynomials are given by,

$$Z_n^m(\rho, \phi) = A \cdot R_n^m(\rho)\cos(m\phi), \tag{2.40}$$

and the odd symmetric Zernike polynomials are expressed as,

$$Z_n^{-m}(\rho, \phi) = A \cdot R_n^m(\rho)\sin(m\phi) \tag{2.41}$$

where $R_n^m(\rho)$ is the radial part of the symmetric Zernike polynomials,

$$R_n^m(\rho) = \sum_{k=0}^{(n-m)/2} \frac{(-1)^k (n-k)!}{k![(n+m)/2-k]![(n-m)/2-k]!}\rho^{n-2k} \tag{2.42}$$

The above expression is valid if $n - m$ is even. But, If $n - m$ is odd $R_n^m(\rho)$ is reduce to,

$$R_n^m(\rho) = 0 \tag{2.43}$$

The coefficient A is determined to satisfy the following normalization condition,

$$\int_0^{2\pi}\int_0^1 Z_j^2 \rho \mathrm{d}\rho \mathrm{d}\phi = \pi \tag{2.44}$$

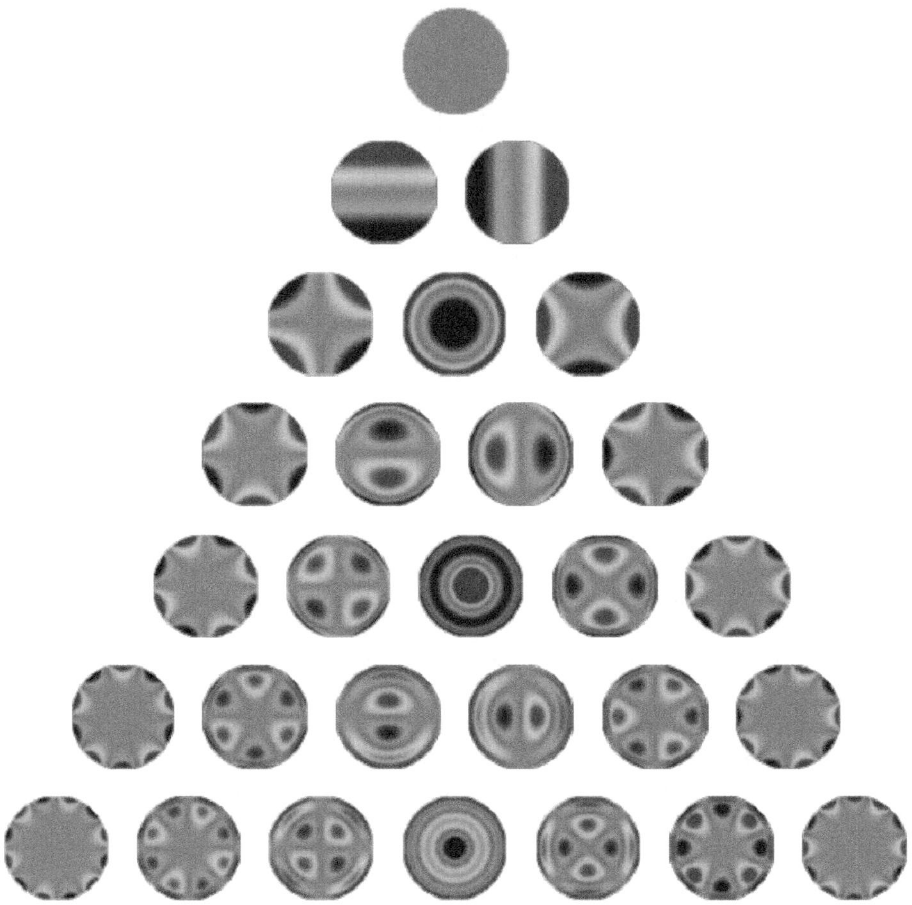

FIGURE 2.36
Symmetric Zernike polynomials according to ANSI Z80.28/ISO 24157 Standard. The index of the Zernike coefficients is counted row by row, commencing from the top with `Z0` for the piston, and progressing from left to right in the figure.

The symmetric Zernike polynomials employed here are consistent with the definitions provided in the ANSI Z80.28/ISO 24157 Standard. These polynomials are normalized to satisfy the condition specified in Eq. (2.44). Their ordering follows the arrangement illustrated in Fig. 2.36.

Symmetric Zernike polynomials according to ANSI Z80.28/ISO 24157 Standard in Table 2.1.

This framework also incorporates the Zernike fringe polynomials, which form a subset of the symmetric Zernike polynomials. The mapping from the symmetric Zernike polynomials to the Zernike fringe polynomials is expressed

TABLE 2.1
Formula of Zernike Polynomials with ANSI Z80.28/ISO 24157 sorting.

Coefficient	Formula
Z0	1
Z1	$2x$
Z2	$2y$
Z3	$\sqrt{6}(2xy)$
Z4	$\sqrt{3}(-1+2y^2+2x^2)$
Z5	$\sqrt{6}(-1y^2+1x^2)$
Z6	$\sqrt{8}(-1y^3+3x^2y)$
Z7	$\sqrt{8}(-2y+3y^3+3x^2y)$
Z8	$\sqrt{8}(-2x+3xy^2+3x^3)$
Z9	$\sqrt{8}(-3xy^2+1x^3)$
Z10	$\sqrt{10}(-4xy^3+4x^3y)$
Z11	$\sqrt{10}(-6xy+8xy^3+8x^3y)$
Z12	$\sqrt{5}(1-6y^2+6y^4-6x^2+12x^2y^2+6x^4)$
Z13	$\sqrt{10}(3y^2-4y^4-3x^2+4x^4)$
Z14	$\sqrt{10}(1y^4-6x^2y^2+1x^4)$
Z15	$\sqrt{12}(1y^5-10x^2y^3+5x^4y)$
Z16	$\sqrt{12}(4y^3-5y^5-12x^2y+10x^2y^3+15x^4y)$
Z17	$\sqrt{12}(3y-12y^3+10y^5-12x^2y+20x^2y^3+10x^4y)$
Z18	$\sqrt{12}(3x-12xy^2+10xy^4-12x^3+20x^3y^2+10x^5)$
Z19	$\sqrt{12}(12xy^2-15xy^4-4x^3-10x^3y^2+5x^5)$
Z20	$\sqrt{12}(5xy^4-10x^3y^2+1x^5)$
Z21	$\sqrt{14}(6xy^5-20x^3y^3+6x^5y)$
Z22	$\sqrt{14}(20xy^3-24xy^5-20x^3y+24x^5y)$
Z23	$\sqrt{14}(12xy-40xy^3+30xy^5-40x^3y+60x^3y^3+30x^5y)$
Z24	$\sqrt{7}(-1+12y^2-29y^4+20y^6+12x^2-59x^2y^2+59x^2y^4)$ $+\sqrt{7}(-29x^4+59x^4y^2+20x^6)$
Z25	$\sqrt{14}(-6y^2+20y^4-15y^6+6x^2-15x^2y^4-20x^4$ $+15x^4y^2+15x^6)$
Z26	$\sqrt{14}(-5y^4+6y^6+30x^2y^2-30x^2y^4-5x^4-30x^4y^2+6x^6)$
Z27	$\sqrt{14}(-1y^6+15x^2y^4-15x^4y^2+1x^6)$

as $Z_n^m \to Z_j$

$$j = \left(1 + \frac{n+|m|}{2}\right)^2 - 2|m| + \frac{1-\text{sgn}(m)}{2} \tag{2.45}$$

Here, sgn(·) denotes the sign function, which returns the sign of its argument.

Fig. 2.37 presents the plots of the first few terms of the Zernike fringe polynomials. The sequence begins with Z_1, representing the piston term at the top, and continues in order from top to bottom and left to right.

Table 2.2 shows the coefficients of Zernike Fringe polynomials.

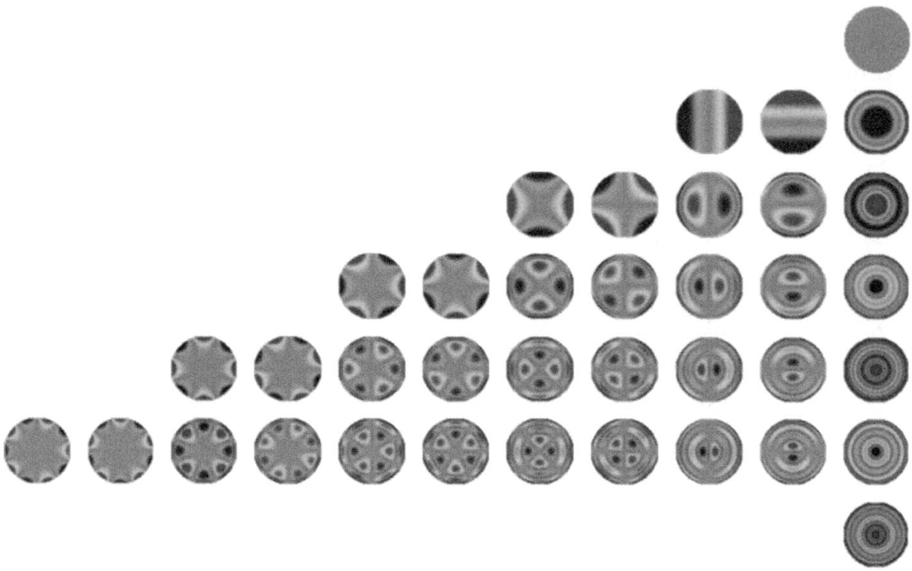

FIGURE 2.37
The Zernike fringe polynomials set is a subset of the symmetric Zernike polynomials. The mapping is given by Eq. (2.45). The following procedure gives the index of Zernike fringe polynomials: starting with $Z1$ which is the piston at the top, and counting from top to bottom and left to right.

2.7 Conclusions

This chapter presents a comprehensive toolkit for geometrical optics, starting from fundamental principles of reflection and refraction governed by the laws of reflection and Snell's law. Fermat's principle of least time serves as the unifying foundation, explaining ray paths in both reflective and refractive systems. Key concepts include the OPL, paraxial approximations, and the vector form of Snell's law. The discussion extends to stigmatic imaging, conic mirrors, cardinal points, pupils, stops, and various ray types (meridional, sagittal, chief, marginal, skew, paraxial). Essential design parameters—focal length, f-number, numerical aperture, field of view—are introduced alongside aberrations (monochromatic and chromatic), spot diagrams, and Zernike polynomials for wavefront analysis. Together, these tools enable precise modeling, and analysis, of optical systems.

TABLE 2.2
Formula of Zernike Fringe Polynomials

Coefficient	Formula
Z1	1
Z2	$2x$
Z3	$2y$
Z4	$-1 + 2y^2 + 2x^2$
Z5	$-1y^2 + 1x^2$
Z6	$2xy$
Z7	$-2x + 3xy^2 + 3x^3$
Z8	$-2y + 3y^3 + 3x^2y$
Z9	$1 - 6y^2 + 6y^4 - 6x^2 + 12x^2y^2 + 6x^4$
Z10	$-3xy^2 + 1x^3$
Z11	$-1y^3 + 3x^2y$
Z12	$3y^2 - 4y^4 - 3x^2 + 4x^4$
Z13	$-6xy + 8xy^3 + 8x^3y$
Z14	$3x - 12xy^2 + 10xy^4 - 12x^3 + 20x^3y^2 + 10x^5$
Z15	$3y - 12y^3 + 10y^5 - 12x^2y + 20x^2y^3 + 10x^4y$
Z16	$-1 + 12y^2 - 30y^4 + 20y^6 + 12x^2 - 60x^2y^2 + 60x^2y^4$ $-30x^4 + 60x^4y^2 + 20x^6$
Z17	$1y^4 - 6x^2y^2 + 1x^4)$
Z18	$-4xy^3 + 4x^3y$
Z19	$12xy^2 - 15xy^4 - 4x^3 - 10x^3y^2 + 5x^5$
Z20	$4y^3 - 5y^5 - 12x^2y + 10x^2y^3 + 15x^4y$
Z21	$-6y^2 + 20y^4 - 15y^6 + 6x^2 - 15x^2y^4 - 20x^4 + 15x^4y^2 + 15x^6$
Z22	$12xy - 40xy^3 + 30xy^5 - 40x^3y + 60x^3y^3 + 30x^5y$
Z23	$-4x + 30xy^2 - 60xy^4 + 35xy^6 + 30x^3 - 120x^3y^2$ $+105x^3y^4 - 60x^5 + 105x^5y^2 + 35x^7$
Z24	$-4y + 30y^3 - 60y^5 + 35y^7 + 30x^2y - 120x^2y^3 + 105x^2y^5$ $-60x^4y + 105x^4y^3 + 35x^6y$

Problems

Problem 1: Fermat's principle and conic sections

Explain why parabolic mirrors are stigmatic for objects at infinity while elliptic mirrors are stigmatic only for two specific finite conjugate points. Use Fermat's principle (minimum optical path length) to justify why these particular conic sections naturally produce point-to-point (stigmatic) imaging, while a spherical mirror generally does not.

Problem 2: Astigmatism vs. field curvature

Both astigmatism and field curvature cause the best focus position to vary with field angle, yet they are considered fundamentally different aberrations.

(a) Describe the key difference in how sagittal and tangential rays behave in each case.

(b) Explain why astigmatism produces line-like blurs oriented in two perpendicular directions, while field curvature does not depend on the azimuthal angle of the ray bundle.

Problem 3: Spherical aberration is absent in mirrors but present in lenses

Spherical aberration is usually the dominant on-axis monochromatic aberration in simple lenses, yet it is absent (to third order) in parabolic mirrors and extremely small in spherical mirrors only for very small apertures.

Using the chapter's discussion of ray bending and surface normals, explain conceptually why reflection at a surface does not (in the paraxial limit) introduce spherical aberration in the same way refraction does.

Part II

Scalar diffraction theory

3

Fundamentals of scalar diffraction theory

This chapter presents the core principles of scalar diffraction theory, a simplified framework for modeling light propagation that treats the electric field as a scalar quantity, ignoring polarization effects. Derived from Maxwell's equations in homogeneous media, the scalar Helmholtz equation governs monochromatic wave behavior. Huygens' principle—that every point on a wavefront emits secondary spherical wavelets—underpins the theory. Combining this with Green's theorem yields Kirchhoff's rigorous diffraction integral. Successive approximations lead to the Huygens-Fresnel form, the Fresnel (near-field) integral with quadratic phase factors, and the Fraunhofer (far-field) integral, which reduces to a Fourier transform of the aperture function. The Fraunhofer result explains the Airy disc pattern for circular apertures and establishes the fundamental diffraction limit in imaging systems, quantified by the Rayleigh criterion.

3.1 Introduction

Geometrical optics simplifies optical design but overlooks interference and diffraction, two fundamental wave phenomena. Diffraction, arising from the wave nature of light, becomes critical when light interacts with optical elements or interfaces with dimensions comparable to the wavelength. This phenomenon is essential for assessing image quality in optical systems [1, 7]. There are two main diffraction theories, the scalar diffraction theory and the vector diffraction theory.

In this chapter, we explore scalar diffraction theory. This theory simplified the electric **E** field as a scalar field U. In contrast to the vector diffraction theory, which accounts for the electric field's vectorial nature and is reserved for specialized applications [11, 14, 2, 47, 48] that we will explore in later chapters.

DOI: 10.1201/9781003726494-5

Scalar diffraction theory, grounded in the Helmholtz equation and the Huygens principle, facilitates the computation of diffraction patterns. The Helmholtz equation governs wave fields, such as the electric field **E**, in a homogeneous medium, while the Huygens principle posits that each point on a wavefront acts as a source of secondary spherical wavelets, forming the new wavefront [7].

This chapter provides a concise overview of scalar diffraction theory without exhaustive derivations, referring readers to authoritative texts for deeper analysis [1, 8, 7]. In subsequent sections, we will derive the Helmholtz equation and explain the Huygens principle, laying the foundation for understanding the key elements of scalar diffraction theory and their diffraction integrals. The diffraction integral is a mathematical construct that describes how light waves evolve as they encounter apertures or obstacles. Mathematically, the diffraction integral is expressed as an integral over the aperture or obstacle, incorporating the complex amplitude of the wavefront at various spatial points. By solving this integral, one can determine the resulting wave pattern or intensity distribution on an observation plane beyond the diffracting element [8].

3.2 Helmholtz equation

The first main concept to formulate scalar diffraction theory is the Helmholtz equation. To get the Helmholtz equation, we need to consider Maxwell's equations in a source-free, linear, isotropic, homogeneous medium (with constant permittivity ϵ and permeability μ),

$$\nabla \cdot \mathbf{E} = 0, \tag{3.1}$$

$$\nabla \cdot \mathbf{B} = 0, \tag{3.2}$$

$$\nabla \times \mathbf{E} = -\frac{\partial \mathbf{B}}{\partial t}, \tag{3.3}$$

$$\nabla \times \mathbf{B} = \mu\epsilon\frac{\partial \mathbf{E}}{\partial t}. \tag{3.4}$$

Now, we assume time-harmonic fields of the form $\mathbf{E}(\mathbf{r}, t) = \mathbf{E}(\mathbf{r})e^{\mathrm{i}\omega t}$ and $\mathbf{B}(\mathbf{r}, t) = \mathbf{B}(\mathbf{r})e^{\mathrm{i}\omega t}$, where ω is the angular frequency. The time derivatives become $\frac{\partial}{\partial t} \to \mathrm{i}\omega$. Substituting into Eqs. (3.1)–3.4, we get,

$$\nabla \cdot \mathbf{E} = 0, \tag{3.5}$$

$$\nabla \cdot \mathbf{B} = 0, \tag{3.6}$$

$$\nabla \times \mathbf{E} = -\mathrm{i}\omega\mathbf{B}, \tag{3.7}$$

$$\nabla \times \mathbf{B} = \mathrm{i}\omega\mu\epsilon\mathbf{E}. \tag{3.8}$$

Getting the curl of Eq. (3.7),

$$\nabla \times (\nabla \times \mathbf{E}) = \nabla \times (-\mathrm{i}\omega \mathbf{B}) = -\mathrm{i}\omega(\nabla \times \mathbf{B}). \tag{3.9}$$

Now, lets substitute Eq. (3.8) into the right-hand side,

$$\nabla \times (\nabla \times \mathbf{E}) = -\mathrm{i}\omega(\mathrm{i}\omega\mu\epsilon\mathbf{E}) = -\omega^2\mu\epsilon\mathbf{E}. \tag{3.10}$$

Using the vector identity $\nabla \times (\nabla \times \mathbf{E}) = \nabla(\nabla \cdot \mathbf{E}) - \nabla^2\mathbf{E}$ and noting from Eq. (3.5) that $\nabla \cdot \mathbf{E} = 0$, we obtain,

$$-\nabla^2\mathbf{E} = -\omega^2\mu\epsilon\mathbf{E}. \tag{3.11}$$

Thus,

$$\nabla^2\mathbf{E} + \omega^2\mu\epsilon\mathbf{E} = 0. \tag{3.12}$$

Define the wave number $k = \omega\sqrt{\mu\epsilon}$, so $\omega^2\mu\epsilon = k^2$. The equation becomes,

$$\nabla^2\mathbf{E} + k^2\mathbf{E} = 0. \tag{3.13}$$

This is the vector Helmholtz equation for $\mathbf{E}$. Each Cartesian component E_x, E_y, E_z satisfies the scalar Helmholtz equation,

$$\nabla^2 E_i + k^2 E_i = 0, \quad (i = x, y, z). \tag{3.14}$$

Example: Obtain the Helmholtz equation of the magnetic field. Taking the curl of Eq. (3.8):

$$\nabla \times (\nabla \times \mathbf{B}) = \nabla \times (\mathrm{i}\omega\mu\epsilon\mathbf{E}) = \mathrm{i}\omega\mu\epsilon(\nabla \times \mathbf{E}). \tag{3.15}$$

Substitute Eq. (3.7) into the right-hand side,

$$\nabla \times (\nabla \times \mathbf{B}) = \mathrm{i}\omega\mu\epsilon(-\mathrm{i}\omega\mathbf{B}) = -\omega^2\mu\epsilon\mathbf{B}. \tag{3.16}$$

Using the vector identity $\nabla \times (\nabla \times \mathbf{B}) = \nabla(\nabla \cdot \mathbf{B}) - \nabla^2\mathbf{B}$ and noting from Eq. (3.6) that $\nabla \cdot \mathbf{B} = 0$, we obtain,

$$-\nabla^2\mathbf{B} = -\omega^2\mu\epsilon\mathbf{B}. \tag{3.17}$$

Thus,

$$\nabla^2\mathbf{B} + \omega^2\mu\epsilon\mathbf{B} = 0. \tag{3.18}$$

With $k^2 = \omega^2\mu\epsilon$, this becomes,

$$\nabla^2\mathbf{B} + k^2\mathbf{B} = 0. \tag{3.19}$$

This is the vector Helmholtz equation for $\mathbf{B}$, with each component satisfying the scalar Helmholtz equation. The scalar version of the Helmholtz equation considers a scalar function u instead of a vector identity like **E** or **B**,

$$\nabla^2 u + k^2 u = 0. \tag{3.20}$$

As previously noted, scalar diffraction theory is fundamentally based on the scalar version of the Helmholtz equation and Huygens' Principle. The scalar form of the Helmholtz equation will be employed alongside all scalar diffraction integrals. By adopting the Helmholtz equation in scalar theory, we assume that the electric fields under consideration conform to Maxwell's equations. Now we will focus on the other fundamental concept in diffraction: Huygens' Principle.

3.3 Huygens' principle

Huygens' Principle is a cornerstone of wave optics, providing an intuitive framework for understanding wave propagation and diffraction. Proposed by Christiaan Huygens in 1678, it posits that every point on a wavefront acts as a source of secondary spherical wavelets, with the new wavefront being the tangent to these wavelets. The Huygens Diffraction Integral, formalized by Kirchhoff, building on Fresnel's contributions, quantifies this principle to compute the wave field at a point [49, 31].

Huygens' Principle explicitly states:

> Every point on a wavefront acts as a source of secondary spherical wavelets that propagate outward at the wave's speed in the medium. The subsequent wavefront is the envelope (tangent surface) of these secondary wavelets.

This principle applies to waves such as light and sound. For a plane wavefront propagating in the positive z-direction, each point emits spherical wavelets, and the new wavefront at time $t + \Delta t$ is the surface tangent to these wavelets. In order to explain this see the diagram of Fig. 3.1. As shown on the (*left*), a point source generates a spherical wavefront. According to Huygens' Principle, every point on a wavefront acts as a secondary source of spherical wavelets, which propagate outward at the speed of the wave in that medium (*center*). The new wavefront is the *envelope* or tangent surface of these secondary wavelets; therefore, wavelets from an aligned series of sources can combine to form a flat (plane) wavefront (*right*).

Once we have formulated both the Helmholtz equation and Huygens' Principle, we are now in a position to derive the primary scalar diffraction integral, known as Kirchhoff's diffraction integral.

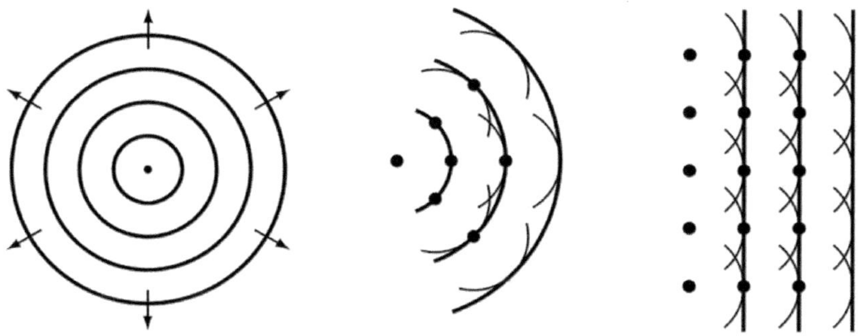

FIGURE 3.1
Diagram explaining Huygens principle. (*left*) It shows that every point source generates a spherical wavefront. Therefore, every point on a wavefront acts as a source of secondary spherical wavelets that propagate outward at the wave's speed in the medium (*center*). The subsequent wavefront is the envelope (tangent surface) of these secondary wavelets, so from an alignment source the envelope can be a flat wavefront (*right*).

3.4 Kirchhoff's diffraction integral

Kirchhoff's diffraction integral is a mathematical framework used to describe the propagation and diffraction of scalar waves from an aperture or obstacle in a wave field. It is derived from the scalar wave equation and provides a way to compute the wave field at a point in space based on the field distribution over a surface, typically an aperture. The integral is the main scalar diffraction integral, since the most important scalar diffraction integrals come from it.

The Kirchhoff diffraction integral expresses the wave field $U(\mathbf{r}_p, t)$ at a point p as an integral over a closed surface s (often an aperture plane and a closing surface) containing the point. Let's start by considering a monochromatic scalar wave field $U(\mathbf{r}_p, t) = u(\mathbf{r}_p)e^{-\mathrm{i}\omega t}$, where $u(\mathbf{r})$ is the spatial component, ω is the angular frequency, and $\mathbf{r}_p$ is the position vector. The Kirchhoff diffraction integral is given by,

$$u(\mathbf{r}_p) = \frac{1}{4\pi}\iint_s \left[u\frac{\partial}{\partial \mathrm{n}}\left(\frac{e^{ikr}}{r}\right) - \frac{e^{ikr}}{r}\frac{\partial u}{\partial \mathrm{n}}\right] \mathrm{d}s, \tag{3.21}$$

where, $u(\mathbf{r}_s)$ is the scalar wave field on the surface s, $r = |\mathbf{r}_p - \mathbf{r}_s|$ is the distance from a point $\mathbf{r}_s$ on the surface to the observation point $\mathbf{r_p}$, k is the wave number ($k = \frac{2\pi}{\lambda}$, with λ being the wavelength), $\frac{\partial}{\partial \mathrm{n}}$ denotes the derivative

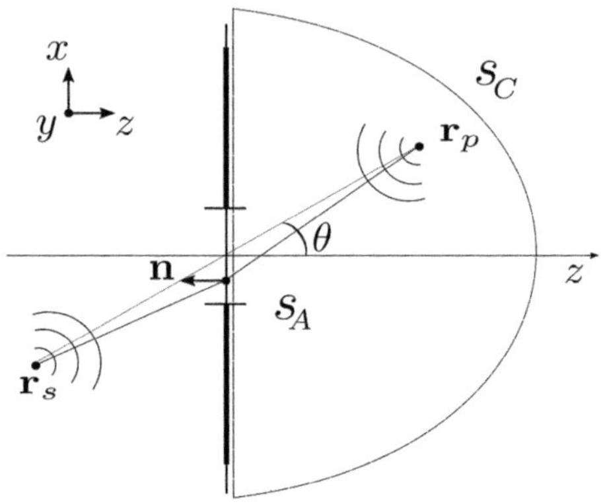

FIGURE 3.2
Diagram of scalar diffraction. The source at $\mathbf{r}_s$ generates a wavefront that goes through the aperture s_A. The intensity of the diffraction pattern at $\mathbf{r}_s$ is given by Kirchhoff's integral, Eq. (3.21). The point $\mathbf{r}_s$ is surrounded by surface s with volume v, where the s_C side of s can be extended to infinity, Eq. (3.35).

along the normal to the surface s, ds is the differential area element on the surface. Here, we consider that the field and its normal derivative are specified on the surface. In the aperture, the field matches the incident wave; outside the aperture, the field and its derivative are often assumed to be zero. These aperture conditions are usually named as Kirchhoff's boundary conditions. Now, it is important to clarify that $u(\mathbf{r}_p)$ is a scalar, but its argument, input, is a vector $\mathbf{r}_p$, which is the position vector. See Fig. 3.2.

Notice that here we only consider the spatial $u(\mathbf{r})$ dependent part of $U(\mathbf{r}, t) = u(\mathbf{r})e^{-i\omega t}$. By focusing on $u(\mathbf{r}_p)$, the diffraction integral simplifies the computation of spatial effects, such as interference and diffraction, across an aperture or obstacle, without needing to account for time-dependent dynamics. This approach is standard in scalar diffraction theory, as it reduces the problem to a spatial one, governed by the Helmholtz equation, which describes the steady-state spatial behavior of the wave field [8].

The proof of the equation tal starts with the scalar wave equation and uses Green's theorem to derive the diffraction integral. We assume the scalar field $u(\mathbf{r})$ satisfies the scalar Helmholtz equation, thus,

$$\nabla^2 u + k^2 u = 0. \tag{3.22}$$

The goal is to express $u(\mathbf{r})$ at a point p inside a volume v enclosed by a surface denoted by s.

We can introduce (due to the Huygens principle) a Green's function $G(\mathbf{r}_p, \mathbf{r}_s)$, which represents the response at point $\mathbf{r}_p$ due to a point source at $\mathbf{r}_s$. For free space, the Green's function is the outgoing spherical wave, thus,

$$G(\mathbf{r}_p, \mathbf{r}_s) = \frac{e^{ik|\mathbf{r}_p - \mathbf{r}_s|}}{|\mathbf{r}_p - \mathbf{r}_s|} = \frac{e^{ikr}}{r}, \tag{3.23}$$

where $r = |\mathbf{r}_p - \mathbf{r}_s|$. The Green's function satisfies,

$$\nabla^2 G + k^2 G = -4\pi\delta(\mathbf{r}_p - \mathbf{r}_s), \tag{3.24}$$

where δ is the Dirac delta function, and the Laplacian is taken with respect to $\mathbf{r}_p$.

Now, consider the volume v enclosed by surface s, with the observation point p at $\mathbf{r}$ inside v. We apply Green's theorem in the form:

$$\iiint_V (\phi\nabla^2\psi - \psi\nabla^2\phi)\,\mathrm{d}v = \iint_s \left(\phi\frac{\partial\psi}{\partial \mathrm{n}} - \psi\frac{\partial\phi}{\partial \mathrm{n}}\right)\mathrm{d}s, \tag{3.25}$$

where $\phi = u(\mathbf{r}_s)$, $\psi = G(\mathbf{r}_p, \mathbf{r}_s)$, and $\frac{\partial}{\partial \mathrm{n}}$ is the outward normal derivative on s. The volume integral is over v, and the surface integral is over s.

Compute the volume integrand,

$$\nabla^2_{\mathbf{r}_s} u = -k^2 u, \quad \nabla^2_{\mathbf{r}_s} G = -k^2 G - 4\pi\delta(\mathbf{r}_p - \mathbf{r}_s), \tag{3.26}$$

where $\nabla^2_{\mathbf{r}_s}$ denotes the Laplacian with respect to $\mathbf{r}_s$. Thus,

$$u\nabla^2_{\mathbf{r}_s} G - G\nabla^2_{\mathbf{r}_s} u = u(-k^2 G - 4\pi\delta(\mathbf{r}_p - \mathbf{r}_s)) - G(-k^2 u). \tag{3.27}$$

The k^2 terms cancel, so the volume integrand becomes,

$$u\nabla^2_{\mathbf{r}_s} G - G\nabla^2_{\mathbf{r}_s} u = -4\pi u\delta(\mathbf{r}_p - \mathbf{r}_s). \tag{3.28}$$

The volume integral is,

$$\iiint_V (u\nabla^2_{\mathbf{r}_s} G - G\nabla^2_{\mathbf{r}_s} u)\,\mathrm{d}v = \iiint_V (-4\pi u\delta(\mathbf{r}_p - \mathbf{r}_s))\,\mathrm{d}v = -4\pi u(\mathbf{r}_p), \tag{3.29}$$

since the delta function picks out the value of u at $\mathbf{r}_s = \mathbf{r}_p$, assuming $\mathbf{r}_p$ is inside v. By Green's theorem, this equals the surface integral,

$$-4\pi u(\mathbf{r}_p) = \iint_s \left(u\frac{\partial G}{\partial \mathrm{n}} - G\frac{\partial u}{\partial \mathrm{n}}\right) ds. \tag{3.30}$$

Thus,

$$u(\mathbf{r}_p) = \frac{1}{4\pi}\iint_s \left(u(\mathbf{r}_s)\frac{\partial G}{\partial \mathrm{n}} - G\frac{\partial u(\mathbf{r}_s)}{\partial \mathrm{n}}\right) ds. \tag{3.31}$$

The minus sign is resolved by the choice of Green's function and normal direction. Using $G = \frac{e^{ikr}}{r}$ with the outward normal, dividing by -4π flips the sign. In diffraction, the normal is often chosen toward the observation point, aligning the sign with wave propagation.

Now, let's substitute the Green's function $G = \frac{e^{ikr}}{r}$. Computing the normal derivative,

$$\frac{\partial G}{\partial \mathrm{n}} = \mathbf{n} \cdot \nabla_{\mathbf{r}_s} \left(\frac{e^{\mathrm{i}kr}}{r} \right), \tag{3.32}$$

where $\mathbf{n}$ is the outward unit normal to s. The gradient is,

$$\nabla_{\mathbf{r}_s} \left(\frac{e^{\mathrm{i}kr}}{r} \right) = \left(\mathrm{i}k - \frac{1}{r} \right) \frac{e^{\mathrm{i}kr}}{r} \frac{\mathbf{r}_p - \mathbf{r}_s}{r}, \tag{3.33}$$

so,

$$\frac{\partial G}{\partial \mathrm{n}} = \left(\mathrm{i}k - \frac{1}{r} \right) \frac{e^{\mathrm{i}kr}}{r} \left(\mathbf{n} \cdot \frac{\mathbf{r}_p - \mathbf{r}_s}{r} \right). \tag{3.34}$$

Thus, the Kirchhoff diffraction integral is

$$u(\mathbf{r}_p) = \frac{1}{4\pi} \iint_S \left[u(\mathbf{r}_s) \frac{\partial}{\partial \mathrm{n}} \left(\frac{e^{\mathrm{i}kr}}{r} \right) - \frac{e^{\mathrm{i}kr}}{r} \frac{\partial u(\mathbf{r}_s)}{\partial \mathrm{n}} \right] ds. \tag{3.35}$$

To apply this to diffraction, consider a common setup where s consists of an aperture plane s_A and a closing surface s_C (e.g., a large hemisphere) enclosing the volume. Kirchhoff's boundary conditions assume that the aperture s_A, U and $\frac{\partial U}{\partial \mathrm{n}}$ match the incident wave. Outside the aperture on s_A and on s_C, $U = 0$ and $\frac{\partial U}{\partial \mathrm{n}} = 0$. Therefore, the contribution from s_C vanishes as the surface extends to infinity; this simplification is also called the Sommerfeld radiation condition.

Thus, the integral reduces to the aperture s_A,

$$u(\mathbf{r}_p) = \frac{1}{4\pi} \iint_{s_A} \left[u(\mathbf{r}_s) \frac{\partial}{\partial \mathrm{n}} \left(\frac{e^{\mathrm{i}kr}}{r} \right) - \frac{e^{\mathrm{i}kr}}{r} \frac{\partial u(\mathbf{r}_s)}{\partial \mathrm{n}} \right] ds. \tag{3.36}$$

In many applications, approximations simplify the integral. For example: Fresnel approximation is valid for small angles and moderate distances, the normal derivative term is approximated, and the obliquity factor $\mathbf{n} \cdot \frac{\mathbf{r}_p - \mathbf{r}_s}{r}$ is simplified. The Fraunhofer approximation, for far-field diffraction, the phase term $e^{\mathrm{i}kr}$ is approximated linearly, leading to Fourier transform-like expressions. Let's explore these approximations in the upcoming sections.

3.5 Huygens diffraction integral

Huygens' diffraction integral is the first approximation to Kirchhoff's formalism. In order to quantify the wave field at a point p, we derive the Huygens diffraction integral using Kirchhoff's formalism, so we recall Eq. (3.21),

$$u(\mathbf{r}_p) = \frac{1}{4\pi} \iint_{s_A} \left[u(\mathbf{r}_s) \frac{\partial}{\partial \mathrm{n}} \left(\frac{e^{ikr}}{r} \right) - \frac{e^{ikr}}{r} \frac{\partial u(\mathbf{r}_s)}{\partial \mathrm{n}} \right] \mathrm{d}s, \tag{3.21}$$

Remember that s is aperture plane at $z = 0$, which it is placed at $\mathbf{r}_s = [x_s, y_s, 0]$. p is the observation point at $\mathbf{r}_p = [x_p, y_p, z_p]$. $r = |\mathbf{r}_p - \mathbf{r}_s| = \sqrt{(x_p - x_s)^2 + (y_p - y_s)^2 + z_p^2}$ is the distance from $\mathbf{r}_s$ to the observation point p. Finally, the unit normal vector $\mathbf{n}$ on s, pointing outward.

Huygens diffraction integral is a special case of the previous equation, where for an open surface s in other it is an aperture. Again, here we assume Kirchhoff's boundary conditions. This implies that $u(\mathbf{r}_s)$ is the incident field in the aperture, and it is zero outside the aperture; also, $\frac{\partial u}{\partial \mathrm{n}}$ is zero outside. Thus, the normal derivative of the Green's function is,

$$\frac{\partial}{\partial n} \left(\frac{e^{ikr}}{r} \right) \approx \frac{e^{ikr}}{r} \left(ik - \frac{1}{r} \right) (\mathbf{n} \cdot \mathbf{r}). \tag{3.37}$$

Also, if we consider that $r \gg \lambda$, which makes $\frac{1}{r}$ term is negligible, thus,

$$\frac{\partial}{\partial n} \left(\frac{e^{ikr}}{r} \right) \approx \mathrm{i}k \frac{e^{ikr}}{r} (\mathbf{n} \cdot \mathbf{r}). \tag{3.38}$$

where $\mathbf{r} = (\mathbf{r}_p - \mathbf{r}_s)/r$.

The term $\mathbf{n} \cdot \mathbf{r} = \cos\theta$, where θ is the angle between $\mathbf{n}$ and $\mathbf{r}$, is called the obliquity factor. Neglecting the term involving $\frac{\partial u}{\partial n}$, the integral simplifies to,

$$u(\mathbf{r}_p) = \frac{1}{\mathrm{i}\lambda} \iint_s u(\mathbf{r}_s) \frac{e^{ikr}}{r} \cos\theta \, \mathrm{d}s, \tag{3.39}$$

where $\frac{1}{\mathrm{i}\lambda} = \frac{k}{i2\pi}$ normalizes amplitude and phase. This sums contributions from secondary wavelets, weighted by the incident field, spherical wave term, obliquity factor, and normalization constant.

Each point on s emits a wavelet with amplitude $u(\mathbf{r}_s)$. The term e^{ikr} accounts for phase delay, and $\frac{1}{r}$ for amplitude decay. Notice that $\cos\theta$ reduces contributions at large angles, explaining forward propagation. Finally, the integral sums complex amplitudes, producing diffraction patterns via interference. It is important to keep in mind that this integral is insufficient for vector fields like electromagnetic waves in some cases, and it only applies when the approximation $(r \gg \lambda)$.

The next important simplification is the Fresnel diffraction integral that we are going to discuss in the next section.

3.6 Fresnel diffraction integral

The Fresnel diffraction integral is a key tool in wave optics for computing near-field diffraction patterns. It builds on the superposition of secondary wavelets from a wavefront with appropriate amplitude, phase, and angular dependence. To derive the Fresnel diffraction integral, we start with the Huygens diffraction integral framework and apply near-field approximations. Recalling Huygens' diffraction integral,

$$u(\mathbf{r}_p) = \frac{1}{\mathrm{i}\lambda} \iint_s u(\mathbf{r}_s) \frac{e^{ikr}}{r} \cos\theta \, \mathrm{d}s, \tag{3.39}$$

where s is the aperture plane at $z = 0$, with points $\mathbf{r}_s = [x_s, y_s, 0]$. p is observation point at $\mathbf{r}_p = [x_p, y_p, z_p]$. The distance r is $r = |\mathbf{r}_p - \mathbf{r}_s| = \sqrt{(x_p - x_s)^2 + (y_p - y_s)^2 + z_p^2}$ is the distance from source $\mathbf{r}_s$ to observation point p. Here, $\cos\theta = \mathbf{n} \cdot \mathbf{r}$, with $\mathbf{r} = \frac{\mathbf{r}_p - \mathbf{r}_s}{r}$, and $\mathrm{d}s = \mathrm{d}x_s \mathrm{d}y_s$.

The Fresnel regime applies for moderate distances z_p, where $z_p \gg \lambda$ but not so large as to reach the far-field (Fraunhofer) regime, which we will explain later in this chapter. Back in track the distance r is given by,

$$r = \sqrt{(x_p - x_s)^2 + (y_p - y_s)^2 + z_p^2}. \tag{3.40}$$

For $z_p \gg \sqrt{(x_p - x_s)^2 + (y_p - y_s)^2}$, we use the binomial approximation, therefore,

$$r = z_p \sqrt{1 + \frac{(x_p - x_s)^2 + (y_p - y_s)^2}{z_p^2}} \approx z_p + \frac{(x_p - x_s)^2 + (y_p - y_s)^2}{2z_p}. \tag{3.41}$$

In the phase term $e^{\mathrm{i}kr}$, this approximation is critical due to phase sensitivity. For the amplitude, $\frac{1}{r} \approx \frac{1}{z_p}$. The obliquity factor becomes,

$$\cos\theta = \frac{z_p}{r} \approx 1, \tag{3.42}$$

since $r \approx z_p$. Substituting,

$$u(\mathbf{r}_p) \approx \frac{1}{\mathrm{i}\lambda z_p} \iint_s u(\mathbf{r}_s) e^{\mathrm{i}k\left(z_p + \frac{(x_p - x_s)^2 + (y_p - y_s)^2}{2z_p}\right)} \mathrm{d}x_s \mathrm{d}y_s. \tag{3.43}$$

Factoring out $e^{\mathrm{i}kz_p}$,

$$u(x_p, y_p, z_p) = \frac{e^{\mathrm{i}kz_p}}{\mathrm{i}\lambda z_p} \iint_s u(x_s, y_s, 0) e^{\mathrm{i}\frac{k}{2z_p}\left[(x_p - x_s)^2 + (y_p - y_s)^2\right]} \mathrm{d}x_s \mathrm{d}y_s \tag{3.44}$$

Therefore, the Fresnel diffraction integral is given by,

$$u(x_p, y_p, z_p) = \frac{e^{ikz_p}}{i\lambda z_p} \iint_s u(x_s, y_s, 0) e^{i\frac{k}{2z_p}\left[(x_p - x_s)^2 + (y_p - y_s)^2\right]} dx_s dy_s. \quad (3.45)$$

This computes the field by summing wavelets with a quadratic phase factor. Each aperture point emits a wavelet with amplitude $u(x_s, y_s, 0)$. Notice that the term $e^{i\frac{k}{2z_p}\left[(x_p - x_s)^2 + (y_p - y_s)^2\right]}$ reflects near-field spherical wavefronts. The factor $\frac{1}{z_p}$ accounts for amplitude decay, so it is called the amplitude scaling factor. Since the integral is the sum of complex amplitudes, it expresses diffraction patterns via interference.

Remember that this integral requires $z_p \gg \sqrt{(x_p - x_s)^2 + (y_p - y_s)^2}$. It also neglects vector properties of electromagnetic waves, since our electric field here is scalar, $u(x_p, y_p, z_p)$. Finally, it assumes the Kirchhoff boundary conditions, which means it assumes zero field outside the aperture, which is less accurate for small apertures.

Example: Diffraction through a rectangular aperture. Consider a rectangular aperture of dimensions $a \times b$ at $z = 0$, illuminated by a plane wave $u(x_s, y_s, 0) = u_0 e^{ikz}$. The aperture extends from $x_s = -\frac{a}{2}$ to $\frac{a}{2}$, and $y_s = -\frac{b}{2}$ to $\frac{b}{2}$. We compute the field at $p(x_p, y_p, z_p)$.

The incident field at $z = 0$ is $u(x_s, y_s, 0) = u_0$. The Fresnel integral becomes,

$$u(x_p, y_p, z_p) = \frac{u_0 e^{ikz_p}}{i\lambda z_p} \int_{-\frac{a}{2}}^{\frac{a}{2}} \int_{-\frac{b}{2}}^{\frac{b}{2}} e^{i\frac{k}{2z_p}\left[(x_p - x_s)^2 + (y_p - y_s)^2\right]} dx_s dy_s. \quad (3.46)$$

The exponent is,

$$(x_p - x_s)^2 + (y_p - y_s)^2 = (x_p^2 + x_s^2 - 2x_p x_s) + (y_p^2 + y_s^2 - 2y_p y_s). \quad (3.47)$$

Since the integrals separate, we write,

$$u(\mathbf{r}_p) = \frac{u_0 e^{ikz_p}}{i\lambda z_p} e^{i\frac{k}{2z_p}(x_p^2 + y_p^2)} \int_{-\frac{a}{2}}^{\frac{a}{2}} e^{i\frac{k}{2z_p}(x_s^2 - 2x_p x_s)} dx_s \int_{-\frac{b}{2}}^{\frac{b}{2}} e^{i\frac{k}{2z_p}(y_s^2 - 2y_p y_s)} dy_s. \quad (3.48)$$

Complete the square in the x_s-integral,

$$x_s^2 - 2x_p x_s = (x_s - x_p)^2 - x_p^2. \quad (3.49)$$

Thus,

$$e^{i\frac{k}{2z_p}(x_s^2 - 2x_p x_s)} = e^{i\frac{k}{2z_p}(x_s - x_p)^2} e^{-i\frac{k}{2z_p}x_p^2}. \quad (3.50)$$

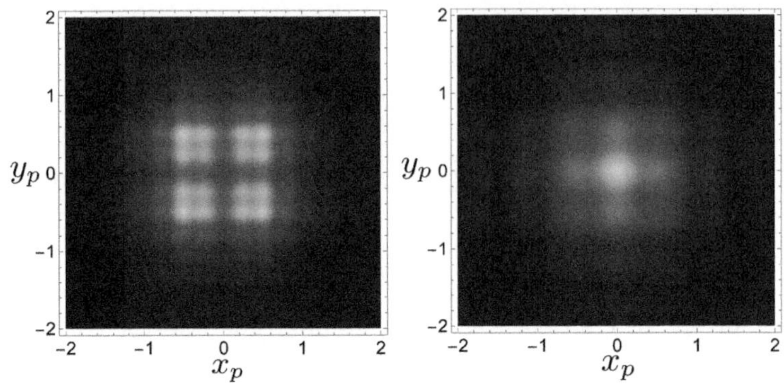

FIGURE 3.3
Fresnel diffraction of a rectangular aperture. For different image planes z_p. The parameters of the plots are $\lambda = 1$, $k = 2\pi/\lambda$, $z_p = 0.5$ (*right*), $z_p = 1$ (*left*). The grayscale has been adjusted to better show the diffraction patterns.

The x_p^2-term cancels with the prefactor. Substituting $u = x_s - x_p$, the integral becomes,

$$\int_{-\frac{a}{2}}^{\frac{a}{2}} e^{\mathrm{i}\frac{k}{2z_p}(x_s-x_p)^2} \mathrm{d}x_s = \int_{-\frac{a}{2}-x_p}^{\frac{a}{2}-x_p} e^{\mathrm{i}\frac{k}{2z_p}u^2} \mathrm{d}u. \tag{3.51}$$

Similarly for the y_s-integral. The intensity is proportional to $|u(x_p, y_p, z_p)|^2$. Fig. 3.3 illustrates the Fresnel diffraction patterns of a rectangular aperture at two distinct image planes z_P. Using simulation parameters $\lambda = 1$ and $k = 2\pi/\lambda$, the plots display the intensity distributions for $z_P = 1$ (*left*) and $z_P = 0.5$ (*right*). The grayscale mapping has been adjusted to enhance the visibility of the specific diffraction fringes and features.

The Fresnel diffraction integral quantifies near-field diffraction by summing secondary wavelets with a quadratic phase. The rectangular aperture example illustrates its application, producing characteristic diffraction patterns. Despite approximations, it is a powerful tool for optical and wave-based systems. Now, let's pay attention to the most famous and widely used diffraction integral, the Fraunhofer diffraction integral

3.7 Fraunhofer diffraction integral

The Fraunhofer diffraction integral applies in the far-field regime, where the observation distance is much larger than the aperture size and wavelength.

We derive it from the Huygens integral. So let's recall it,

$$u(\mathbf{r}_p) = \frac{1}{\mathrm{i}\lambda} \iint_s u(\mathbf{r}_s) \frac{e^{\mathrm{i}kr}}{r} \cos\theta \, \mathrm{d}s. \tag{3.21}$$

It maybe repetitive but let me remember the reader that s is the aperture plane at $z = 0$, with points $\mathbf{r}_s = [x_s, y_s, 0]$. p is observation point at $\mathbf{r}_p = [x_p, y_p, z_p]$. $r = |\mathbf{r}_p - \mathbf{r}_s| = \sqrt{(x_p - x_s)^2 + (y_p - y_s)^2 + z_p^2}$ is the distance from source $\mathbf{r}_s$ to observation point p. Here, $\cos\theta = \mathbf{n} \cdot \mathbf{r}$, with $\mathbf{r} = \frac{\mathbf{r}_p - \mathbf{r}_s}{r}$, and $\mathrm{d}s = \mathrm{d}x_s \mathrm{d}y_s$.

In the far-field ($z_p \gg \sqrt{x_s^2 + y_s^2}$, $z_p \gg \lambda$), we approximate the distance r,

$$r = \sqrt{(x_p - x_s)^2 + (y_p - y_s)^2 + z_p^2}. \tag{3.52}$$

For the amplitude, $\frac{1}{r} \approx \frac{1}{z_p}$. For the phase, we use,

$$r \approx z_p + \frac{x_p^2 + y_p^2}{2z_p} - \frac{x_p x_s + y_p y_s}{z_p}. \tag{3.53}$$

The quadratic term $\frac{x_s^2 + y_s^2}{2z_p}$ is negligible in the far-field. Thus,

$$e^{\mathrm{i}kr} \approx e^{\mathrm{i}k\left(z_p + \frac{x_p^2 + y_p^2}{2z_p}\right)} e^{-\mathrm{i}k\frac{x_p x_s + y_p y_s}{z_p}}. \tag{3.54}$$

The obliquity factor is,

$$\cos\theta = \frac{z_p}{r} \approx 1. \tag{3.55}$$

Substituting, we get the Fraunhofer diffraction integral,

$$u(\mathbf{r}_p) = \frac{e^{\mathrm{i}k\left(z_p + \frac{x_p^2 + y_p^2}{2z_p}\right)}}{\mathrm{i}\lambda z_p} \iint_s u(x_s, y_s, 0) e^{-\mathrm{i}\frac{k}{z_p}(x_p x_s + y_p y_s)} \mathrm{d}x_s \mathrm{d}y_s. \tag{3.56}$$

Fraunhofer diffraction integral is a Fourier transform of the aperture field, scaled by a phase factor and a constant.

Once again, like in other diffraction integrals, each aperture point emits a wavelet with amplitude $u(x_s, y_s, 0)$. The term $e^{-\mathrm{i}\frac{k}{z_p}(x_p x_s + y_p y_s)}$ represents plane wave propagation in the far-field. The integral is a 2D Fourier transform, mapping the aperture field to the diffraction pattern. The superposition produces characteristic far-field patterns, since it is an integral, it sums (superposition).

The integral is used to analyze far-field diffraction patterns, for example: slits, gratings, and apertures. It is widely used in designed optical systems, such as telescopes and microscopes.

Remember that the far-field approximation requires $z_p \gg \sqrt{x_s^2 + y_s^2}$, so this integral is only valid in those conditions. Like the other scalar integral, all the vector properties of electromagnetic waves are fully ignored. Once again, the integral is computed only across the aperture, since the Kirchhoff boundary conditions are assumed, the field is zero outside the aperture.

Example: Single slit diffraction. Consider a rectangular slit of width a along x_s, infinite along y_s, at $z = 0$, illuminated by a plane wave $u(x_s, y_s, 0) = u_0$. The aperture extends from $x_s = -\frac{a}{2}$ to $\frac{a}{2}$. The integral becomes,

$$u(x_p, y_p, z_p) = \frac{u_0 e^{\mathrm{i}k\left(z_p + \frac{x_p^2 + y_p^2}{2z_p}\right)}}{\mathrm{i}\lambda z_p} \int_{-\infty}^{\infty} e^{-\mathrm{i}\frac{k y_p y_s}{z_p}} \mathrm{d}y_s \int_{-\frac{a}{2}}^{\frac{a}{2}} e^{-\mathrm{i}\frac{k x_p x_s}{z_p}} \mathrm{d}x_s. \tag{3.57}$$

The y_s-integral gives a delta function, so we focus on,

$$u(x_p, z_p) \propto \int_{-\frac{a}{2}}^{\frac{a}{2}} e^{-\mathrm{i}\frac{k x_p x_s}{z_p}} \mathrm{d}x_s. \tag{3.58}$$

Let $\beta = \frac{k x_p a}{2 z_p}$, then,

$$u(x_p, z_p) \propto \int_{-\frac{a}{2}}^{\frac{a}{2}} e^{-\mathrm{i}\frac{2\beta x_s}{a}} \mathrm{d}x_s = a\frac{\sin\beta}{\beta}. \tag{3.59}$$

Intensity is,

$$I \propto \left|\frac{\sin\beta}{\beta}\right|^2, \quad \beta = \frac{\pi a x_p}{\lambda z_p}. \tag{3.60}$$

Minima occur at $\beta = n\pi$, or $x_p = \frac{n\lambda z_p}{a}$.

Example: Fraunhofer diffraction of a rectangular aperture. Fig. 3.4 depicts the Fraunhofer diffraction patterns produced by a rectangular aperture with dimensions $a = 1 \times 10^{-3}$ mm and $b = 0.5 \times 10^{-3}$ mm. The intensity distributions are calculated at a fixed image plane $z_P = 1$ mm for two different wavelengths: $\lambda = 555$ nm (left) and $\lambda = 633$ nm (right). To ensure the secondary maxima and fine fringe structures are clearly visible, the grayscale contrast has been modified.

Example: Double slit diffraction. Consider two slits of width a, separated by distance d, centered at $x_s = \pm\frac{d}{2}$. The aperture function is,

$$u(x_s, y_s, 0) = u_0 \left[\mathrm{rect}\left(\frac{x_s - \frac{d}{2}}{a}\right) + \mathrm{rect}\left(\frac{x_s + \frac{d}{2}}{a}\right)\right]. \tag{3.61}$$

The integral splits,

$$u(x_p, z_p) \propto \int_{-\frac{a}{2}+\frac{d}{2}}^{\frac{a}{2}+\frac{d}{2}} e^{-\mathrm{i}\frac{k x_p x_s}{z_p}} \mathrm{d}x_s + \int_{-\frac{a}{2}-\frac{d}{2}}^{\frac{a}{2}-\frac{d}{2}} e^{-\mathrm{i}\frac{k x_p x_s}{z_p}} \mathrm{d}x_s. \tag{3.62}$$

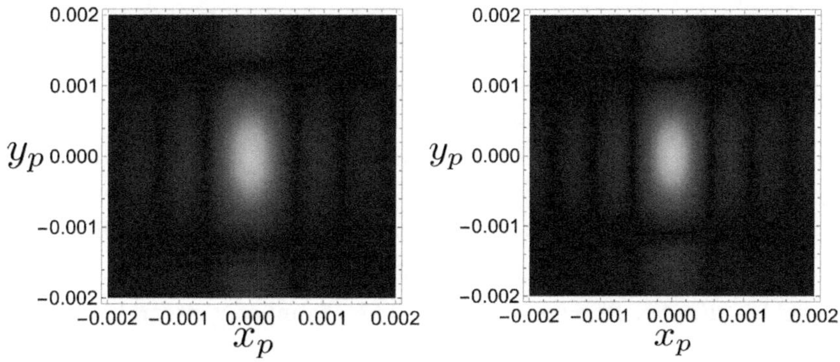

FIGURE 3.4
Fraunhofer diffraction of a rectangular aperture. For different image planes z_p. The parameters of the plots are $\lambda = 633$nm (*right*) $\lambda = 555$nm (*left*), $k = 2\pi/\lambda$, $z_p = 1$mm, $a = 1\times^{-3}$mm, $b = 0.5\times^{-3}$mm. The grayscale has been adjusted to better show the diffraction patterns.

This yields,

$$u(x_p, z_p) \propto \frac{\sin\beta}{\beta}\cos\left(\frac{kx_p d}{2z_p}\right), \quad \beta = \frac{\pi a x_p}{\lambda z_p}. \tag{3.63}$$

Intensity is,

$$I \propto \left(\frac{\sin\beta}{\beta}\right)^2 \cos^2\left(\frac{\pi d x_p}{\lambda z_p}\right). \tag{3.64}$$

Interference maxima occur at $x_p = \frac{m\lambda z_p}{d}$.

The Fraunhofer diffraction integral quantifies far-field diffraction as a Fourier transform of the aperture field. The examples illustrate its application to classic diffraction patterns, showing characteristic features like minima and interference fringes. Despite its approximations, it is essential for optical and wave-based systems.

3.8 Airy disc

The Airy disc, named after George Biddell Airy, is the far-field diffraction pattern produced by a circular aperture. It is central to wave optics, limiting the resolution of optical systems like telescopes and microscopes [35, 50].

The Airy disc arises in the far-field, modeled by the Fraunhofer diffraction integral, recalling it,

$$u(x_p, y_p, z_p) = \frac{e^{ik\left(z_p + \frac{x_p^2 + y_p^2}{2z_p}\right)}}{i\lambda z_p} \iint_s u(x_s, y_s, 0) e^{-i\frac{k}{z_p}(x_p x_s + y_p y_s)} dx_s dy_s. \quad (3.56)$$

For a circular aperture of radius a, we use polar coordinates: $(x_s, y_s) = (\rho\cos\phi, \rho\sin\phi)$, $(x_p, y_p) = (r_p\cos\theta, r_p\sin\theta)$, where $r_p = \sqrt{x_p^2 + y_p^2}$. The incident field is $u(x_s, y_s, 0) = u_0$ for $\rho \leq a$. The Fraunhofer integral becomes,

$$u(r_p, z_p) = \frac{u_0 e^{ik\left(z_p + \frac{r_p^2}{2z_p}\right)}}{i\lambda z_p} \int_0^a \int_0^{2\pi} e^{-i\frac{kr_p\rho\cos(\phi-\theta)}{z_p}} \rho d\phi d\rho. \quad (3.65)$$

The ϕ-integral yields a Bessel function of the first kind,

$$\int_0^{2\pi} e^{-i\frac{kr_p\rho\cos(\phi-\theta)}{z_p}} d\phi = 2\pi J_0\left(\frac{kr_p\rho}{z_p}\right). \quad (3.66)$$

Thus,

$$u(r_p, z_p) = \frac{u_0 e^{ik\left(z_p + \frac{r_p^2}{2z_p}\right)}}{i\lambda z_p} 2\pi \int_0^a J_0\left(\frac{kr_p\rho}{z_p}\right) \rho d\rho. \quad (3.67)$$

Substitute $u = \frac{kr_p\rho}{z_p}$, so $\rho = \frac{z_p u}{kr_p}$, $d\rho = \frac{z_p}{kr_p} du$,

$$\int_0^a J_0\left(\frac{kr_p\rho}{z_p}\right) \rho d\rho = \frac{z_p^2}{k^2 r_p^2} \int_0^{\frac{kar_p}{z_p}} J_0(u) u du. \quad (3.68)$$

Using the identity $\int u J_0(u) du = u J_1(u)$, we evaluate [16],

$$\int_0^{\frac{kar_p}{z_p}} J_0(u) u du = [u J_1(u)]_0^{\frac{kar_p}{z_p}} = \frac{kar_p}{z_p} J_1\left(\frac{kar_p}{z_p}\right). \quad (3.69)$$

Thus,

$$u(r_p, z_p) = \frac{u_0 e^{ik\left(z_p + \frac{r_p^2}{2z_p}\right)}}{i\lambda z_p} \frac{2\pi z_p^2}{k^2 r_p^2} \cdot \frac{kar_p}{z_p} J_1\left(\frac{kar_p}{z_p}\right). \quad (3.70)$$

Simplify,

$$u(r_p, z_p) = u_0 \frac{\pi a^2}{i\lambda z_p} e^{ik\left(z_p + \frac{r_p^2}{2z_p}\right)} \frac{2J_1\left(\frac{kar_p}{z_p}\right)}{\frac{kar_p}{z_p}}. \quad (3.71)$$

The intensity of the Airy disc is,

$$I \propto \left| \frac{2\mathrm{J}_1\left(\frac{kar_p}{z_p}\right)}{\frac{kar_p}{z_p}} \right|^2. \tag{3.72}$$

The first minimum occurs at $\frac{kar_p}{z_p} \approx 3.8317$, or,

$$r_p/z_p \approx \frac{3.8317\lambda}{2\pi a} = \frac{1.22\lambda}{2a}. \tag{3.73}$$

The Airy disc consists of a bright central maximum surrounded by concentric rings of decreasing intensity. The bright central disc contains most of the light energy. The concentric rings, see Fig. 3.5, are the result of interference, with the first minimum at $r_p/z_p \approx \frac{1.22\lambda}{2a}$.

A diffraction-limited imaging system is characterized by its ability to convert a diverging spherical wave, originating from a point-source object, into a perfectly spherical wave once again [23]. This newly formed spherical wave converges towards an ideal point within the image plane. The location of this ideal image point is directly related to the position of the original object point through a consistent scaling factor, known as magnification. For the system to be considered ideal, this magnification factor must remain uniform across all points within the image field of interest [21, 51, 26].

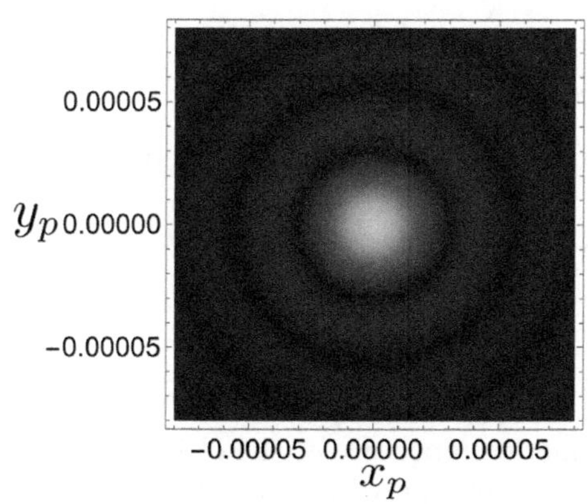

FIGURE 3.5
The Airy disk is the central lobe of the diffraction pattern. It the lobe presented in yellow and within the figure. The grayscale has been adjusted to better show the diffraction pattern of the concentric rings. The x_p-y_p is the image plane and its units are in milliliters. The parameters of this example are $\lambda = 500$nm, $a = 1$cm, $z_p = 1$m.

It is critical in understanding the resolution limits of optical systems like telescopes, microscopes, and cameras, as the size of the Airy disk determines the smallest resolvable detail, governed by the diffraction limit. It sets a minimum resolvable distance between two points. Additionally, the Airy disk's size is wavelength-dependent, with longer wavelengths producing larger disks, reducing resolution in certain spectral ranges. Optical aberrations, such as spherical or chromatic aberration, can further distort the Airy disk, degrading image quality.

Finally, it is good to remember that the Airy disk, since it comes from the Fraunhofer diffraction integral, has all the approximations that the Fraunhofer integral has by design. So it comes from the far-Field approximation, then it requires $z_p \gg a^2/\lambda$. It neglects the polarization effects of electromagnetic waves. That means that the diffraction limit on scalar theory that the Airy disk imposes may not be the same when the vector behavior of the electric field is considered. It is less accurate near edges, since it assumes the Kirchhoff boundary conditions (the field is zero outside the aperture).

The Airy disc limits a telescope's ability to resolve two point sources. The Rayleigh criterion states that two Airy discs are resolvable if the maximum of one falls on the first minimum of the other, giving the angular resolution,

$$\theta \approx \frac{1.22\lambda}{D} \tag{3.74}$$

where $D = 2a$ is the aperture diameter.

Example: For a telescope with aperture diameter $D = 100\,\text{mm}$, wavelength $\lambda = 550\,\text{nm} = 5.5 \times 10^{-4}\,\text{mm}$,

$$\theta \approx \frac{1.22 \times 5.5 \times 10^{-4}}{100} = 6.71 \times 10^{-6}\,\text{rad} \approx 1.38\,\text{arcseconds}. \tag{3.75}$$

Two stars separated by less than 1.38 arcseconds appear as a single blurred image.

In a microscope, the Airy disc determines the minimum resolvable distance between two points, approximated as,

$$d \approx \frac{1.22\lambda}{2\text{NA}} \tag{3.76}$$

where NA is the numerical aperture of the objective lens.

Example: For a microscope with $\text{NA} = 1.4$, $\lambda = 550\,\text{nm} = 0.55\,\mu\text{m}$,

$$d \approx \frac{1.22 \times 0.55}{2 \times 1.4} \approx 0.24\,\mu\text{m}. \tag{3.77}$$

Objects closer than 0.24 μm cannot be resolved as distinct points.

3.9 Conclusion

This chapter has presented the foundational framework of scalar diffraction theory, establishing the mathematical and physical tools required to predict how coherent light propagates through apertures and obstacles when feature sizes approach the wavelength scale. Beginning from Maxwell's equations in source-free media, we derived the scalar Helmholtz equation that governs monochromatic wave fields, then combined it with Huygens' principle – the concept that every point on a wavefront acts as a source of secondary spherical wavelets – to arrive at Kirchhoff's rigorous diffraction integral via Green's theorem.

Practical computation, however, demands controlled approximations. The Huygens–Fresnel integral simplifies Kirchhoff's formula by neglecting backward propagation and the derivative term. The Fresnel diffraction integral further introduces the paraxial quadratic-phase approximation, accurately describing near-field patterns where curved wavefronts produce characteristic interference fringes. Finally, in the far-field Fraunhofer regime, the diffraction field reduces to the Fourier transform of the aperture function – a result of profound importance that links spatial aperture geometry directly to angular intensity distributions.

A central result of far-field Fraunhofer approximation is the Airy disc, which is the diffraction pattern of a circular aperture. The Airy disc, with its central maximum radius $1.22\lambda/(2a)$ and the associated Rayleigh resolution limit $1.22\lambda/D$, sets the ultimate performance bound for diffraction-limited imaging systems. Scalar theory thus provides both predictive power and deep insight into resolution limits that no purely geometrical model can capture, forming the essential bridge between ray optics and full wave optics.

Problems

Problem 1: Why the far-field pattern is a Fourier transform

The chapter states that in the Fraunhofer regime the diffracted field is essentially the Fourier transform of the aperture transmittance function. Explain conceptually why the far-field diffraction pattern takes this specific mathematical form (a Fourier transform). In particular, discuss what physical feature of plane-wave propagation and interference in the far zone leads to the linear phase term $-\mathrm{i}k(x_p x_s + y_p y_s)/z_p$ appearing in the exponent — and why this term naturally produces a Fourier kernel rather than, say, a derivative or a convolution.

Problem 2: The dropped quadratic phase term

In deriving the Fraunhofer integral from the more general expression, the chapter approximates the propagation distance

$$r \approx z_p + \frac{x_p^2 + y_p^2}{2z_p} - \frac{x_p x_s + y_p y_s}{z_p}$$

and explicitly neglects the quadratic term in the aperture coordinates $(x_s^2 + y_s^2)/(2z_p)$ inside the phase factor. Explain physically why this quadratic phase term can be ignored in the true far field ($z_p \gg$ aperture size$^2/\lambda$), but would cause serious errors if kept when the observation distance is only moderately large. What kind of wavefront curvature or interference behavior does retaining that term actually describe?

Problem 3: Scaling laws of the Airy pattern with aperture size

The chapter derives the Airy disc intensity pattern for a circular aperture of radius a and shows that the radius of the first dark ring is

$$r_p/z_p \approx \frac{1.22\lambda}{2a}.$$

Consider two otherwise identical optical systems (same wavelength λ, same observation distance z) but with different circular aperture diameters: one with diameter $D = 2a$ and another with diameter $4a$. Qualitatively compare (and explain why) the following features of their Fraunhofer diffraction patterns will differ:

(a) the angular size of the central bright disc,

(b) the angular spacing between successive bright rings,

(c) the fraction of total power contained inside the central maximum.

Problem 4: Fraunhofer pattern of an "infinite" versus a finite aperture

The chapter emphasizes that the Fraunhofer integral is computed only over the finite aperture (with Kirchhoff boundary conditions assuming zero field outside). Imagine a hypothetical situation in which the aperture function is constant (plane-wave illumination) over an infinitely extended plane with no obstacles or edges whatsoever.

(a) What would the mathematical result of the Fraunhofer integral be in this idealized case?

(b) Physically interpret this result in terms of wave propagation.

(c) Explain why — even when using a very large laboratory aperture or telescope pupil — we never actually observe this idealized mathematical outcome in experiments, and how the finite pupil size fundamentally modifies the observed far-field pattern (refer to the chapter's discussion of the Airy disc and resolution limit).

Part III

Stigmatic vector diffraction theory

4

Stigmatic vector diffraction theory: Richards-Wolfs Integrals

This chapter introduces stigmatic vector diffraction theory via the Richards–Wolf diffraction integrals, providing a rigorous vectorial description of electromagnetic fields near the focus of high-numerical-aperture aplanatic systems. It begins with the fundamentals of polarization, deriving the transversality of plane waves and reviewing linear, circular, and elliptical states, together with Fresnel coefficients at interfaces. The angular spectrum representation is presented, linking propagating and evanescent waves to Fourier optics and geometrical optics (far-field) through the stationary phase method. The Richards–Wolf integrals are then derived, expressing focused electric and magnetic fields by integrating vectorial contributions over a spherical reference wavefront (stigmatic), incorporating apodization and polarization transformations. Key properties—linearity, scaling, rotational symmetry, transversality, and intensity—are analyzed. Polarization effects are explored in detail for uniform linear, radial, and azimuthal incident fields, revealing distinct focal structures, including strong longitudinal components and needle-like spots in radial polarization.

4.1 Introduction

This chapter delves into the fundamental concepts of polarization and its interplay with vector diffraction, providing a comprehensive foundation for vector based optical phenomena [7]. We begin by exploring the basics of polarization, establishing key principles that govern the behavior of light's electric field orientation. Building on this, we introduce the angular spectrum representation, a powerful method for analyzing wave propagation in free space.

DOI: 10.1201/9781003726494-7

The method of the stationary phase is then discussed, offering insights to introduce vector diffraction. Central to the chapter is an in-depth examination of the Richards-Wolf diffraction integrals, which describe diffraction considering the vector behavior of the electric field. We explore the properties of these integrals, highlighting their mathematical elegance and physical significance.

Finally, we investigate polarization effects on the Richards-Wolf integrals, revealing how polarization influences the structure of focused fields and their applications in modern optics, in a particular case in photolithography. This chapter aims to equip readers with the theoretical tools and conceptual understanding necessary to navigate the intricate relationship between polarization and diffraction, via Richards-Wolf diffraction integrals.

4.2 Polarization

We have mentioned before that diffraction can be analyzed with the scalar and vector diffraction theories. The vector theory is more accurate than the scalar theory since the electric field is a vector identity. It considers the polarization state of the electric field. For a monochromatic plane wave propagating in a homogeneous, isotropic medium free of charge, the electric field can be expressed as [52, 53],

$$\mathbf{E}(\mathbf{r},t)=\mathbf{E}_0e^{\mathrm{i}(\mathbf{k}\cdot\mathbf{r}-\omega t)} \tag{4.1}$$

where $\mathbf{E}_0$ is the complex amplitude, $\mathbf{k}$ is the propagation vector ($|\mathbf{k}|=k=\frac{2\pi}{\lambda}$), $\omega=kc$ is the angular frequency ($c=\frac{1}{\sqrt{\mu\epsilon}}$), and $\mathbf{r}$ is the position vector.

If we apply the divergence in Eq. (4.1), must be zero since homogeneous medium free of charge,

$$\nabla\cdot\mathbf{E}=\mathrm{i}\mathbf{k}\cdot\mathbf{E}_0e^{\mathrm{i}(\mathbf{k}\cdot\mathbf{r}-\omega t)}=0\implies\mathbf{k}\cdot\mathbf{E}_0=0 \tag{4.2}$$

The electric field is transverse to the propagation direction. Now, lets apply the curl operator on $\mathbf{E}$ to find $\mathbf{B}$, via Maxwell's equations (Faraday's law, Eq. (3.7)),

$$\nabla\times\mathbf{E}=\mathrm{i}\mathbf{k}\times\mathbf{E}_0e^{\mathrm{i}(\mathbf{k}\cdot\mathbf{r}-\omega t)}=-\frac{\partial\mathbf{B}}{\partial t} \tag{4.3}$$

We can assume that, $\mathbf{B}$ is also a plane wave,

$$\mathbf{B}(\mathbf{r},t)=\mathbf{B}_0e^{\mathrm{i}(\mathbf{k}\cdot\mathbf{r}-\omega t)}, \tag{4.4}$$

computing the time derivative on the last expression,

$$-\frac{\partial\mathbf{B}}{\partial t}=\mathrm{i}\omega\mathbf{B}_0e^{\mathrm{i}(\mathbf{k}\cdot\mathbf{r}-\omega t)}. \tag{4.5}$$

Equating expressions Eq. (4.3) and Eq. (4.5),

$$\mathrm{i}\mathbf{k} \times \mathbf{E}_0 = \mathrm{i}\omega\mathbf{B}_0 \implies \mathbf{B}_0 = \frac{1}{\omega}\mathbf{k} \times \mathbf{E}_0 = \frac{\mathbf{k} \times \mathbf{E}_0}{kc} = \frac{1}{c}\hat{\mathbf{k}} \times \mathbf{E}_0, \tag{4.6}$$

where $\hat{\mathbf{k}} = \mathbf{k}/c$. Now, lets verify $\nabla \cdot \mathbf{B} = 0$,

$$\mathbf{k} \cdot \mathbf{B}_0 = \mathbf{k} \cdot \left(\frac{1}{c}\hat{\mathbf{k}} \times \mathbf{E}_0\right) = \frac{1}{c}\hat{\mathbf{k}} \cdot (\hat{\mathbf{k}} \times \mathbf{E}_0) = 0. \tag{4.7}$$

Verify curl of $\mathbf{B}$, Ampere's law, Eq. (3.4),

$$\nabla \times \mathbf{B} = \mathrm{i}\mathbf{k} \times \mathbf{B}_0 e^{\mathrm{i}(\mathbf{k}\cdot\mathbf{r}-\omega t)} = \mu\epsilon\frac{\partial \mathbf{E}}{\partial t} = -\mathrm{i}\omega\mu\epsilon\mathbf{E}_0 e^{\mathrm{i}(\mathbf{k}\cdot\mathbf{r}-\omega t)} \tag{4.8}$$

Applying the property of the double cross product, in the last expression,

$$\mathbf{k} \times \mathbf{B}_0 = \mathbf{k} \times \left(\frac{1}{c}\hat{\mathbf{k}} \times \mathbf{E}_0\right) = \frac{1}{c}[\mathbf{k}(\hat{\mathbf{k}} \cdot \mathbf{E}_0) - \mathbf{E}_0(\mathbf{k} \cdot \hat{\mathbf{k}})] = -\frac{k}{c}\mathbf{E}_0 \tag{4.9}$$

$$-\omega\mu\epsilon\mathbf{E}_0 = -\frac{\omega}{c^2}\mathbf{E}_0 = -\frac{k}{c}\mathbf{E}_0 \tag{4.10}$$

This is consistent, confirming the plane wave solution.

$$\mathbf{E}(\mathbf{r}, t) = \mathbf{E}_0 e^{\mathrm{i}(\mathbf{k}\cdot\mathbf{r}-\omega t)}, \quad \mathbf{B}(\mathbf{r}, t) = \frac{1}{c}(\hat{\mathbf{k}} \times \mathbf{E}_0)e^{\mathrm{i}(\mathbf{k}\cdot\mathbf{r}-\omega t)} \tag{4.11}$$

From this we have set the basis for the study of polarization. The electric $\mathbf{E}(\mathbf{r}, t)$ and magnetic field $\mathbf{B}(\mathbf{r}, t)$ are orthogonal between them. Both of them had the same frequency w and they are orthogonal respect for a propagation vector $\mathbf{k}$.

The Poynting vector $\mathbf{s}$ represents the energy flux (power per unit area) of an electromagnetic wave. It is defined as,

$$\mathbf{s} = \frac{1}{\mu}\mathbf{E} \times \mathbf{B} \tag{4.12}$$

where μ is the permeability of the medium. Notice that $\mathbf{s}$ is parallel to $\mathbf{k}$. Notice that $\mathbf{k}$ has the direction of the rays traced in geometrical optics, in chapter 2, therefore $\mathbf{s}$ is parallel to $\mathbf{k}$. A representation of monochromatic plane wave propagating in a homogeneous, isotropic medium free of charge is presented in Figure (4.1). From the picture it is clear that the electric field and magnetic field are perpendicular to each other and perpendicular to to propagation vector. The magnitude of the oscillations along the propagation axis. Since this monochromatic is represented by Eq. (4.1).

Polarization of light refers to the orientation of the oscillations of the electric/magnetic field component of an electromagnetic wave. The polarization is based on Eqs. (4.1), (4.11), and (4.12). These vector quantities describe the

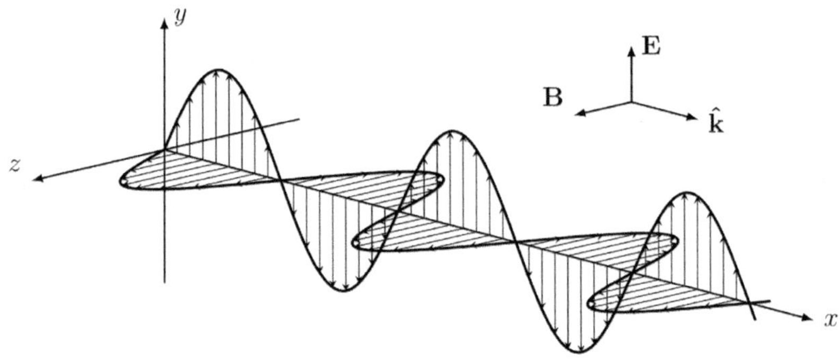

FIGURE 4.1
A representation of monochromatic plane wave propagating in a homogeneous, isotropic medium free of charge. From the picture it is clear that the electric field and magnetic field are perpendicular to each other and perpendicular to to propagation vector **s**.

propagation of electromagnetic waves in a homogeneous, isotropic, charge-free medium. How these vectors interact explains in detail the polarization phenomena of electromagnetic waves. For convenience, polarization is usually described only in terms of the electric field **E**, since, by the given equations, obtaining **B** is straightforward.

In an unpolarized light wave, the electric field oscillates in all possible directions perpendicular to the direction of propagation. When light is polarized. There are several ways in which light can be polarized; the most common cases are elliptical, linear, and circular polarizations [53]. This polarization has a huge impact on the effect of vector diffraction patterns, and a linear combination of them can lead to other polarization states. So let's review them case by case. First, we focus on elliptical polarization since it is the most general case, being linear and circular polarization special cases of it.

4.2.1 Elliptical polarization

Elliptical polarization is the most general state of polarization of an electromagnetic wave. It occurs when the tip of the electric field vector traces out an *ellipse* in the plane perpendicular to the direction of propagation. Linear and circular polarization are special cases [52, 53].

Consider a plane wave propagating along the z-axis,

$$\mathbf{E}(z,t) = E_x \cos(\omega t - kz)\hat{\mathbf{x}} + E_y \cos(\omega t - kz + \delta)\hat{\mathbf{y}}, \tag{4.13}$$

where E_x and E_y are the amplitudes of the components, δ is the phase difference between them, ω is the angular frequency, and k is the wavenumber.

The geometry of the ellipse is characterized by its axial ratio (AR), defined as the ratio of the major to minor axes. Another important quantity is the ellipticity angle χ, defined as

$$\chi = \tan^{-1}\left(\frac{b}{a}\right), \tag{4.14}$$

where a and b denote the semi-major and semi-minor axes, respectively. When $\chi = 0°$ the polarization is linear, while $\chi = \pm 45°$ corresponds to circular polarization. In addition, the *handedness* of the polarization is determined by the sense of rotation of the electric field vector: if the vector rotates counterclockwise (as seen by an observer facing the incoming wave), the light is said to be left-handed, while clockwise rotation corresponds to right-handed polarization.

The electric field of an elliptically polarized wave can be expressed in terms of its two orthogonal components as, for convenience let's set $z = 0$, then we have,

$$E_x(t) = E_x \cos(\omega t), \qquad E_y(t) = E_y \cos(\omega t + \delta), \tag{4.15}$$

where E_x and E_y are the field amplitudes along the x- and y-directions, and δ is the phase difference between them. By eliminating the time dependence, the relation between E_x and E_y can be cast into the Cartesian equation of an ellipse,

$$\left(\frac{E_x(t)}{E_x}\right)^2 + \left(\frac{E_y(t)}{E_y}\right)^2 - 2\frac{E_x(t)}{E_x}\frac{E_y(t)}{E_y}\cos\delta = \sin^2\delta, \tag{4.16}$$

which confirms that the tip of the electric field vector traces out an ellipse in the transverse plane.

The semi-major and semi-minor axes of this ellipse, denoted by a and b, can be obtained from the eigenvalues of the quadratic form above. The squared lengths of these axes are given by

$$a^2, b^2 = \tfrac{1}{2}\left(E_x^2 + E_y^2 \pm \sqrt{(E_x^2 - E_y^2)^2 + 4E_x^2E_y^2\cos^2\delta}\right). \tag{4.17}$$

The larger value corresponds to the semi-major axis a, while the smaller value corresponds to the semi-minor axis b. Explicitly,

$$a = \sqrt{\tfrac{1}{2}\left(E_x^2 + E_y^2 + \sqrt{(E_x^2 - E_y^2)^2 + 4E_x^2E_y^2\cos^2\delta}\right)}, \tag{4.18}$$

$$b = \sqrt{\tfrac{1}{2}\left(E_x^2 + E_y^2 - \sqrt{(E_x^2 - E_y^2)^2 + 4E_x^2E_y^2\cos^2\delta}\right)}. \tag{4.19}$$

These expressions reduce to the expected special cases. When the phase difference is $\delta = 0$ or π, the ellipse collapses into a straight line and the polarization is linear, which corresponds to $b = 0$. When the amplitudes are equal $E_x = E_y$ and the phase difference is $\delta = \pm\frac{\pi}{2}$, the ellipse becomes a circle with $a = b$, describing circular polarization. In all other situations the polarization remains elliptical with unequal semi-axes.

FIGURE 4.2
In this diagram the magnitude of E_y is bigger than the magnitude of E_x forming a ellipticity polarization pattern. For the magnetic field there is a similar diagram which is orthogonal to the one presented in figure tal. he projection of the polarization is placed at coordinate system and its shape is in an ellipse.

A diagram of the elliptical polarization is shown at Fig. 4.2. In this diagram the magnitude of E_y is bigger than the magnitude of E_x forming a ellipticity polarization pattern. For the magnetic field there is a similar diagram which is orthogonal to the one presented in Fig. 4.2. Notice that a coordinate system we have drawn the shape of the projection polarization of the polarization, the shape in an ellipse as explained in the previews paragraphs.

4.2.2 Linear polarization

Linear polarization represents the simplest form of polarization, in which the electric field vector oscillates along a fixed straight line in the plane perpendicular to the direction of wave propagation. This occurs when the two orthogonal components of the electric field are either in phase or out of phase by exactly π. In this case, the tip of the electric field vector does not trace out an ellipse but instead moves back and forth along a straight line, whose orientation depends on the relative amplitudes of the two field components.

Mathematically, if the electric field is written as

$$\mathbf{E}(z,t) = E_x \cos(\omega t - kz)\hat{\mathbf{x}} + E_y \cos(\omega t - kz + \delta)\hat{\mathbf{y}}. \tag{4.20}$$

The linear polarization occurs when $\delta = 0$ or $\delta = \pi$. The resulting field can always be expressed as a single cosine function along some fixed direction in the x-y plane, given by

$$\mathbf{E}(z,t) = E_0 \cos(\omega t - kz)\hat{\mathbf{u}}, \tag{4.21}$$

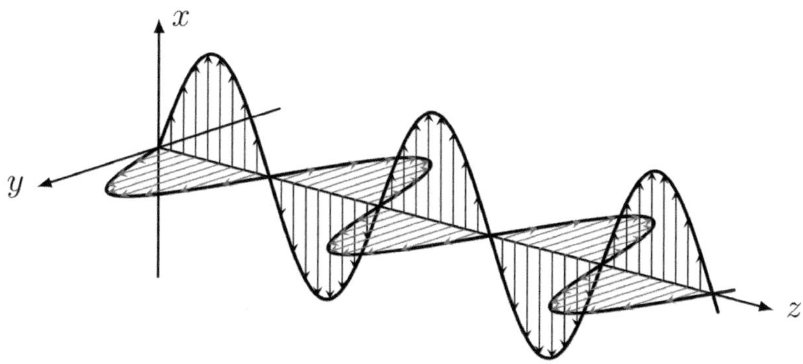

FIGURE 4.3
A diagram of the linear polarization. The axis of propagation is along the z-axis. The two orthogonal components of the electric field are either in phase or out of phase by exactly π.

where $\hat{\mathbf{u}}$ is a unit vector defining the polarization direction and E_0 is the resultant amplitude. The angle of this polarization direction with respect to the x-axis depends on the ratio E_y/E_x of the two field amplitudes. It is also important to note that when one of the components vanishes—for example, $E_y = 0$, the ellipse degenerates into a straight line, corresponding also to linear polarization.

Therefore, linear polarization can be interpreted as a special case of elliptical polarization, in which the ellipse degenerates into a line because the semi-minor axis vanishes. The orientation of this line is fixed, and the electric field oscillates back and forth along it without any rotation of the tip of the vector. Linearly polarized light can be achieved by passing unpolarized light through a polarizing filter, which selectively transmits light waves oscillating in a specific orientation while blocking others. See Fig. 4.3 for a graphical representation of a monochromatic linear polarization beam.

4.2.3 Circular polarization

Circular polarization is a particular case of elliptical polarization in which the semi-major and semi-minor axes of the polarization ellipse are equal, so that the tip of the electric field vector traces out a circle in the plane perpendicular to the direction of propagation. In this state, the magnitude of the electric field remains constant while its direction rotates uniformly in time.

Mathematically, circular polarization occurs when the amplitudes of the two orthogonal components are equal, $E_x = E_y$, and the phase difference between them is $\delta = \pm\frac{\pi}{2}$. Under these conditions, the electric field can be

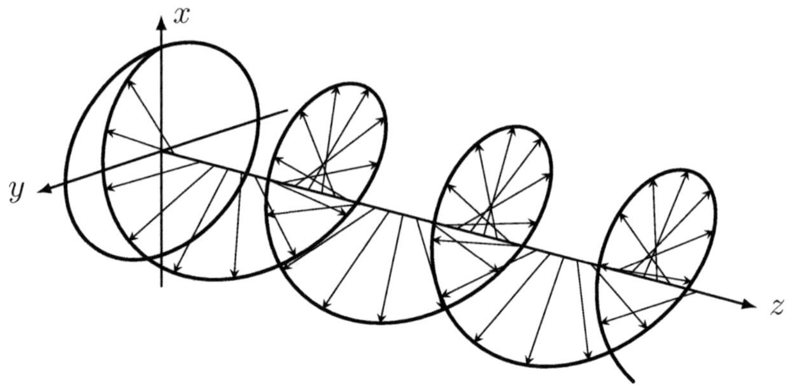

FIGURE 4.4
Graphical representation of the circular polarization.

written as,

$$\mathbf{E}(z,t) = E_0 \cos(\omega t - kz)\hat{\mathbf{x}} + E_0 \cos\big(\omega t - kz \pm \tfrac{\pi}{2}\big)\,\hat{\mathbf{y}}, \qquad (4.22)$$

which describes a vector of constant length rotating with angular frequency ω. The sign of the phase difference determines the sense of rotation: a positive phase shift of $+\frac{\pi}{2}$ corresponds to left-handed circular polarization, while a negative phase shift of $-\frac{\pi}{2}$ gives right-handed circular polarization, as viewed by an observer looking into the incoming wave.

Circular polarization is therefore distinguished by its perfect rotational symmetry in the transverse plane. Unlike linear polarization, where the electric field oscillates back and forth along a fixed axis, here the field vector maintains constant magnitude and continuously rotates, describing a circle as time evolves.

A diagram of the circular polarization is shown in Fig. 4.4. The projection of the polarization is shown in the coordinate system. The projection indeed is a circle.

4.2.4 Fresnel coefficients

When an electromagnetic wave encounters the boundary between two media with different refractive indices, its electric field components parallel and perpendicular to the plane of incidence experience different transmission and reflection amplitudes. These effects are described by the Fresnel coefficients, denoted r^s, r^p for reflection and t^s, t^p for transmission, where the subscripts s and p refer to the electric field components polarized perpendicular (s) and parallel (p) to the plane of incidence [52, 53].

If the incident light is linearly polarized but oriented at an arbitrary angle with respect to the plane of incidence, its electric field can be decomposed into s and p components. Upon reflection or transmission, each of these components is scaled by its corresponding Fresnel coefficient.

Both reflection and transmission coefficients are determined by the Fresnel equations, which are derived from Maxwell's equations describing the behavior of electromagnetic waves at boundaries between media with different refractive indices. These equations provide explicit formulas for calculating the reflection and transmission coefficients based on the angle of incidence, the refractive indices of the two media, and the polarization state of the incident light. The Fresnel equations are [26],

$$r^s(\theta_1) = \left[\frac{n_1 \cos\theta_1 - n_2 \cos\theta_2}{n_1 \cos\theta_1 + n_2 \cos\theta_2}\right]^2 \tag{4.23}$$

and

$$r^p(\theta_1) = \left[\frac{n_1 \cos\theta_2 - n_2 \cos\theta_1}{n_1 \cos\theta_2 + n_2 \cos\theta_1}\right]^2 \tag{4.24}$$

where θ_1 and θ_2 are the incident and refractive angles, respectively, $r^s(\theta_1)$ and $r^p(\theta_1)$ are the reflectance of light with polarization vertical and parallel to the reflection plane, respectively, as illustrated in Fig. 4.5. In the figure, the refraction index of the first medium is n_1 the refraction index of the second medium is n_2. $\mathbf{n}$ is the normal vector of the surface. $\mathbf{s}_1$ is a unit vector with the direction of the incident ray and $\mathbf{s}_2$ is a unit vector with the direction of the refracted ray. S and P are vertical and parallel to the reflection plane, respectively. The S direction is going out of the page.

Due to the principle of energy conservation, the transmitted amount of light can be determined by subtracting the portion of incident light that is reflected,

$$t^s(\theta_1) = 1 - r^s(\theta_1) \tag{4.25}$$

and

$$t^p(\theta_1) = 1 - r^p(\theta_1) \tag{4.26}$$

$t^s(\theta_1)$ and $t^p(\theta_1)$ are the reflectances of light with polarization vertical and parallel to the reflection plane.

Since r^s and r^p (or t^s and t^p) generally differ in both magnitude and phase, the recombination of the two components leads to a change in the polarization state. For example, a linearly polarized wave incident at an oblique angle may emerge as elliptically polarized, because the two orthogonal components acquire different phase shifts. In the special case where the phase difference reaches $\pm\frac{\pi}{2}$ and the amplitudes become equal, the outgoing wave may even be circularly polarized.

An important situation arises at the Brewster angle, where the reflection coefficient for the p component vanishes, $r^p = 0$. In this case, the reflected

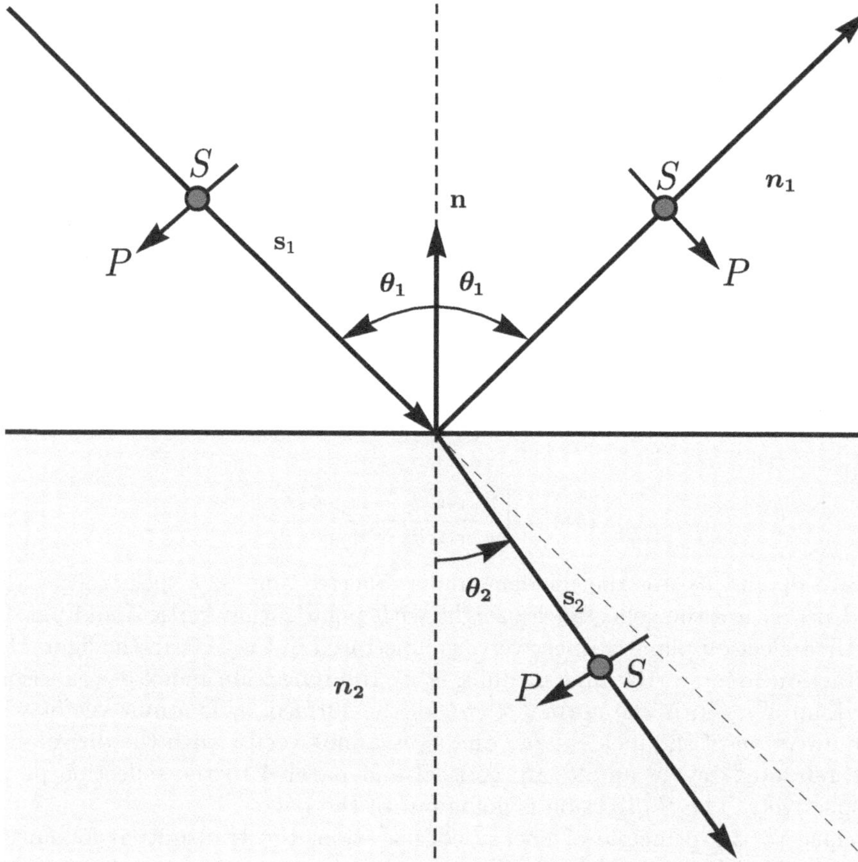

FIGURE 4.5
Polarized light incident on an optical interface. S and P are the directions vertical and parallel to the reflection plane, respectively. The S direction is going out of the page. θ_1 and θ_2 are the incident and refractive angles, respectively. $\mathbf{n}$ is the normal vector of the surface. $\mathbf{s}_1$ is a unit vector with the direction of the indecent ray and $\mathbf{s}_2$ is a unit vector with the direction of the refracted ray. The refraction indexes are n_1 and n_2 for the first and second medium, respectively.

wave is purely s-polarized, regardless of the incident polarization state. More generally, by carefully choosing the angle of incidence and the refractive indices of the media, the Fresnel coefficients can be exploited to convert one state of polarization into another, making them a fundamental tool in polarization optics.

We have now revisited the fundamental concepts of polarization, providing the reader with a structured overview of linear, circular, and elliptical polarization, as well as the role of Fresnel coefficients in modifying polarization states upon reflection and transmission. While this discussion offers a foundational understanding, the interested reader is encouraged to consult the recommended references for a more comprehensive and in-depth treatment of polarization phenomena, including advanced topics such as depolarization, polarization conversion, and the influence of complex media. Having established this groundwork, we now shift our focus to the next essential concept in vector diffraction: the angular spectrum representation. This formalism provides a powerful framework for decomposing arbitrary wavefields into plane-wave components, enabling rigorous treatment of wave propagation, diffraction, and focusing.

4.3 Angular spectrum representation

In this section, we derive the key result that both Fourier optics [8] and geometrical optics [23] naturally arise from the angular spectrum representation.

The angular spectrum representation is a mathematical technique to describe optical fields in homogeneous media. Optical fields are described as a superposition of plane waves and evanescent waves, both of which are physically intuitive solutions of Maxwell's equations. Furthermore, in the paraxial limit, the angular spectrum representation becomes identical with the framework of Fourier optics, which extends its importance even further.

By the angular spectrum representation we understand the series expansion of an arbitrary field in terms of plane (and evanescent) waves with variable amplitudes and propagation directions.

In this plane we can evaluate the two-dimensional Fourier transform of the field $\mathbf{E}(x, y, z)$ as,

$$\hat{\mathbf{E}}(k_x, k_y; z) = \frac{1}{4\pi^2} \iint_{-\infty}^{\infty} \mathbf{E}(x, y, z) e^{-\mathrm{i}[k_x x + k_y y]} \mathrm{d}x \mathrm{d}y, \tag{4.27}$$

where x, y are the Cartesian transverse coordinates and k_x, k_y the corresponding spatial frequencies or reciprocal coordinates. Similarly, the inverse Fourier transform reads as

$$\mathbf{E}(x, y, z) = \frac{1}{4\pi^2} \iint_{-\infty}^{\infty} \hat{\mathbf{E}}(k_x, k_y; z) e^{\mathrm{i}[k_x x + k_y y]} \mathrm{d}k_x \mathrm{d}k_y. \tag{4.28}$$

It is also important at to assume that $\mathbf{E}(x, y, z)$ must obey the Helmholtz equation, since $\mathbf{E}(x, y, z)$ it should be a solution to of Maxwell's equations (see Chapter 3),

$$(\nabla^2 + k^2)\mathbf{E}(\mathbf{r}) = 0, \tag{4.29}$$

where k is the wavenumber, it is defined as $k = (w/c)n$, n is refraction index, c is the speed of light in the medium. Thus, replacing (4.29) in (4.28),

$$\frac{1}{4\pi^2}\iint_{-\infty}^{\infty}(\nabla^2 + k^2)\hat{\mathbf{E}}(k_x, k_y; z)e^{\mathrm{i}[k_x x + k_y y]}\mathrm{d}k_x\mathrm{d}k_y = 0. \tag{4.30}$$

Expanding, and computing the Laplacian operator,

$$\frac{1}{4\pi^2}\iint_{-\infty}^{\infty}\left[(k^2 - k_x^2 - k_y^2)\hat{\mathbf{E}}(k_x, k_y; z) + \frac{\partial^2\hat{\mathbf{E}}(k_x, k_y; z)}{\partial z^2}\right]e^{\mathrm{i}[k_x x + k_y y]}\mathrm{d}k_x\mathrm{d}k_y = 0. \tag{4.31}$$

Since, the integral above is zero, then the integrand must be zero,

$$k_z^2\hat{\mathbf{E}}(k_x, k_y; z) + \frac{\partial^2\hat{\mathbf{E}}(k_x, k_y; z)}{\partial z^2} = 0, \tag{4.32}$$

where, we have defined k_z as,

$$k_z \equiv \sqrt{k^2 - k_x^2 - k_y^2} \tag{4.33}$$

The general solution of differential equation given by Eq. (4.32) is,

$$\hat{\mathbf{E}}(k_x, k_y; z) = \hat{\mathbf{E}}_1(k_x, k_y)e^{\pm \mathrm{i}k_z z} + \hat{\mathbf{E}}_2(k_x, k_y)e^{\mp \mathrm{i}k_z z}, \tag{4.34}$$

where $\hat{\mathbf{E}}_1(k_x, k_y)$ and $\hat{\mathbf{E}}_2(k_x, k_y)$ are vector amplitudes. We do not consider evenest waves, thus, we can eliminate $\hat{\mathbf{E}}_2(k_x, k_y)$,

$$\hat{\mathbf{E}}(k_x, k_y; z) = \hat{\mathbf{E}}_1(k_x, k_y)e^{\pm \mathrm{i}k_z z}. \tag{4.35}$$

Replacing Eq. (4.35) in Eq. (4.28),

$$\mathbf{E}(x, y, z) = \frac{1}{4\pi^2}\iint_{-\infty}^{\infty}\hat{\mathbf{E}}_1(k_x, k_y)e^{\mathrm{i}[k_x x + k_y y \pm k_z z]}\mathrm{d}k_x\mathrm{d}k_y. \tag{4.36}$$

So, from the last expression if $z = 0$,

$$\begin{aligned}\mathbf{E}(x, y, 0) &= \frac{1}{4\pi^2}\iint_{-\infty}^{\infty}\hat{\mathbf{E}}_1(k_x, k_y)e^{\mathrm{i}[k_x x + k_y y]}\mathrm{d}k_x\mathrm{d}k_y &(4.37)\\ &= \frac{1}{4\pi^2}\iint_{-\infty}^{\infty}\hat{\mathbf{E}}(k_x, k_y; 0)e^{\mathrm{i}[k_x x + k_y y]}\mathrm{d}k_x\mathrm{d}k_y &(4.38)\end{aligned}$$

Thus, that means that the solution of differential equation is the Fourier transform in two dimensions,

$$\hat{\mathbf{E}}(k_x, k_y; 0) = \hat{\mathbf{E}}_1(k_x, k_y), \tag{4.39}$$

replacing Eq. (4.39) in Eq. (4.35),

$$\hat{\mathbf{E}}(k_x, k_y; z) = \hat{\mathbf{E}}(k_x, k_y; 0)e^{\pm \mathrm{i}k_z z}. \tag{4.40}$$

Observe that in the Fourier space, the propagation of light from point 0 to point z therefore, it is just given by the multiplication of $\hat{\mathbf{E}}(k_x, k_y; 0)$ with $e^{\pm \mathrm{i}k_z z}$. To recover $\mathbf{E}(x, y, z)$, we can compute the Fourier inverse transform,

$$\mathbf{E}(x, y, z) = \frac{1}{4\pi^2} \iint_{-\infty}^{\infty} \hat{\mathbf{E}}(k_x, k_y; 0)e^{\mathrm{i}[k_x x + k_y y \pm k_z z]} \mathrm{d}k_x \mathrm{d}k_y. \tag{4.41}$$

The last expression is known as the angular spectrum representation. We can apply the same treatment to the $\mathbf{B}(x, y, z)$ and get the following expression,

$$\mathbf{B}(x, y, z) = \frac{1}{4\pi^2} \iint_{-\infty}^{\infty} \hat{\mathbf{B}}(k_x, k_y; 0)e^{\mathrm{i}[k_x x + k_y y \pm k_z z]} \mathrm{d}k_x \mathrm{d}k_y. \tag{4.42}$$

For a purely dielectric medium without losses, the refractive index n is a real and positive quantity. The wavenumber k_z is then either real or imaginary, causing the factor $\exp(\pm \mathrm{i}k_z z)$ to become either oscillatory or exponentially decaying. For a given pair (k_x, k_y), we find two distinct characteristic solutions: plane waves and evanescent waves. The angular spectrum is thus a superposition of these two types of waves. Plane waves are oscillatory functions in the z-direction and are subject to the condition $k_x^2 + k_y^2 \leq k^2$. In contrast, when $k_x^2 + k_y^2 > k^2$, we encounter evanescent waves, which exhibit exponential decay along the z-axis.

4.3.1 Far-fields in the angular spectrum representation

We had mentioned before that angular representation for far-fields becomes identical with the frame-work of Fourier optics, let's explore that condition. Let a localized field distribution exist in the plane $z = 0$. The angular spectrum representation offers a formal description of how this field propagates and how it is transferred to other planes, such as $z = z_0$, see Fig. 4.6. Specifically, one might be interested in the field distribution in the asymptotic limit $z \to \infty$, which corresponds to the Fraunhofer regime (far-field). Alternatively, one could inquire about the field distribution in the image plane when a specified far-field is focused.

$$\mathbf{E}(x, y, z) = \frac{1}{4\pi^2} \iint_{-\infty}^{\infty} \hat{\mathbf{E}}(k_x, k_y; 0)e^{\mathrm{i}[k_x x + k_y y \pm k_z z]} \mathrm{d}k_x \mathrm{d}k_y \tag{4.43}$$

We are interested in the asymptotic far-zone approximation of this field. To this end, let $\mathbf{s}$ denote the dimensionless unit vector pointing in the direction

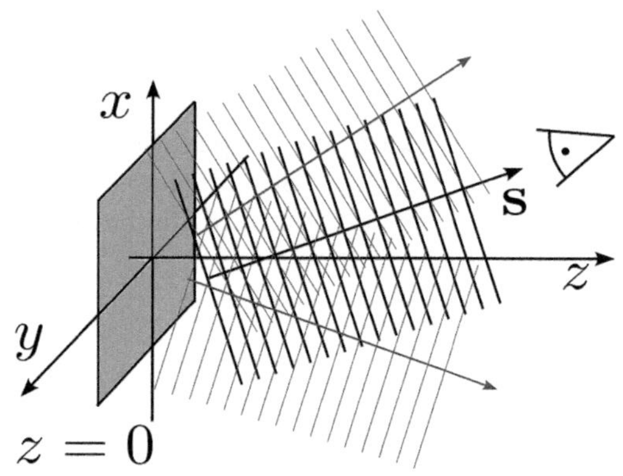

FIGURE 4.6
Illustration of the far-field approximation. According to the angular spectrum representation, According to the angular spectrum approach, each point at the source plane ($z = 0$) emits plane waves toward every possible direction. In the far field, however, a distant detector only *sees* the single plane wave traveling straight toward it. All other plane-wave components interfere destructively and cancel out.

of the image plane, which is assumed to be located at a large distance from the distribution origin plane,

$$\mathbf{s} = [s_x, s_y, s_z] = \left[\frac{x}{\mathrm{r}}, \frac{y}{\mathrm{r}}, \frac{z}{\mathrm{r}}\right] \tag{4.44}$$

where $\mathrm{r} = (x^2 + y^2 + z^2)^{1/2}$ is the distance from a point of image plane from the origin. The far-field approximation $\mathbf{E}_\infty$ is determined by evaluating the asymptotic limit $r \to \infty$ of Eq. (4.41).

$$\mathbf{E}_\infty(s_x, s_y) = \lim_{kr\to\infty} \iint_{(k_x^2+k_y^2)<k^2} \hat{\mathbf{E}}(k_x, k_y; 0) e^{\mathrm{i}kr\left[\frac{k_x}{k} s_x + \frac{k_y}{k} s_y \pm \frac{k_z}{k} s_z\right]} \mathrm{d}k_x \mathrm{d}k_y. \tag{4.45}$$

Here, $s_z = \sqrt{1 - s_x^2 - s_y^2}$. Since evanescent waves decay exponentially, they do not contribute to the field in the far zone, and the integration domain can therefore be restricted to the propagating region $s_x^2 + s_y^2 \leq 1$. The asymptotic behavior of the resulting double integral in the limit $k\mathrm{r} \to \infty$ is then obtained by applying the method of stationary phase.

4.3.2 Stationary phase

The method of stationary phase is an asymptotic technique used to evaluate highly oscillatory integrals in the limit of a large parameter, such as $k\mathrm{r} \to \infty$

[54, 55]. The central idea is that, for rapidly oscillating integrands, contributions from most regions cancel due to destructive interference, and the dominant contribution arises from points where the phase of the integrand is stationary, i.e., where its gradient vanishes. In the present context, this method allows one to approximate the far-field behavior of the angular spectrum integral by evaluating the contribution of the stationary point corresponding to the propagation direction.

Without going into details, and applying the stationary phase method over $\mathbf{E}_\infty$ the result is expressed as [54],

$$\mathbf{E}_\infty(s_x, s_y) = -2\pi\mathrm{i}ks_z\hat{\mathbf{E}}(ks_x, ks_y; 0)\frac{e^{\mathrm{i}k\mathrm{r}}}{\mathrm{r}} \tag{4.46}$$

This equation shows that the far-field distribution is entirely determined by the Fourier spectrum of the fields $\hat{\mathbf{E}}(ks_x, ks_y; 0)$ in the object plane, with the substitutions $k_x \to ks_x$ and $k_y \to ks_y$. This correspondence arises because, in the far-field, the spatial frequency components of the object plane map directly onto the angular directions defined by the unit vector $\mathbf{s}$, which therefore satisfies,

$$\mathbf{s} = [s_x, s_y, s_z] = \left[\frac{k_x}{k}, \frac{k_y}{k}, \frac{k_z}{k}\right] \tag{4.47}$$

This result implies that, in the far-field, only a single plane wave from the angular spectrum at $z = 0$, with wavevector $\mathbf{k} = (k_x, k_y, k_z)$, contributes to the field at a point in the direction of the unit vector $\mathbf{s}$. Consequently, the far-zone field can be interpreted as a superposition of rays, each corresponding to a specific plane wave component of the original angular spectrum, in agreement with the principles of geometrical optics. Evaluating this correspondence allows one to determine $\mathbf{E}_\infty(s_x, s_y)$ explicitly,

$$\hat{\mathbf{E}}(k_x, k_y; 0) = \frac{-\mathrm{i}\mathrm{r}e^{\mathrm{i}k\mathrm{r}}}{z\pi k_z}\mathbf{E}_\infty\left(\frac{k_x}{k}, \frac{k_y}{k}\right) \tag{4.48}$$

The preceding expression can be directly substituted into the angular spectrum representation (4.41), yielding,

$$\mathbf{E}(x, y, z) = \frac{-\mathrm{i}\mathrm{r}e^{\mathrm{i}k\mathrm{r}}}{2\pi}\iint_{(k_x^2+k_y^2)<k^2}\mathbf{E}_\infty\left(\frac{k_x}{k}, \frac{k_y}{k}\right)e^{\mathrm{i}[k_x x + k_y y + \pm k_z z]}\frac{1}{k_z}\mathrm{d}k_x\mathrm{d}k_y \tag{4.49}$$

Observe that in the absence of evanescent components, the field $\mathbf{E}$ in the plane $z = 0$ and its far-field $\mathbf{E}_\infty$ constitute, to a very good approximation, a Fourier-transform pair. The only deviation arises from the factor k_z in the angular spectrum. Under the common approximation $k_z \approx k$, this correspondence becomes exact, establishing the formal limit of Fourier optics.

The preceding equation is one of the most important results in this book and the central result of this chapter. It states that the field distribution **E** in the image plane is obtained as an inverse Fourier transform over k_x and k_y of the far-field $\mathbf{E}_\infty$, with a normalization factor of $1/k_z$. Eq. (4.49) is the basis for Richards-Wolfs diffraction integrals, stigmatic vector diffraction theory, and their generalization to arbitrary wavefronts, freeform vector diffraction theory.

4.4 Richards-Wolfs diffraction integrals

The Richards-Wolf diffraction integrals, developed by B. Richards and E. Wolf in 1959, provide a vectorial framework for computing the electromagnetic field near the focus of a high numerical aperture (NA) optical system, such as a microscope or telescope, imaging a point source [56, 57, 2]. Unlike scalar diffraction theories (Fraunhofer or Fresnel), they account for the vector nature of light, including polarization effects. For that, B. Richards and E. Wolf considered Eq. (4.49), where $\mathbf{E}_\infty$ is a spherical wavefront. Stigmatic optical system are the only ones that can generate spherical waterfront, therefore here we call Richards-Wolfs diffraction integrals as the basis of stigmatic vector diffraction theory.

So, let's consider an aplanatic optical system with focal length f, imaging a point source at infinity to a focal point at the origin of the image space (*left* of Fig. 4.7). An aplanatic optical system is a system that is stigmatic and follows the Abbe sine condition. The incident field is a plane wave with electric

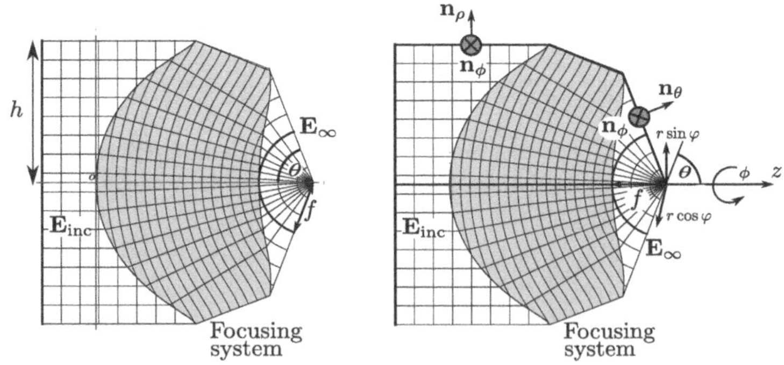

FIGURE 4.7
The sine condition of geometrical optics. The refraction of light rays at an aplanatic lens is determined by a spherical surface with radius f. (b) The intensity law of geometrical optics. The energy carried along a ray must stay constant.

field $\mathbf{E}_{\text{inc}}(\theta, \phi)$ on a reference sphere of radius f, centered at the focus. The angles θ and ϕ are the polar and azimuthal angles in spherical coordinates as shown in *right* of Fig. 4.7). $\theta \leq \theta_{\text{max}}$, where θ_{max} is the maximum aperture angle, related to numerical aperture, $\text{NA} = n \sin \theta_{\text{max}}$.

The Richards-Wolf integrals compute the electric field $\mathbf{E}(\mathbf{r})$ at a point $\mathbf{r}$ near the focus by integrating contributions from the reference sphere, accounting for its polarization $\mathbf{E}_\infty$. Let us express the angular representation in terms of a converging sphere, let the observation point in cylindrical coordinates be,

$$\mathbf{r} = [r \cos \varphi, r \sin \varphi, z]. \tag{4.50}$$

So, we can express the electric field in the image space as $\mathbf{E}$ as $\mathbf{E}(r, \varphi, z)$. Now, let's pay attention on the unit vectors in the direction of the ray, coming from the spherical wavefront,

$$\mathbf{s} = [\sin \theta \cos \phi, \sin \theta \sin \phi, \cos \theta] \tag{4.51}$$

Therefore, the components for the propagation vector are $\mathbf{k} = k\mathbf{s}$,

$$k_x = k \sin \theta \cos \phi, \quad k_y = k \sin \theta \sin \phi, \quad k_z = k \cos \theta \tag{4.52}$$

Remember that the propagation vector must be perpendicular to $\mathbf{E}_\infty$. The argument of the exponential of the integral can be written as $\mathbf{k} \cdot \mathbf{r}$. Thus computing $\mathbf{k} \cdot \mathbf{r}$ with Eq. (4.50) and Eq. (4.52),

$$\begin{aligned}
\mathbf{k} \cdot \mathbf{r} &= xk \sin \theta \cos \phi + yk \sin \theta \sin \phi + zk \cos \theta \\
&= kr \cos \varphi \sin \theta \cos \phi + kr \sin \varphi \sin \theta \sin \phi + zk \cos \theta \\
&= rk \sin \theta \cos(\phi - \varphi) + zk \cos \theta,
\end{aligned} \tag{4.53}$$

The focusing wavefront is spherical as it can be seen in the right side of Fig. 4.7. The easiest way to express the coming rays from the spherical wavefront and their electric field $\mathbf{E}_\infty(\theta, \phi)$ is in terms of spherical coordinates. Thus, lets express spherical coordinates unit vectors $\mathbf{n}_r$, $\mathbf{n}_\phi$, $\mathbf{n}_\theta$ in terms of the Cartesian unit vectors $\mathbf{n}_x$, $\mathbf{n}_y$, $\mathbf{n}_z$,

$$\begin{aligned}
\mathbf{n}_r &= \cos \phi \mathbf{n}_x + \sin \phi \mathbf{n}_y, \\
\mathbf{n}_\phi &= -\sin \phi \mathbf{n}_x + \cos \phi \mathbf{n}_y, \\
\mathbf{n}_\theta &= \cos \theta \cos \phi \mathbf{n}_x + \cos \theta \sin \phi \mathbf{n}_y - \sin \theta \mathbf{n}_z.
\end{aligned} \tag{4.54}$$

Therefore, the refracted electric field on the reference sphere $\mathbf{E}_\infty(\theta, \phi)$, in terms of the spherical coordinate terms θ, and ϕ, can be expressed in terms of the incident vector $\mathbf{E}_{\text{inc}}$, multiplied by the corresponding transmission Fresnel coefficients t^s and t^p,

$$\boxed{\mathbf{E}_\infty = q(\theta) \left[t^s [\mathbf{E}_{\text{inc}} \cdot \mathbf{n}_\phi] \mathbf{n}_\phi + t^p [\mathbf{E}_{\text{inc}} \cdot \mathbf{n}_r] \mathbf{n}_\theta \right].} \tag{4.55}$$

So, replacing the unit vectors, Eq. (4.54), in Eq. (4.55),

$$\mathbf{E}_\infty(\theta,\phi) = t^s(\theta)q(\theta)\left[\mathbf{E}_{\text{inc}}\cdot\begin{bmatrix}-\sin\phi\\ \cos\phi\\ 0\end{bmatrix}\right]\begin{bmatrix}-\sin\phi\\ \cos\phi\\ 0\end{bmatrix} + t^p(\theta)q(\theta)\left[\mathbf{E}_{\text{inc}}\cdot\begin{bmatrix}\cos\phi\\ \sin\phi\\ 0\end{bmatrix}\right]\begin{bmatrix}\cos\phi\cos\theta\\ \sin\phi\cos\theta\\ -\sin\theta\end{bmatrix}. \tag{4.56}$$

The factor $q(\theta) = \sqrt{\cos\theta}$ in $\mathbf{E}_\infty$ is included in the integral's apodization term of an aplanatic system, ensuring energy conservation. We are going to discuss this factor in more detail in the following chapters for non-aplanatic systems. For the moment, let's keep it as $\sqrt{\cos\theta}$ [58, 48].

If we ignore the Fresnel coefficients t^s and t^p and set them to one, $t^s \to 1$ and $t^p \to 1$. Then, we can relate the refracted field $\mathbf{E}_\infty$ with the incident field $\mathbf{E}_{\text{inc}}$ via a transformation matrix $\mathbf{M}$, accounting for the change in polarization due to refraction,

$$\mathbf{E}_\infty(\theta,\phi) = q(\theta)\mathbf{M}\mathbf{E}_{\text{inc}}(\theta,\phi) \tag{4.57}$$

The matrix $\mathbf{M}$ projects the incident field onto the spherical coordinate basis $(\hat{\mathbf{s}},\hat{\theta},\hat{\phi})$,

$$\mathbf{M} = \begin{pmatrix}(\cos\theta-1)\cos^2\phi+1 & (\cos\theta-1)\sin\phi\cos\phi & -\sin\theta\cos\phi\\ (\cos\theta-1)\sin\phi\cos\phi & (\cos\theta-1)\sin^2\phi+1 & -\sin\theta\sin\phi\\ -\sin\theta\cos\phi & -\sin\theta\sin\phi & \cos\theta\end{pmatrix} \tag{4.58}$$

For most of the examples in this book, we are going to set $t^s \to 1$ and $t^p \to 1$. But in some of them, we are going to use the transformation matrix $\mathbf{M}$ and in others Eq. (4.55). Both systems are equivalent. Now, we know that we can expresses $\mathbf{E}_\infty(\theta,\phi)$ in terms of the spherical coordinate terms θ, and ϕ.

Let's focus now on the differentials $\mathrm{d}k_x$ and $\mathrm{d}k_y$. Applying the Jacobean matrix, used in Chapter 1, we can get the differentials,

$$\mathbf{Jcb} = \begin{bmatrix}\dfrac{\partial k_x}{\partial\theta} & \dfrac{\partial k_x}{\partial\phi}\\ \dfrac{\partial k_y}{\partial\theta} & \dfrac{\partial k_y}{\partial\phi}\end{bmatrix} \tag{4.59}$$

Computing the determinant of the Jacobian,

$$\begin{aligned}\det(\mathbf{Jcb}) &= \frac{\partial}{\partial\theta}(k\sin\theta\cos\phi)\frac{\partial}{\partial\phi}(k\sin\theta\cos\phi)\\ &\quad - \frac{\partial}{\partial\theta}(k\sin\theta\sin\phi)\frac{\partial}{\partial\phi}(k\sin\theta\sin\phi)\\ &= (k^2\sin\theta\cos\theta\cos^2\phi + k^2\sin\theta\cos\theta\sin^2\phi)\mathrm{d}\theta\mathrm{d}\phi\\ &= k^2\sin\theta\cos\theta\mathrm{d}\theta\mathrm{d}\phi. \end{aligned}\tag{4.60}$$

Then, diving by $k_z = k\cos\theta$,

$$\frac{1}{k_z}\mathrm{d}k_x\mathrm{d}k_y = k\sin\theta\mathrm{d}\theta\mathrm{d}\phi \tag{4.61}$$

Then, replacing Eq. (4.50), (4.52), (4.61) in Eq. (4.49)

$$\mathbf{E}(r,\varphi,z) = \frac{-\mathrm{i}fe^{\mathrm{i}kf}}{2\pi}\int_0^{\theta_{\max}}\int_0^{2\pi}\mathbf{E}_\infty(\theta,\phi)e^{\mathrm{i}kz\cos\theta}e^{\mathrm{i}kr\sin\theta\cos(\phi-\varphi)}k\sin\theta\mathrm{d}\theta\mathrm{d}\phi. \tag{4.62}$$

where, r→ f turns to be the effective focal length. The factor $-\frac{\mathrm{i}f}{2\pi}$ normalizes the integral based on the lens's geometry. Eq. (4.62) is called the Richards-Wolf diffraction integrals, it is in plural since they are actually three integrals, since $\mathbf{E}(r,\varphi,z)$ has three different components, x, y, z. They effectively describe the diffraction pattern given by the vectorial electric field $\mathbf{E}_\infty(\theta,\phi)$. It is one of the most important equations of the book. Since it considers the vectorial behavior of $\mathbf{E}_\infty(\theta,\phi)$ is most accurate than its predecessors of scalar diffraction theory, since by nature the electric field is a vector an not an scalar element. The biggest limitation of this expression is that it is valid only for spherical wavefronts. Thus, system that are stigmatic that can be studied with this equation, but all other system which no produce a stigmatic wavefront they are out of the scope of it. A typical example can be a miss aligned system, it is out the scope expression Eq. (4.62). However it is remarkable the elegance in which B. Richards and E. Wolf reach it and it deeply helps us to understood the diffraction limit of light.

Eq. (4.62) is too large to study its properties, so we can rewrite it in a simplified notation where, the electric field at $\mathbf{r}$ in this new notation is given by,

$$\mathbf{E}(\mathbf{r}) = -\frac{\mathrm{i}kfe^{\mathrm{i}kf}}{2\pi}\iint_\Omega \mathbf{E}_\infty(\theta,\phi)e^{\mathrm{i}k\mathbf{s}\cdot\mathbf{r}}\sin\theta\,\mathrm{d}\phi\mathrm{d}\theta. \tag{4.63}$$

where, Ω is the solid angle defined by $\theta\in[0,\theta_{\max}]$, $\phi\in[0,2\pi)$. Remember that here $\mathbf{s}\cdot\mathbf{r} = x\sin\theta\cos\phi + y\sin\theta\sin\phi + z\cos\theta$.

From Eq. (4.1) and Eq. (4.11), we know that the magnetic field $\mathbf{E}(\mathbf{r})$ is perpendicular to the electric field and the propagation vector $\mathbf{k}$. Therefore, the magnetic field $\mathbf{B}(\mathbf{r})$ is computed applying Eq. (4.11) over Eq. (4.62),

$$\mathbf{B}(\mathbf{r}) = -\frac{\mathrm{i}kfe^{\mathrm{i}kf}}{2\pi c}\iint_\Omega(\hat{\mathbf{k}}\times\mathbf{E}_\infty(\theta,\phi))e^{\mathrm{i}\mathbf{k}\cdot\mathbf{r}}\sin\theta\,d\theta\,d\phi \tag{4.64}$$

Eq. (4.63) and (4.64) represent the electric and magnetic fields in the stigmatic vector diffraction theory. In this compact notation, they are more convenient for exploring the properties of the fields. We will examine these properties in the next section.

4.5 Properties of the Richards-Wolf integrals

Eqs. (4.63) and (4.64) describe the electric and magnetic fields as formulated in the Richards–Wolf integrals. These expressions characterize the spatial distribution of the diffraction pattern generated by an aplanatic optical system, given a specific polarization state of the incident field. In this section, we will examine some of their key properties in detail.

4.5.1 Linearity

The Richards–Wolf integrals are linear, meaning that the superposition principle applies: the field resulting from a combination of sources is the sum of the fields produced by each source individually.

For incident fields $\mathbf{E}_{\text{inc}} = a\mathbf{E}_{i1} + b\mathbf{E}_{i2}$, the resulting field is:

$$\mathbf{E}(\mathbf{r}) = a\mathbf{E}_1(\mathbf{r}) + b\mathbf{E}_2(\mathbf{r}) \tag{4.65}$$

Therefore,

$$\begin{aligned}
\mathbf{E}(\mathbf{r}) &= -\frac{\mathrm{i}kfe^{\mathrm{i}kf}}{2\pi}\iint_\Omega \mathbf{M}(a\mathbf{E}_{i1} + b\mathbf{E}_{i2})e^{\mathrm{i}k\mathbf{s}\cdot\mathbf{r}}q(\theta)\sin\theta\,\mathrm{d}\phi\mathrm{d}\theta && (4.66)\\
&= a\left(-\frac{\mathrm{i}kfe^{\mathrm{i}kf}}{2\pi}\iint_\Omega \mathbf{M}\mathbf{E}_{i1}e^{\mathrm{i}k\mathbf{s}\cdot\mathbf{r}}q(\theta)\sin\theta\,\mathrm{d}\phi\mathrm{d}\theta\right) && (4.67)\\
&+ b\left(-\frac{\mathrm{i}kfe^{\mathrm{i}kf}}{2\pi}\iint_\Omega \mathbf{M}\mathbf{E}_{i2}e^{\mathrm{i}k\mathbf{s}\cdot\mathbf{r}}q(\theta)\sin\theta\,\mathrm{d}\phi\mathrm{d}\theta\right) && (4.68)\\
&= a\mathbf{E}_1(\mathbf{r}) + b\mathbf{E}_2(\mathbf{r}) && (4.69)
\end{aligned}$$

The linearity of the integral and matrix $\mathbf{M}$ ensures the property.

This property greatly simplifies the analysis of complex polarization states and composite optical systems. For example a elliptical polarization can be obtained by the sum of two different linear polarizations. Then, the diffraction pattern of elliptical polarized field can be easy computed with by the sum of its components.

4.5.2 Scaling

The Richards–Wolf integrals exhibit scaling behavior: if the amplitude of the input field is multiplied by a constant factor, the resulting diffraction field is scaled by the same factor.

If the spatial coordinates are scaled, $\mathbf{r}' = a\mathbf{r}$, and the incident field is scaled in amplitude, $\mathbf{E}'_\infty = b\mathbf{E}_\infty$, then:

$$\mathbf{E}'(\mathbf{r}') = b\mathbf{E}(\mathbf{r}/a) \tag{4.70}$$

For the scaled field, we have,

$$\mathbf{E}'(\mathbf{r}') = -\frac{ikfe^{ikf}}{2\pi}\iint_{\Omega}\mathbf{M}(b\mathbf{E}_{\infty})e^{ik\mathbf{s}\cdot(a\mathbf{r})}q(\theta)\sin\theta\,\mathrm{d}\phi\mathrm{d}\theta \tag{4.71}$$

$$= b\left(-\frac{ikfe^{ikf}}{2\pi}\iint_{\Omega}\mathbf{M}\mathbf{E}_{\infty}e^{ik(a\mathbf{s})\cdot\mathbf{r}}q(\theta)\sin\theta\,\mathrm{d}\phi\mathrm{d}\theta\right) \tag{4.72}$$

Let $\mathbf{s}' = a\mathbf{s}$, but since $\mathbf{s}$ is a unit vector, we adjust the phase term,

$$\mathbf{s}\cdot(a\mathbf{r}) = (a\mathbf{s})\cdot\mathbf{r} \tag{4.73}$$

Thus,

$$\mathbf{E}'(\mathbf{r}') = b\mathbf{E}(\mathbf{r}/a) \tag{4.74}$$

This shows the field scales with amplitude and inversely with spatial coordinates. This property allows straightforward control of field intensity without altering the spatial distribution or polarization characteristics.

4.5.3 Rotational symmetry

For systems with circularly symmetric apertures and uniform polarization, the Richards-Wolf integrals produce a vectorial field that exhibits rotational symmetry about the optical axis. In such cases, the intensity and polarization distributions are invariant under rotations around the axis, simplifying both analytical and numerical analysis. This symmetry is particularly useful for understanding focal spot shapes, energy distribution, and the behavior of radially or azimuthally polarized fields.

For a rotationally symmetric incident field, radially or azimuthally polarized, the intensity $I(\mathbf{r})$ is invariant under rotation about the optical axis. If we rotate the observation point by $\varphi \to \varphi + \Delta\varphi$, the phase becomes $\cos(\phi - \varphi - \Delta\varphi)$. Since $\mathbf{E}_{\infty}(\theta, \phi)$ is symmetric, the integral's magnitude $|\mathbf{E}|^2$ remains unchanged after a change of variables, confirming rotational symmetry of the intensity. We will dig more into this property in the on the polarization effects on Richards-Wolf Integrals,

4.5.4 Transversality

The fields described by the Richards–Wolf integrals satisfy the transversality condition, meaning that both the electric and magnetic fields are divergence-free in free space, as we have seen in previews sections. Each plane wave component satisfies,

$$\mathbf{E}_{\infty}(\theta, \phi)\cdot\mathbf{k} = 0 \tag{4.75}$$

This ensures that the field vectors remain perpendicular to the local propagation direction, preserving the physically correct polarization structure of the diffraction pattern.

4.5.5 Intensity

The Richards-Wolf integrals model the electric field distribution in the focal plane, determining the intensity pattern. The intensity of the focused field is given by the squared magnitude of the vectorial electric field,

$$I(\mathbf{r}) = |\mathbf{E}(\mathbf{r})|^2 = |E_x(\mathbf{r})|^2 + |E_y(\mathbf{r})|^2 + |E_z(\mathbf{r})|^2. \tag{4.76}$$

This ensures the resulting intensity distribution reflects contributions from both transverse and longitudinal components. For each of the components of $I(\mathbf{r})$ the full width at half maximum (FWHM) can be computed. FWHM measures the width of a peak in an intensity distribution $I(\mathbf{r})$ at half its maximum value.

1. Find the maximum intensity $I_{\text{max}} = I(x_0)$.
2. Identify points x_1, x_2 where $I(x_1) = I(x_2) = I_{\text{max}}/2$.
3. Compute FWHM $= |x_2 - x_1|$.

The generalization of the FWHM to intensity distribution with a vector $I(\mathbf{r})$ as an argument is straightforward and left for the reader as an exercise.

4.6 Polarization effects on Richards-Wolf integrals

The Richards–Wolf integrals depend strongly on the polarization state of the incident field $\mathbf{E}_{\text{inc}}$. Here we mean the spatially varying polarization. Spatially varying polarization means that the polarization state of light (linear, circular, elliptical) changes from point to point across the wave front's transverse plane. For the purpose of the book we will only consider linear polarization from point to point. Circular and elliptical polarization can easy extended form the linear polarization but are consider outside the scope of the book.

In this section, we focus on the most commonly studied cases of spatially varying polarization: linear, radial, and azimuthal polarizations. Each of these states leads to distinct focal field distributions, illustrating how polarization plays a central role Richards–Wolf integrals.

Here we have used the term linear twice for the local state and the spatially varying polarization. As mentioned before, all the local state will be considered a linear as in Fig. 4.3. So from now on, when we mention linear polarization state, then it means spatially varying polarization. Linear spatially varying polarization considers that the polarization is linear along the wavefront, and its protection over is local xy plane has the same direction as see in *left* of Fig. 4.8. In other means linear spatially varying polarization means the light's polarization is everywhere linear (not circular or elliptical), but the orientation (angle) of that linear polarization changes from point to point across the beam.

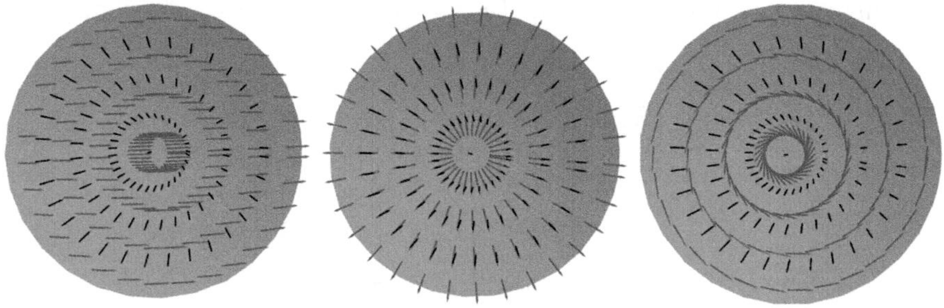

FIGURE 4.8
Comparison of three types of linearly polarized spatially varying polarization states. In black is the normal vector **k** and in light gray is the polarization vector $\mathbf{E}_\infty$. At the *left* of the figure is the uniform linear polarization, constant orientation across the beam). *Center:* radial polarization, polarization vectors pointing radially outward from the beam axis, like wheel spikes, and *right*: azimuthal polarization, polarization vectors oriented tangentially around the beam center, following concentric circles.

Radial spatially varying polarization is presented in *center* of Fig. 4.8. It means the light is linearly polarized everywhere, with the polarization direction always pointing radially outward from the beam center (like spokes of a wheel). Finally, in the *right* of Fig. 4.8 we have azimuthal spatially varying polarization. It means the light is linearly polarized everywhere, with the polarization direction always oriented tangentially around the beam center (like concentric circles or the direction of rotation in a clock face).

The linear, radial, and azimuthal polarization configurations are the one will be used along the book. Because they are the most common in literature but it is important to remark that there is a whole jungle of possible polarization configurations. There are infinitely many spatially varying polarization states, since the polarization can vary continuously (in direction, ellipticity, phase difference) from point to point across the beam. When spatially varying polarization that covers all the possible polarization state is considered a full Poincaré beam.

4.6.1 Linearly polarized plane wave

Consider a plane wave polarized along $\hat{\mathbf{x}}$. The incidence electric field can be written as $\mathbf{E}_{\text{inc}} = E_0\hat{\mathbf{x}}$. From Eq. (4.55) we know that, given an incident electric $\mathbf{E}_{\text{inc}}$, we can express the refracted field $\mathbf{E}_\infty$ as,

$$\mathbf{E}_\infty = q(\theta)\left[t^s[\mathbf{E}_{\text{inc}} \cdot \mathbf{n}_\phi]\mathbf{n}_\phi + t^p[\mathbf{E}_{\text{inc}} \cdot \mathbf{n}_r]\mathbf{n}_\theta\right]. \tag{4.77}$$

Replacing the unit vectors $\mathbf{n}_\phi$, $\mathbf{n}_\theta$, (Eq. (4.53)) in the last expression,

$$\mathbf{E}_\infty(\theta,\phi) = t^s(\theta)q(\theta)\left[\mathbf{E}_{\text{inc}}\cdot\begin{bmatrix}-\sin\phi\\ \cos\phi\\ 0\end{bmatrix}\right]\begin{bmatrix}-\sin\phi\\ \cos\phi\\ 0\end{bmatrix} + t^p(\theta)q(\theta)\left[\mathbf{E}_{\text{inc}}\cdot\begin{bmatrix}\cos\phi\\ \sin\phi\\ 0\end{bmatrix}\right]\begin{bmatrix}\cos\phi\cos\theta\\ \sin\phi\cos\theta\\ -\sin\theta\end{bmatrix}. \tag{4.78}$$

We assume that Fresnel coefficients are one, which implies that $t^s(\theta) \to 1$, $t^p(\theta) \to 1$.Thereforee,

$$\mathbf{E}_\infty(\theta,\phi) = E_0q(\theta)\left[\begin{bmatrix}1\\0\\0\end{bmatrix}\cdot\begin{bmatrix}-\sin\phi\\ \cos\phi\\ 0\end{bmatrix}\right]\begin{bmatrix}-\sin\phi\\ \cos\phi\\ 0\end{bmatrix} + E_0q(\theta)\left[\begin{bmatrix}1\\0\\0\end{bmatrix}\cdot\begin{bmatrix}\cos\phi\\ \sin\phi\\ 0\end{bmatrix}\right]\begin{bmatrix}\cos\phi\cos\theta\\ \sin\phi\cos\theta\\ -\sin\theta\end{bmatrix}. \tag{4.79}$$

Computing the dot products,

$$\mathbf{E}_\infty(\theta,\phi) = -E_0q(\theta)\sin\phi\begin{bmatrix}-\sin\phi\\ \cos\phi\\ 0\end{bmatrix} + E_0q(\theta)\cos\phi\begin{bmatrix}\cos\phi\cos\theta\\ \sin\phi\cos\theta\\ -\sin\theta\end{bmatrix} \tag{4.80}$$

Simplifying,

$$\mathbf{E}_\infty(\theta,\phi) = -E_0q(\theta)\begin{bmatrix}-\sin\phi^2\\ \cos\phi\sin\phi\\ 0\end{bmatrix} + E_0q(\theta)\begin{bmatrix}\cos\phi^2\cos\theta\\ \cos\phi\sin\phi\cos\theta\\ -\sin\theta\cos\phi\end{bmatrix} \tag{4.81}$$

The refracted field $\mathbf{E}_\infty$ of the linear polarization state can be written as,

$$\mathbf{E}_\infty = E_0q(\theta)\begin{bmatrix}\cos\theta\cos^2\phi + \sin^2\phi\\ \cos\theta\cos\phi\sin\phi - \cos\phi\sin\phi\\ -\sin\theta\cos\phi\end{bmatrix}. \tag{4.82}$$

Applying some basic trigonometry, we remove the quadratic terms,

$$\mathbf{E}_\infty = E_0q(\theta)\begin{bmatrix}(1/2)(1+\cos\theta) - (1/2)(1-\cos\theta)\cos(2\phi)\\ \cos\theta\cos\phi\sin\phi - \cos\phi\sin\phi\\ -\sin\theta\cos\phi\end{bmatrix}. \tag{4.83}$$

Substituting the refracted field $\mathbf{E}_\infty$ in the Richards-Wolf integral, Eq. (4.83) in Eq. (4.62), in x,

$$E_x = -\frac{ikfe^{ikf}E_0}{2\pi}\iint_\Omega [(1/2)(1+\cos\theta) - (1/2)(1-\cos\theta)\cos(2\phi)] \times e^{ikz\cos\theta}e^{ikr\sin\theta\cos(\phi-\varphi)}\sin\theta\, d\phi d\theta, \quad (4.84)$$

in y,

$$E_y = -\frac{ikfe^{ikf}E_0}{2\pi}\iint_\Omega ((\cos\theta - 1)\cos\phi\sin\phi) \times e^{ikz\cos\theta}e^{ikr\sin\theta\cos(\phi-\varphi)}q(\theta)\sin\theta\, d\phi d\theta, \quad (4.85)$$

and z,

$$E_z = \frac{ikfe^{ikf}E_0}{2\pi}\iint_\Omega \sin\theta\cos\phi e^{ikz\cos\theta}e^{ikr\sin\theta\cos(\phi-\varphi)}q(\theta)\sin\theta\, d\phi d\theta. \quad (4.86)$$

Now the components E_x, E_y, and E_z, can be simplified due to their cylindrical symmetry. From Bessel functions of the first kind, we know that [16],

$$\int_0^{2\pi} \cos(n\phi)e^{ix\cos(\phi-\varphi)}d\phi = 2\pi(i)^n J_n(x)\cos(n\varphi), \quad (4.87)$$

and,

$$\int_0^{2\pi} \sin(n\phi)e^{ix\cos(\phi-\varphi)}d\phi = 2\pi(i)^n J_n(x)\sin(n\varphi). \quad (4.88)$$

By introducing Bessel functions of the first kind, Eq. 4.87 and Eq. 4.88, Eq. 4.84, Eq. 4.85, and Eq. 4.86 in x, we have,

$$E_x = -\frac{ikfe^{ikf}}{2kr}\int_0^{\theta_{\max}} q(\theta)e^{ikz\cos\theta}\Big((1+\cos\theta)J_0(kr\sin\theta) + (-1+\cos\theta)\sin\theta\cos(2\varphi)J_2(kr\sin\theta)\Big)\sin\theta d\theta. \quad (4.89)$$

E_y turns to,

$$E_y = -ikfe^{ikf}\int_0^{\theta_{\max}} q(\theta)(1-\cos\theta)\cos\varphi\sin\theta\sin\varphi e^{ikz\cos\theta}J_2(kr\sin\theta)d\theta. \quad (4.90)$$

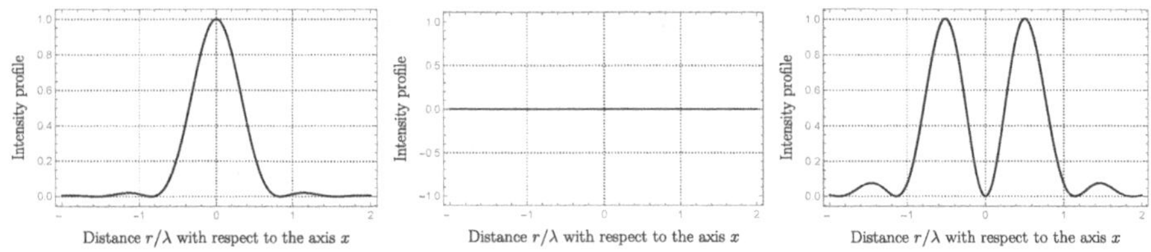

FIGURE 4.9
Normalized intensity profiles of the linear polarization beam. In the *left*, *center* and *right* are the $\hat{\mathbf{x}}$ component, the $\hat{\mathbf{y}}$ component, and the $\hat{\mathbf{z}}$ component of the normalized intensity profiles of $\mathbf{E}_\infty(\theta,\phi)$ in the horizontal axis.

Finally, for E_z is given by,

$$E_z = -kfe^{\mathrm{i}kf} \int_0^{\theta_{\max}} \sin\theta^2 q(\theta)\cos\varphi e^{\mathrm{i}kz\cos\theta} \mathrm{J}_1(kr\sin\theta)\mathrm{d}\theta. \tag{4.91}$$

By applying the Bessel function properties, we have reduced the double integral over the solid angle Ω to a single integral over θ. The normalized density distribution is presented in Fig. 4.9. The normalized density distribution has been computed by taking $\theta_{\max}$, $f = 1\text{mm}$, $\lambda = 1$.

The distributions plotted in Fig. 4.9 are along of the x in the image plane. In order to show-up the pattern of the distribution along the image plane it is necessary to compute it along the xy plane as it is show in Fig. 4.10.

The x-component, E_x, is the dominant component, shows an elongated lobe intensity distribution stretched along the incident polarization direction (x-axis), this is the main contributor to the total intensity PSF. Much weaker, the y-component appears as four symmetric lobes in the xy plane. It creates depolarization and contributes to a butterfly-like or four-lobed total intensity pattern. Finally the z-component appears as two symmetric lobes along the y-direction (perpendicular to polarization), zero on the x-axis.

4.6.2 Radially polarized beam

Now consider a radially polarized beam, $\mathbf{E}_{\text{inc}} = E_0(\cos\phi\hat{\mathbf{x}} + \sin\phi\hat{\mathbf{y}})$. We recall Eq. (4.55), for the refracted field $\mathbf{E}_\infty$,

$$\mathbf{E}_\infty = q(\theta)\left[t^s[\mathbf{E}_{\text{inc}}\cdot\mathbf{n}_\phi]\mathbf{n}_\phi + t^p[\mathbf{E}_{\text{inc}}\cdot\mathbf{n}_r]\mathbf{n}_\theta\right] \tag{4.92}$$

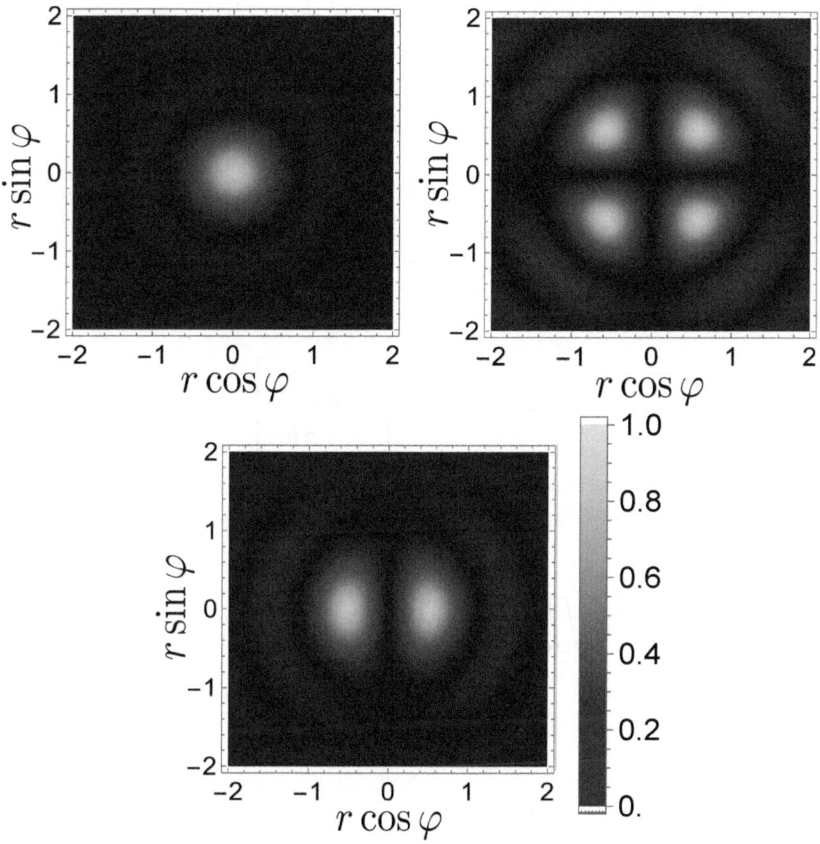

FIGURE 4.10
Normalized intensity profiles of the linear polarization beam in the xy plane in terms of wavelengths. In the *left*, *center* and *right* are the $\hat{\mathbf{x}}$ component, the $\hat{\mathbf{y}}$ component, and the $\hat{\mathbf{z}}$ component of the normalized intensity profiles of $\mathbf{E}_\infty(\theta, \phi)$.

where the unit vectors are $\mathbf{n}_\phi$, $\mathbf{n}_\theta$ from Eq. (4.53). Thus,

$$\mathbf{E}_\infty(\theta,\phi) = t^s(\theta)q(\theta)\left[\mathbf{E}_{\text{inc}} \cdot \begin{bmatrix} -\sin\phi \\ \cos\phi \\ 0 \end{bmatrix}\right] \begin{bmatrix} -\sin\phi \\ \cos\phi \\ 0 \end{bmatrix} + t^p(\theta)q(\theta)\left[\mathbf{E}_{\text{inc}} \cdot \begin{bmatrix} \cos\phi \\ \sin\phi \\ 0 \end{bmatrix}\right] \begin{bmatrix} \cos\phi\cos\theta \\ \sin\phi\cos\theta \\ -\sin\theta \end{bmatrix}. \quad (4.93)$$

Replacing the incident field $\mathbf{E}_{\text{inc}}$ and setting $t^s(\theta) \to 1$ and $t^p(\theta) \to 1$,

$$\mathbf{E}_\infty(\theta,\phi) = E_0 q(\theta) \left[\begin{bmatrix}\cos\phi\\ \sin\phi\\ 0\end{bmatrix} \cdot \begin{bmatrix}-\sin\phi\\ \cos\phi\\ 0\end{bmatrix}\right] \begin{bmatrix}-\sin\phi\\ \cos\phi\\ 0\end{bmatrix} + E_0 q(\theta) \left[\begin{bmatrix}\cos\phi\\ \sin\phi\\ 0\end{bmatrix} \cdot \begin{bmatrix}\cos\phi\\ \sin\phi\\ 0\end{bmatrix}\right] \begin{bmatrix}\cos\phi\cos\theta\\ \sin\phi\cos\theta\\ -\sin\theta\end{bmatrix} \tag{4.94}$$

Computing the dot products and simplifying,

$$\mathbf{E}_\infty = E_0 q(\theta) \begin{bmatrix}\cos\theta\cos\phi\\ \cos\theta\sin\phi\\ -\sin\theta\end{bmatrix}. \tag{4.95}$$

Substituting the refracted field $\mathbf{E}_\infty$, in the Richards-Wolf integral, Eq. (4.62). The field components are,

$$E_x = -\frac{ikfe^{ikf}E_0}{2\pi} \iint_\Omega \cos\theta\cos\phi e^{ikz\cos\theta} e^{ikr\sin\theta\cos(\phi-\varphi)} q(\theta)\sin\theta \, d\phi d\theta, \tag{4.96}$$

E_y turns to,

$$E_y = -\frac{ikfe^{ikf}E_0}{2\pi} \iint_\Omega \cos\theta\sin\phi e^{ikz\cos\theta} e^{ikr\sin\theta\cos(\phi-\varphi)} \sin\theta q(\theta) \, d\phi d\theta, \tag{4.97}$$

and,

$$E_z = \frac{ikfe^{ikf}E_0}{2\pi} \iint_\Omega e^{ikz\cos\theta} e^{ikr\sin\theta\cos(\phi-\varphi)} q(\theta)\sin^2\theta \, d\phi d\theta. \tag{4.98}$$

E_x, E_y, and E_z can be simplified using Eq (4.87) and Eq (4.88). Just as we did in the previous subsection. The outcome for E_x is,

$$E_x = kfe^{ikf} \int_0^{\theta_{\max}} \sin\theta\cos\theta q(\theta)\cos\varphi e^{ikz\cos\theta} J_1(kr\sin\theta) d\theta, \tag{4.99}$$

E_y is given by,

$$E_y = kfe^{ikf} \int_0^{\theta_{\max}} \sin\theta\cos\theta q(\theta)\sin\varphi e^{ikz\cos\theta} J_1(kr\sin\theta) d\theta, \tag{4.100}$$

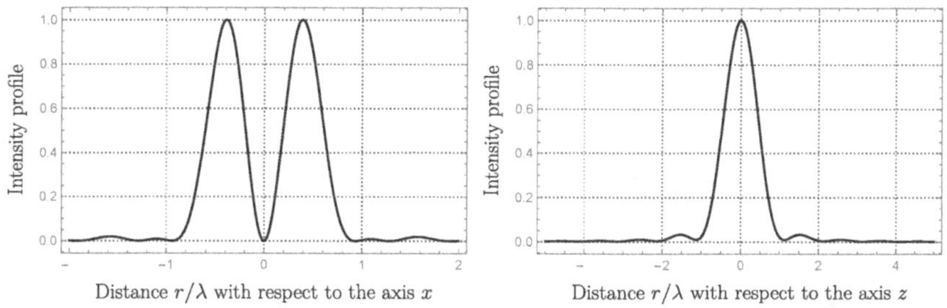

FIGURE 4.11
Normalized intensity profile of a radially polarized beam. In the *left* is the profile along z and in the *right* is a long r.

and E_z is expressed as,

$$E_z = \mathrm{i}kfe^{\mathrm{i}kf} \int_0^{\theta_{\max}} \sin\theta^2 q(\theta) e^{\mathrm{i}kz\cos\theta} \mathrm{J}_0(kr\sin\theta)\mathrm{d}\theta. \tag{4.101}$$

Eqs. (4.96), (4.97), (4.98) are equivalent to equations Eqs. (4.99), (4.100), (4.101), we had just written them in a more elegant way to reduced them to one integral instead of two.

Radially polarized beam is radially symmetric, therefore the x and y components can be reduced to,

$$E_r = kfe^{\mathrm{i}kf} \int_0^{\theta_{\max}} \sin\theta\cos\theta q(\theta) e^{\mathrm{i}kz\cos\theta} \mathrm{J}_1(kr\sin\theta)\mathrm{d}\theta, \tag{4.102}$$

We plot the normalized intensity profile along z and in the *right* is a long r. For this plot $\theta_{\max} =$ and $f = 1$mm. The plot is displayed in Fig. 4.11.

In the rz meridional plane (where r is the radial distance from the axis), the total intensity distribution forms a sharp, nearly circular focal spot on the axis, significantly smaller than that obtained with linearly polarized light and exhibiting very weak sidelobes. The radial electric field component E_r is zero on the optical axis and displays a characteristic doughnut-shaped ring in the focal plane ($z = 0$), appearing in the $r - z$ plane as two symmetric lobes centered around the focus and decaying along the propagation direction with diffraction rings.

In contrast, the longitudinal electric field component E_z becomes the dominant feature at high numerical apertures (typically $NA > 0.8$) [59]. E_z reaches its maximum intensity precisely on the optical axis ($r = 0$) at the focus ($z = 0$), often comparable to or even exceeding the transverse components, and appears in the rz plane as a single strong, elongated central lobe, this effect is called *optical needle* in literature [47, 60, 61, 62]. This on-axis longitudinal field, combined with the ring-like E_r distribution, produces a highly confined, bright focal spot with axial symmetry, making radially polarized beams particularly valuable for applications such as super-resolution microscopy [63, 64, 65], optical trapping [66], and material processing. The reader is invited to read more about these applications in the following references [67, 68, 69, 70, 71]. We will continue with the azimuthal polarized field.

4.6.3 Azimuthal polarized field

Finally, the we are going to consider azimuthal polarized field, this field is also widely used. For a azimuthal polarized beam, we have $\mathbf{E}_{\text{inc}} = E_0(-\sin\phi\hat{\mathbf{x}} + \cos\phi\hat{\mathbf{y}})$. The refracted field is given by Eq. (4.55), recalling it,

$$\mathbf{E}_\infty = q(\theta)\left[t^s[\mathbf{E}_{\text{inc}}\cdot\mathbf{n}_\phi]\mathbf{n}_\phi + t^p[\mathbf{E}_{\text{inc}}\cdot\mathbf{n}_r]\mathbf{n}_\theta\right] \tag{4.103}$$

replacing the unit vectors are $\mathbf{n}_\phi$, $\mathbf{n}_\theta$ (from Eq. (4.53)),

$$\begin{aligned}\mathbf{E}_\infty(\theta,\phi) = t^s(\theta)q(\theta)&\left[\mathbf{E}_{\text{inc}}\cdot\begin{bmatrix}-\sin\phi\\ \cos\phi\\ 0\end{bmatrix}\right]\begin{bmatrix}-\sin\phi\\ \cos\phi\\ 0\end{bmatrix}\\ &+t^p(\theta)\left[\mathbf{E}_{\text{inc}}\cdot\begin{bmatrix}\cos\phi\\ \sin\phi\\ 0\end{bmatrix}\right]\begin{bmatrix}\cos\phi\cos\theta\\ \sin\phi\cos\theta\\ -\sin\theta\end{bmatrix}\end{aligned} \tag{4.104}$$

Substituting the $\mathbf{E}_{\text{inc}}$ and setting the Fresnel coefficients to one, $t^s(\theta)\to 1$, $t^p(\theta)\to 1$,

$$\begin{aligned}\mathbf{E}_\infty(\theta,\phi) = E_0q(\theta)&\left[\begin{bmatrix}-\sin\phi\\ \cos\phi\\ 0\end{bmatrix}\cdot\begin{bmatrix}-\sin\phi\\ \cos\phi\\ 0\end{bmatrix}\right]\begin{bmatrix}-\sin\phi\\ \cos\phi\\ 0\end{bmatrix}\\ &+E_0q(\theta)\left[\begin{bmatrix}-\sin\phi\\ \cos\phi\\ 0\end{bmatrix}\cdot\begin{bmatrix}\cos\phi\\ \sin\phi\\ 0\end{bmatrix}\right]\begin{bmatrix}\cos\phi\cos\theta\\ \sin\phi\cos\theta\\ -\sin\theta\end{bmatrix}\end{aligned} \tag{4.105}$$

Simplifying we get that $\mathbf{E}_\infty$ for the azimuthal polarization as

$$\mathbf{E}_\infty = E_0q(\theta)\begin{bmatrix}-\sin\phi\\ \cos\phi\\ 0\end{bmatrix} \tag{4.106}$$

The field components of $\mathbf{E}_\infty$ for the azimuthal polarization are, in x,

$$E_x = \frac{ikfe^{ikf}E_0}{2\pi}\iint_\Omega \sin\phi e^{ikz\cos\theta}e^{ikr\sin\theta\cos(\phi-\varphi)}q(\theta)\sin\theta\,\mathrm{d}\phi\mathrm{d}\theta, \tag{4.107}$$

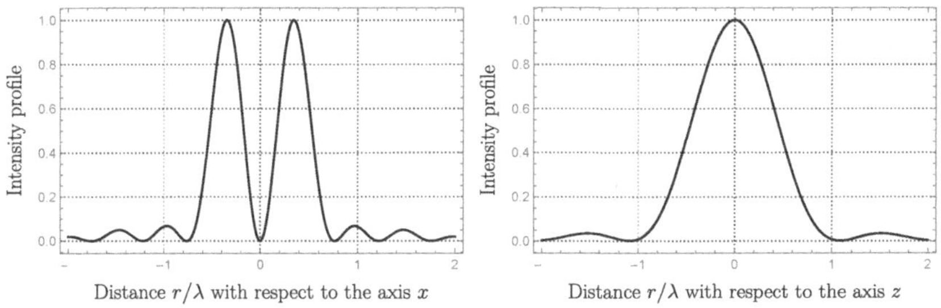

FIGURE 4.12
Normalized intensity profile of azimuthal polarized field. In the *left* is the profile along r and in the *right* is a long z at $r = 0.3$.

in y,

$$E_y = -\frac{\mathrm{i}kfe^{\mathrm{i}kf}E_0}{2\pi}\iint_\Omega \cos\phi e^{\mathrm{i}kz\cos\theta}e^{\mathrm{i}kr\sin\theta\cos(\phi-\varphi)}q(\theta)\sin\theta\,\mathrm{d}\phi\mathrm{d}\theta. \quad (4.108)$$

Finally, in z,

$$E_z = 0. \quad (4.109)$$

In the focal plane, the field can be simplified using Eq. (4.87) and (4.88),

$$E_x = -kfe^{\mathrm{i}kf}E_0\int_0^{\theta_{\max}} q(\theta)\sin\varphi e^{\mathrm{i}kz\cos\theta}\mathrm{J}_1(kr\sin\theta)\sin\theta\mathrm{d}\theta \quad (4.110)$$

and,

$$E_y = kfe^{\mathrm{i}kf}E_0\int_0^{\theta_{\max}} q(\theta)\cos\varphi e^{\mathrm{i}kz\cos\theta}\mathrm{J}_1(kr\sin\theta)\sin\theta\mathrm{d}\theta \quad (4.111)$$

Fig. 4.12 shows the normalized intensity profile of azimuthal polarized field. On the *left* is the profile along r and on the *right* is a long z at $r = 0.3$. For this plot $\theta_{\max} =$ and $f = 1\mathrm{mm}$.

The interest in azimuthally polarized beams is due to their unique hollow-core intensity distribution at the focus. In the field of optical manipulation, these beams enable to trapping of particles and generate optical tweezers [72, 73]. Also they can be applied to electron acceleration and scanning microscopy. The reader is invited to deep on this topic [74, 75, 76, 77].

4.7 Conclusions

This chapter has established the foundational framework of stigmatic vector diffraction theory through the Richards-Wolf integrals, providing a rigorous vectorial description of electromagnetic fields near the focus of high-numerical-aperture aplanatic systems. Beginning with the fundamentals of polarization, we reviewed the vector nature of the electric field in monochromatic plane waves, deriving the transversality condition and exploring linear, circular, and elliptical polarization states as special cases of the general elliptical form. The role of Fresnel coefficients in altering polarization upon refraction was highlighted, underscoring their importance in realistic optical interfaces.

The angular spectrum representation was then introduced as a tool for decomposing arbitrary fields into propagating plane waves and evanescent components, naturally linking to Fourier optics in the paraxial limit and geometrical optics in the far field via the method of stationary phase. This formalism culminated in the derivation of the Richards-Wolf integrals, which elegantly compute the focused electric and magnetic fields by integrating vectorial contributions over a spherical reference wavefront, incorporating apodization and polarization transformations. Key properties of these integrals—linearity, scaling, rotational symmetry for circular apertures, transversality, and intensity distribution—were examined, revealing their physical robustness. Polarization effects were demonstrated through explicit calculations for linear, radial, and azimuthal incident fields, illustrating diverse focal structures, including longitudinal components and symmetry-dependent spot shapes with potential applications in photolithography and beyond.

Ultimately, the Richards-Wolf framework surpasses scalar diffraction theories by faithfully capturing the vectorial behavior of light, offering profound insights into polarization-dependent focusing and the diffraction limit in stigmatic systems. This equips readers with essential theoretical tools for advanced optics.

4.8 Problems

Problem 1: Basic polarization properties

A monochromatic plane wave propagates in the $+z$ direction. The electric field is given by

$$\mathbf{E} = \frac{E_0}{\sqrt{2}}\big(\hat{\mathbf{x}} + \mathrm{i}\hat{\mathbf{y}}\big)e^{\mathrm{i}(kz-\omega t)}.$$

(a) Determine whether this is linear, circular, or elliptical polarization.

(b) What is the handedness (left- or right-circular)?

(c) Write the time-dependent real electric field vector $\mathbf{E}(z,t)$.

(d) At $z = 0$, sketch how the tip of the $\mathbf{E}$ vector behaves as time progresses.

Problem 2: Transversality and Poynting vector

Prove that for a monochromatic plane wave

$$\mathbf{E} = \mathbf{E}_0 \, e^{\mathrm{i}(\mathbf{k}\cdot\mathbf{r}-\omega t)}$$

propagating in a charge-free homogeneous isotropic medium, the following relations hold:

(a) $\mathbf{k} \cdot \mathbf{E}_0 = 0$

(b) $\mathbf{k} \cdot \mathbf{B}_0 = 0$

(c) $\mathbf{B}_0 = \frac{1}{c} \hat{\mathbf{k}} \times \mathbf{E}_0$

(d) The time-averaged Poynting vector $\langle \mathbf{s} \rangle$ is parallel to $\mathbf{k}$.

Problem 3: Elliptical → special cases

Consider an elliptically polarized wave propagating along z with

$$E_x(t) = 3\cos(\omega t), \qquad E_y(t) = 2\cos(\omega t + \delta).$$

(a) For which values of δ do we obtain strictly linear polarization?

(b) For which values of δ and relative amplitude do we obtain circular polarization?

(c) Compute the axial ratio $\mathrm{AR} = a/b$ and ellipticity angle χ when $\delta = \pi/4$.

Problem 4: Fresnel coefficients - Brewster angle

Unpolarized light is incident from air ($n_1 \approx 1$) onto glass ($n_2 = 1.52$) at angle θ_i.

(a) Calculate the Brewster angle θ_B.

(b) At the Brewster angle, what is the polarization state of the reflected light?

(c) What happens to the p-component of the reflected field?

(d) Qualitatively explain why the reflected light can be used to produce linearly polarized light.

Problem 5: Angular spectrum - propagating vs. evanescent waves

The angular spectrum of a field in the plane $z = 0$ is given. The free-space wavenumber is $k = 2\pi/\lambda$.

(a) For which spatial frequencies (k_x, k_y) do we have purely propagating plane waves?

(b) For which region do we obtain evanescent waves?

(c) How does the amplitude of evanescent waves behave as a function of z?

(d) Explain why evanescent waves do *not* contribute to the far-field.

Problem 6: Richards-Wolf integral - conceptual understanding

(a) Explain in your own words why the scalar diffraction theories (Fresnel, Fraunhofer) are insufficient for high-NA focusing systems.

(b) What physical assumptions are made when using the Richards-Wolf integrals? (name at least four important ones)

(c) What kind of optical systems can be rigorously described by the classical Richards-Wolf formulation?

Problem 7: Properties of Richards-Wolf integrals

(a) Explain what is meant by the *linearity* property of the Richards-Wolf integrals. Why is this property very useful when dealing with complex polarization states?

(b) What does the *scaling property* tell us? Write the mathematical relation between input amplitude scaling and coordinate scaling.

(c) For which two polarization types discussed in the chapter is the intensity distribution in the focal plane rotationally symmetric about the optical axis?

(d) Why is there generally a strong longitudinal electric field component (E_z) when using radially polarized light at high NA, but almost none when using azimuthally polarized light?

Problem 8: Richards-Wolf integrals—conceptual understanding
Show (at least schematically) how the x- and y-components of $\mathbf{E}_\infty(\theta, \phi)$ look like in spherical coordinates for

(a) uniform x-linear polarization

(b) radial polarization

(c) azimuthal polarization

(You may sketch the polarization arrows on a meridional plane or write the dominant vector components.)

5

The apodization function on the Richards-Wolf equations

This chapter examines how the apodization function shapes the focused light field in high-numerical-aperture optical systems when described by the Richards-Wolf diffraction integrals. Apodization acts as an amplitude weighting across the pupil that preserves energy conservation when a plane wave is transformed into a converging spherical wavefront. It strongly influences the point-spread function by controlling sidelobe levels and overall sharpness, which matters greatly for applications such as microscopy and lithography. This chapter derives and compares apodization profiles for several well-known stigmatic focusing systems. For the aplanatic lens the weighting follows the square root of the cosine of the convergence angle. The parabolic mirror produces a different, stronger weighting that favors larger angles. The Cartesian oval yields yet another profile, depending on the refractive indices of the object and image spaces. Using the Richards-Wolf framework with uniform radially polarized illumination, the chapter shows that the parabolic mirror delivers the tightest focal spot compared with the aplanatic lens and the Cartesian oval under identical maximum convergence angles.

5.1 Introduction

The apodization factor is a fundamental concept in optical systems, describing the modulation of the amplitude of the electromagnetic field across the pupil or aperture of a focusing element, such as a lens or mirror. This factor, often denoted as $q(\theta)$, accounts for the geometric and energetic transformations that occur when an incident plane wave is focused into a spherical wavefront. The apodization factor ensures energy conservation and optimizes the PSF by

DOI: 10.1201/9781003726494-8

reducing diffraction artifacts like sidelobes, which is critical in high numerical aperture systems used in applications such as microscopy and lithography.

In the context of Richards-Wolf integrals, the apodization factor is important for accurately modeling the vectorial diffraction of light near the focal point of a high-NA system. These integrals, studied in detail in last chapter, describe the focused field $\mathbf{E}_\infty$ as a function of the incident field $\mathbf{E}_{\text{inc}}$, incorporating the apodization factor to scale the amplitude based on the system's geometry. For example, in an aplanatic lens, the apodization factor is $\sqrt{\cos\theta}$, where θ is the polar angle, ensuring uniform energy distribution across the spherical wavefront. This factor is derived from geometric analysis, such as the relationship $h = fg(\theta)$, where h is the incidence height and f is the focal length, making it specific to the optical system's design.

The significance of the apodization factor in Richards-Wolf integrals lies in its role in maintaining energy conservation and enabling precise field calculations. By weighting the pupil function, it ensures that the energy of the incident beam matches that on the reference sphere after focusing, as seen in systems like parabolic mirrors or elliptical mirrors with more complex forms. Additionally, the apodization factor allows optical designers to tailor the PSF, enhancing image quality by suppressing unwanted diffraction effects, which is vital for applications requiring high resolution and contrast, such as astronomical imaging or semiconductor manufacturing.

In this chapter we are going to dig into the topics studied where apodization functions in the most common stigmatic system are implemented in the Richards-Wolf integrals: aplanatic lens, elliptic mirror and the parabolic mirror. Then we are going to compare the point spread function using the Richards-Wolf integrals.

5.2 Apodization

Within geometrical optics, when a point source in the object space is perfectly focused by an optical system to a single image point, then the object-image pair is called a stigmatic pair and the image point is free of aberrations [23]. If we consider a given ray incident at a height h on a perfectly focusing system, there must exist a function $g(\theta)$ such that [78],

$$h = fg(\theta), \tag{5.1}$$

where f is the focal length of the system.

This function is entirely dependent on the focusing system and enables the connection between the incident geometry and the focusing geometry. Specifically, the purpose of this geometry transfer is to describe the focused field $\mathbf{E}_\infty$ as a function of the incident field $\mathbf{E}_{\text{inc}}$. The principle linking these two fields is energy conservation, according to which the energy contained in

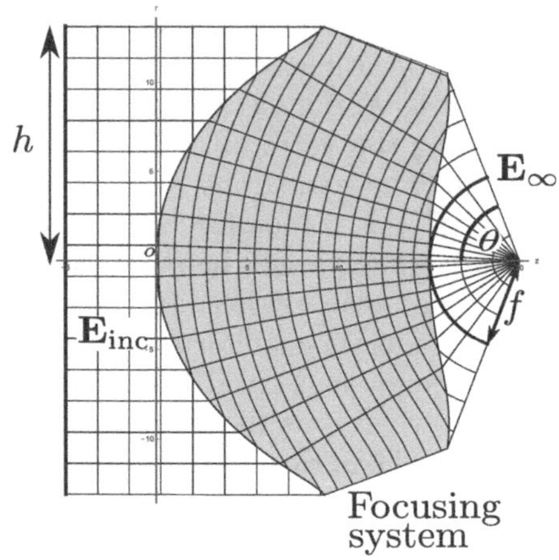

FIGURE 5.1
Diagram showing the parameters related to the apodization function. h is the height of a given ray incident, f is the focal length of the system, and θ is the converging angle of the ray at the image plane.

the bundle of incident beams must be equal to that on the surface of the reference sphere (after focusing).

Considering a three-dimensional and axisymmetric beam (with one plane defined by Fig. 5.1), the energy conservation condition can be expressed with the following integrals,

$$\int |\mathbf{E}_{\text{inc}}|^2 2\pi h \mathrm{d}h = \int |\mathbf{E}_{\infty}|^2 2\pi f^2 \sin\theta \mathrm{d}\theta. \tag{5.2}$$

Since it must hold for any integration domain, no matter how small [12].

$$|\mathbf{E}_{\text{inc}}|^2 2\pi h \mathrm{d}h = |\mathbf{E}_{\infty}|^2 2\pi f^2 \sin\theta \mathrm{d}\theta. \tag{5.3}$$

By using the change of reference from Eq. (2.6), we obtain,

$$|\mathbf{E}_{\text{inc}}|^2 g(\theta) g'(\theta) = |\mathbf{E}_{\infty}|^2 \sin\theta \mathrm{d}\theta, \tag{5.4}$$

where $g'(\theta)$ is the derivative, with respect to θ, of the function $g(\theta)$. This equality is used to define the apodization factor $g(\theta)$, an important concept in focusing, defined as,

$$|\mathbf{E}_{\infty}| = |\mathbf{E}_{\text{inc}}| q(\theta). \tag{5.5}$$

As can be observed, this apodization factor is, in a sense, a weighting factor that adjusts the focused plane wave spectrum based on the description of the focusing system. Indeed, it is noted that the function $g(\theta)$ is related to the relation $h(\theta)$ as,

$$q(\theta) = \frac{1}{f}\sqrt{\frac{h(\theta)h'(\theta)}{\sin\theta}}. \tag{5.6}$$

By knowing the focusing system and through geometric analysis of the focusing angle as a function of the incidence height h of a given ray, it is possible to define this apodization function to relate the amplitude of the known incident field in a plane to the amplitude of the focused field on the reference sphere after focusing.

The apodization factor relates the amplitude of the incident light beam with the focusing beam. If the incident angle is small, the value of the apodization factor $q(\theta) \rightarrow 1$, which indicates the energy density remains unchanged. Now, if the apodization factor $q(\theta)$ is greater or less than 1, the energy after the transmission will be compressed or stretched, respectively.

5.3 Apodization of the aplanatic lens

We have seen from last chapter that the apodization function is given by $q(\theta) = \sqrt{\cos\theta}$. But we have not mentioned where it comes from, for that we need to explore first the geometry an aplanatic lens and with this, the aplanatic condition, or the so called Abbe since condition.

The geometry of such an aplanatic lens is presented in Fig. 5.2. The first surface of the lens is characterized by (z_a, r_a), the axial thickness of the lens is τ and the second surface is given by (z_b, r_b). The lens has a refraction index n and the refraction indices of the object and image spaces are n_o and n_i, respectively. Earlier work has shown that the only independent variable of the functions z_a, z_b and r_b is the radius r_a [23]. The distance from the object to the first surface is $z_o \rightarrow -\infty$ and the distance from the second surface to the image or focus is z_i (back focal length). Notice that the coordinate system is placed in the vertex of the first surface.

The parameters z_b and r_b can be expressed as,

$$z_b = z_a + \vartheta\wp_z, \qquad r_b = r_a + \vartheta\wp_r, \tag{5.7}$$

where, $\vartheta \equiv \sqrt{(z_b - z_a)^2 + (r_b - r_a)^2}$ and $\wp_z$ and $\wp_r$ are the cosine directors of the ray traveling since the lens. To be considered aplanatic, the lens needs to satisfied the Abbe sine condition, which is given by the following expression,

$$h = f\sin\theta, \tag{5.8}$$

where, h is the height at the entrance of the input ray. f is the effective focal length and θ is the output angle of the departing ray. Using (5.7), one can obtain the definition of the first surface to be in accordance with the

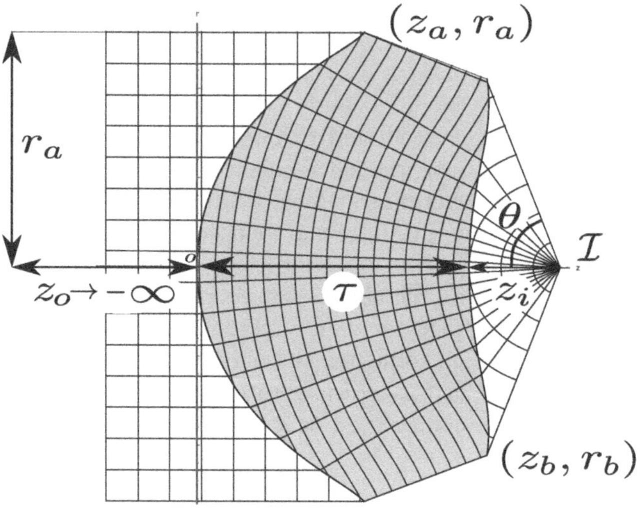

FIGURE 5.2
Diagram of an aplanatic lens. The geometry of the aplanatic lens is defined by its two surfaces. The first (front) surface is described by the coordinates (z_a, r_a), while the second (rear) surface is given by (z_b, r_b). The lens has a constant axial thickness τ and a refractive index n. The refractive indices of the surrounding media are n_o in the object space and n_i in the image space. In this configuration, the object is located at infinity ($z_o \to -\infty$), so the incident rays are collimated (parallel to the optical axis). The distance from the vertex of the second surface to the focal point (image position) is denoted z_i and is commonly referred to as the back focal length. The coordinate system is placed with its origin at the vertex of the first surface, with the positive z-axis pointing toward the image space.

condition. Then, the surface sag will be expressed in term of the parameters of the stigmatic lens and can be solved [43]. We obtain the relation for r_a following,

$$r_a = f \frac{r_b}{\sqrt{r_b^2 + (z_b - z_i - \tau)^2}}. \tag{5.9}$$

From [43], it is clear that (5.9) can be considered seen has $r_a = h$. Therefore,

$$q(\theta) = \frac{1}{f}\sqrt{\frac{r_a}{\sin\theta}\frac{\partial r_a}{\partial \theta}}, \tag{5.10}$$

replacing Eq. (5.8) in previews expression,

$$q(\theta) = \frac{1}{f}\sqrt{f^2 \frac{\sin\theta}{\sin\theta}\cos\theta}, \tag{5.11}$$

simplifying,

$$q(\theta) = \sqrt{\cos\theta}. \tag{5.12}$$

The last expression is the apodization function of the aplanatic lens. This expression was used along the previews chapter, due historical reasons. When initially B. Richards and E. Wolf in 1959 published their vector diffraction integrals, they were only considering an aplanatic lens [79].

5.4 Apodization of the parabolic mirror

To get the apodization function for the parabolic mirror (Fig. 5.3), we need first study the elliptic mirror and its stigmatic properties. The we shall get the expression of the parabola and from it, we can obtain the apodization using Eq. (5.6).

So consider first the elliptic mirror, a stigmatic mirror for a point object at z_o and point image at z_i. The Fermat principle predicts that the light follows the least time path. If the refraction index of the medium is constant, the least time path from and object placed in z_o is a straight line to the image placed in z_i. This line, we can call the optical axis if z_o and z_i are aligned; and the ray that follows it is the axial ray. Hence, if all the rays follow the least time path, then the optical path of a non-axial ray is same as the path of the axial

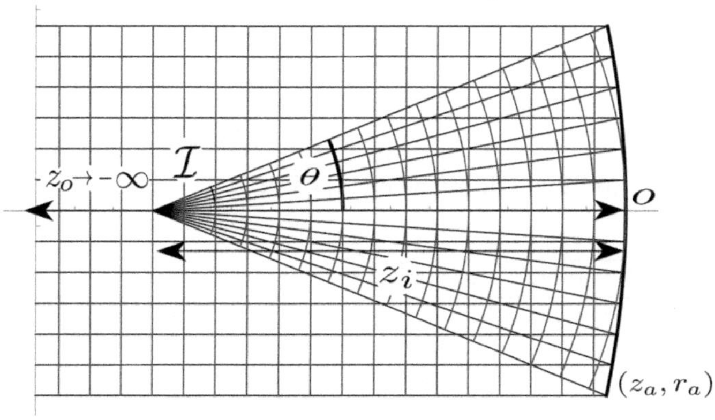

FIGURE 5.3
A parabolic mirror is an stigmatic mirror for a point object at z_o and point image at z_i, when $z_o = -\infty$.

ray, The optical path of the axial ray is,

$$-z_o - z_i, \tag{5.13}$$

while the optical path of the non-axial ray is,

$$\sqrt{(z_a - z_i)^2 + r_a^2} + \sqrt{r_a^2 + (z_a - z_o)^2}, \tag{5.14}$$

where r_a is the radius of the surface z_a, and it works as the independent variable. The idea is how must be z_a such that the mirror is stigmatic. Tacking on count what we have deduced in the first paragraph of this chapter we have,

$$-z_o - z_i = \sqrt{(z_a - z_i)^2 + r_a^2} + \sqrt{r_a^2 + (z_a - z_o)^2}, \tag{5.15}$$

where in the last expression we have $-z_o - z_i$ in the left side because the coordinate system is placed at the vertex of the mirror.

Solving for z_a,

$$z_a = \frac{1}{2}\left[z_o + z_i \pm \frac{\sqrt{z_o z_i (z_o + z_i)^2 (z_o z_i - r_a^2)}}{z_o z_i}\right], \tag{5.16}$$

from the last expression, from the $\pm$ we take the positive sign, in order to have an elliptic mirror, hence,

$$z_a = \frac{1}{2}\left[z_o + z_i + \frac{\sqrt{z_o z_i (z_o + z_i)^2 (z_o z_i - r_a^2)}}{z_o z_i}\right]. \tag{5.17}$$

The last expression give us the shape of a mirror that for a point object and a point image located at z_o and z_i respectively the system is stigmatic.

If the object is very far away from the image $z_o \to \infty$ and we want a stigmatic mirror we should manipulate Eq. (5.15) in order to get the following expression,

$$1 = \frac{\sqrt{r_a^2 + (z_a - z_o)^2} + z_o}{-\sqrt{(z_a - z_i)^2 + r_a^2} - z_i}, \tag{5.18}$$

then, we can compute the limit when $z_o \to \infty$,

$$\lim_{z_o \to -\infty}\left[\frac{\sqrt{r_a^2 + (z_a - z_o)^2} + z_o}{-\sqrt{(z_a - z_i)^2 + r_a^2} - z_i}\right] = -\frac{z_a}{\sqrt{(z_i - z_a)^2 + r_a^2} + z_i}, \tag{5.19}$$

substituting equation (5.19) in equation (5.18) we have,

$$1 = -\frac{z_a}{\sqrt{(z_i - z_a)^2 + r_a^2} + z_i}, \tag{5.20}$$

and finally, solving for z_a,

$$z_a = \frac{r_a^2}{4z_i}. \tag{5.21}$$

It is clear that the last expression is a parabola. An example of the parabolic mirror can be found in Fig. 5.3.

The polar angle θ can be obtained the division of r_a, the height, and the $z_i - z_a$ the width,

$$\tan(\theta) = \frac{r_a}{z_i - z_a}, \tag{5.22}$$

Replacing Eq. (5.21) in the last expression,

$$\tan(\theta) = \frac{r_a}{z_i - \frac{r_a^2}{4z_i}}. \tag{5.23}$$

Solving for r_a

$$r_a = 2z_i \tan(\theta/2), \tag{5.24}$$

setting $r_a = h$ and $z_i = f$,

$$h = 2f \tan(\theta/2), \tag{5.25}$$

By substituting last expression in Eq. (5.6), the apodization factor of the parabolic mirror can be expressed as,

$$q(\theta) = \sec^2(\theta/2). \tag{5.26}$$

The last expression is the apodization of the parabolic mirror, a system that does not follow the Abbe sine condition. In fact the parabolic mirror by its shape suffers heavy from coma aberration. Aplanatic systems (that follow the abbe sine condition) have controlled the coma.

5.5 Cartesian oval

The Cartesian oval is one well-known stigmatic refractive surface and had been successfully used in illumination [80], refocusing problems [81, 82] or even angular position converter in fiber optics [83]. The first person to formally analyze the oval was Ibn Shal over a millennium ago and it has since been extensively studied even for negative index of refraction [84, 85, 86, 87, 88, 89, 90, 91, 92, 93, 94].

Here we need to obtain closed-form relations for the pupil apodization function of a Cartesian oval. For that, we derive an analytical expression of the Cartesian oval for a far object and a finite object whose form is convenient to obtain the apodization. This derivation is partially based on the general stigmatic surfaces [95]. Let's explore first this general stigmatic surfaces.

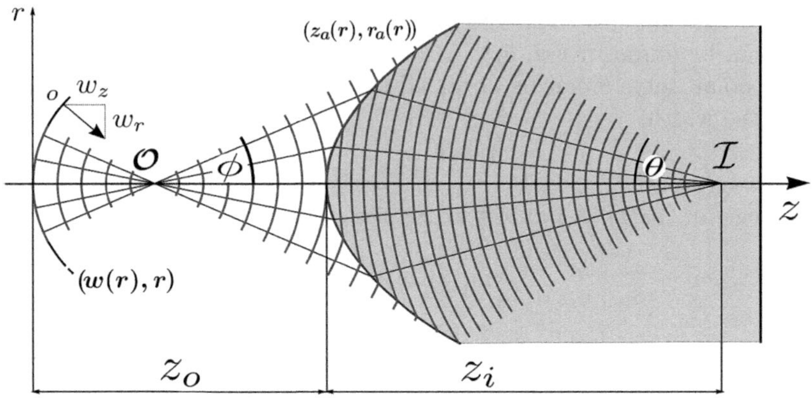

FIGURE 5.4
Diagram of a stigmatic surface. The wavefront $w(r)$ is located at the origin of the coordinate system. The refractive surface is defined by the unknown function $\{z_a, r_a\}$. The separation between the wavefront w and the refractive surface $\{z_a, r_a\}$ is z_o, and from the vertex of the refractive surface to the image point is z_i. The quantities w_r and w_z are the cosine directors of the rays in r and z directions, respectively.

Fig. 5.4 shows the geometry of the problem. The input wavefront is located at the origin of the coordinate system, and it is described by the function $w(r)$, where r is the radial variable. The distance between the wavefront w and the refractive surface (z_a, r_a) is z_o. The distance from the vertex of the refractive surface to the image point is z_i. The refractive indices at the object and image spaces are n_o and n_i, respectively. The problem can be stated as follows, given the wavefront $w(r)$, the distances z_o and z_i, and the refractive indices, how must be refractive surface (z_a, r_a) such that the rays of the wavefront converge into a single point image. This problem is a generalization of the classical problem of the stigmatic Cartesian ovals where only an axial image point is considered.

According to the Fermat principle, to obtain the stigmatic condition between the wavefront w and the image point, all the rays that emerge from the wavefront must have the same OPL. Consequently,

$$n_i\sqrt{(z_a - z_i - z_o)^2 + r_a^2} + n_o\sqrt{(r_a - r)^2 + (z_a - w)^2} = n_i z_i + n_o z_o \equiv K, \tag{5.27}$$

where K is the constant value of the OPL. The sum of the two square roots is the OPL of a non-axial ray, and $n_i z_i + n_o z_o$ is the OPL of the axial ray.

According to the Malus-Dupin theorem, the rays of the wavefront w are perpendicular to the wavefront [**?**]. Then, the rays propagate along the normal vector of w. With this in mind, the cosine directors of the input rays are

$$w_r = -\frac{w'}{\sqrt{w'^2 + 1}}, \quad w_z = \frac{1}{\sqrt{w'^2 + 1}}, \tag{5.28}$$

where $w' \equiv \mathrm{d}w/\mathrm{d}r$ is the derivative of w with respect to r, and w_r and w_z are the cosine directors in r and z directions, respectively. Now, we can set the distance between the wavefront w and the refractive surface as (z_a, r_a)

$$\vartheta = \sqrt{(r_a - r)^2 + (z_a - w)^2}. \tag{5.29}$$

With Eqs. (5.28) and (5.36) we can express the refracting surface in a parametric form as follows

$$r_a = r + \vartheta w_r, \quad z_a = w + \vartheta w_z. \tag{5.30}$$

We have now the structure of the solution. We only need to express ϑ in terms of the known parameters. For this we replace Eqs. (5.36) and (5.30) in Eq.(5.35), namely

$$n_i\sqrt{(-z_i - z_o + \vartheta w_z + w)^2 + (\vartheta w_r + r)^2} + \vartheta n_o = K. \tag{5.31}$$

Solving for ϑ we get

$$\begin{aligned} \vartheta &= \frac{+n_i^2(rw_r - w_z z_\tau) + kn_o}{n_o^2 - n_i^2} \\ &+ \frac{\sqrt{(n_o^2 - n_i^2)(n_i^2(r^2 + z_\tau^2) - k^2) + (n_i^2(rw_r - w_z z_\tau) + kn_o)^2}}{n_o^2 - n_i^2}, \end{aligned} \tag{5.32}$$

where

$$z_\tau \equiv -w + z_i + z_o. \tag{5.33}$$

Eqs. (5.30) and (5.32) describe in a mathematical closed-form how must be the refractive surface such the rays coming from a given wavefront converge geometrically into a single image point. We remark that this wavefront can be spherical too, and we end with the shape of the Cartesian Oval.

One last observation is that the angles θ and ϕ are defined as,

$$\tan\theta = \frac{r_a}{z_o + z_i - z_a}, \quad \tan\phi = \frac{r - r_a}{z_a - w} \tag{5.34}$$

5.5.1 Stigmatic Cartesian oval with object at infinity

The Cartesian oval of an object at infinity is a special case of the general Cartesian oval presented earlier where $w = 0$ and $z_o \to -\infty$. In that case, the radial derivation of the wavefront is null (i.e. $w' = 0$) resulting in the expected cosine directors $w_r = 0$ and $w_z = 1$ as presented in Fig. 5.5. To obtain the shape for this particular Cartesian oval, the OPL of any given ray should be constant and takes the form,

$$n_i z_i = n_i\vartheta(\theta) + n_o[z_i - \vartheta(\theta)\cos\theta], \tag{5.35}$$

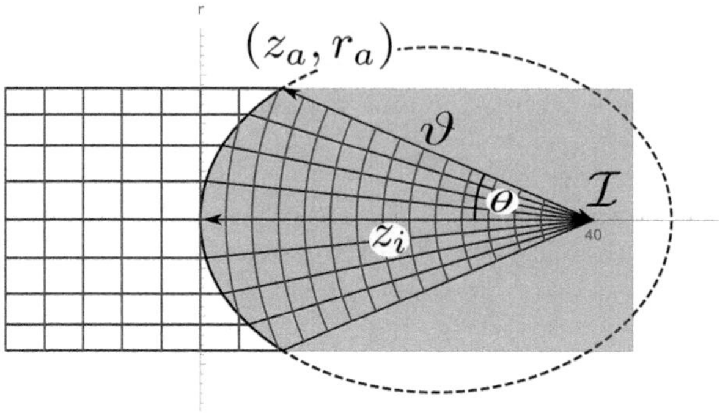

FIGURE 5.5
The diagram of the Cartesian oval of a distant object. The Cartesian oval is given by (z_a, r_a). The distance between the Cartesian oval and the image is z_i. The coordinate system is placed at the waterfront. The output ray has an angle θ concerning the optical axis.

where $\vartheta(\theta)$ is the distance from the refractive surface to the image point $(\mathcal{I})$. The left side of Eq.(5.35) is the OPL of an axial-ray, and the right side is the OPL of the refracted ray. Then, solving for $\vartheta(\theta)$ results in,

$$\vartheta(\theta) = \frac{(n_i - n_o)z_i}{n_i - n_o \cos\theta}. \tag{5.36}$$

Since, $\sin\theta = r_a/\vartheta(\theta)$, we can express the radial height of the Cartesian oval as,

$$r_a = \frac{(n_i - n_o)z_i}{n_i - n_o \cos\theta} \sin\theta. \tag{5.37}$$

Notice that we naturally get the radial height r_a from the OPL (i.e., Eq. (5.35)) which corresponds to the well-known equation of the ellipse with eccentricity $\epsilon = n_o/n_i$ when the coordinates are defined in accordance to the furthermost focus. Eq. (5.37) is not only the analytic closed-form, but also it is intuitive, compact and, it is valid for positive and negative refractive indices since OPL is valid for both cases.

The apodization function of a stigmatic Cartesian oval with an object at infinity is quite easy to obtain. Replacing Eq. (5.37) in Eq. (5.6), the apodization function of the Cartesian oval is given by,

$$q(\theta) = \sqrt{\frac{(n_i - n_o)^2 (n_i \cos\theta - n_o)}{(n_i - n_o \cos\theta)^3}}. \tag{5.38}$$

Eq. (5.38) describes the pupil apodization function of a Cartesian oval for a faraway object. Thus, it expresses how the density energy compresses and stretches along with the exit pupil. This closed-form expression was only possible using the parameterization presented in presented in this section, since none of the presented forms of the Cartesian ovals in Refs. [84, 85, 86, 87, 88, 89, 90, 91, 92, 93, 94] is useful to get Eq. (5.38).

Observe that Eq. (5.38) does not depend on the distance z_i and it only depends on the refractive indices and the focusing angle θ. One observation is that we obtain $q(\theta \to 0) = 1$ showing that Eq.(5.38) is well normalized. The maximum accessible angle θ_{max} can be computed with $\frac{\mathrm{d}}{\mathrm{d}\theta} r_a = 0$. Ignoring the trivial case when $n_0/n_i = 1$, the relation for the angle is,

$$\theta_{max} = \cos^{-1}(n_o/n_i). \tag{5.39}$$

One interesting note about Eq. (5.39) is valid when both n_o, n_i are positive and when n_o is positive and n_i negative or less than minus one.

5.6 Apodization influence on the diffraction integrals

In the last chapter we extensively studied the Richards-Wolf integrals [79]. In this section we are going to compare several apodization factors over the mentioned diffraction integrals. For this comparison we are going to use the radially polarized light of a symmetric stigmatic optical system can be computed using the following integrals, [79, 96],

$$\begin{bmatrix} E_r \\ E_z \end{bmatrix} = E_0 \int_0^{\theta_{\max}} q(\theta) l_0(\theta) \sin\theta \, \mathrm{e}^{-ikz\cos\theta} \begin{bmatrix} \mathrm{i}\cos\theta \, \mathrm{J}_1(kr\sin\theta) \\ \sin\theta \, \mathrm{J}_0(kr\sin\theta) \end{bmatrix} \mathrm{d}\theta, \tag{5.40}$$

where $l_0(\theta)$ is the illumination pattern and $q(\theta)$ is the apodization function which is different for every optical system. Notice that in the previews expression we use to have the term Eq. (5.40), now we absorbed inside the constant $kfe^{ikf}E_0 \to E_0$, since we are plotting the normalized intensity profile. For now on, this will be the case through the book.

Fig. 5.6 presents the comparison of the apodization aplanatic lens (gray), the parabolic mirror (black), for the Cartesian oval for a far object (gray dashed). We can see that the apodization function of both Cartesian ovals behaves like an aplanatic lens system, with the distinction that the apodization function of the Cartesian oval for a far object decays faster.

Fig. 5.7 presents the normalized intensity of the diffraction of the Cartesian oval, the aplanatic lens and the parabolic mirror under the same conditions. For this comparison, we didn't consider the Fresnel transmission and reflection coefficients. The illumination pattern used for this example is a uniform

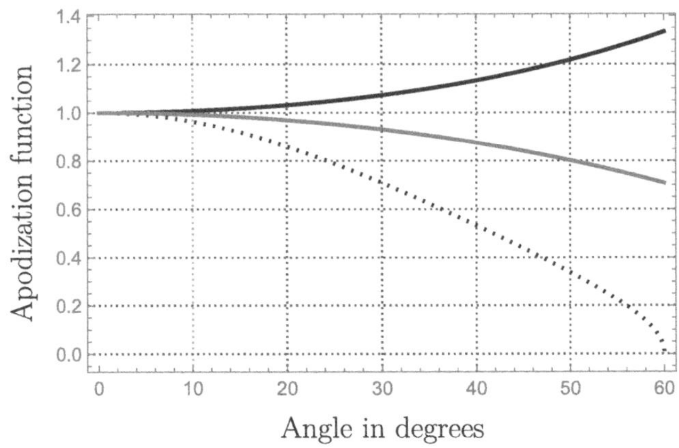

FIGURE 5.6
Pupil apodization function of several optical systems. The curves corresponding to the apodization function of a parabolic mirror, an aplanatic lens, Cartesian oval for a far object are in black, gray, and dashed gray, respectively. The specifications of the Cartesian oval are $n_o = 1$, $n_i = 2$, thus $\theta_{max} = 60°$.

illumination following $l(\theta) = 1$ with radial polarization. The specifications of the design for the Cartesian oval are $n_o = 1$, $z_i = 5$mm and $n_i = \pm 2$ imposing a maximum focusing angle $\theta_{\max} = 60°$ for the positive index oval. In such conditions, for the oval, the FWHM for the radial dimension is 1.33λ and 2.07λ for the on-axis component for the positive refraction index. As it can be seen in Fig. 5.7, the aplanatic lens and the parabolic mirror focus are sharper

FIGURE 5.7
Normalized intensity of the diffraction of the Cartesian oval with a radially polarized uniform illumination (i.e. $l_0(\alpha) = 1$) with $\alpha_{\max} = 60°$. (a) positive refraction index of $n_i = 2$, considering $n_o = 1$ and a focal length $z_i = 5$mm, (b) and (c) are the normalized intensity of the aplanatic lens and the parabolic mirror in the same conditions.

than the one obtained by the positive index Cartesian oval. The FWHM for the aplanatic lens is 0.901λ and 1.94λ and the parabolic mirror are 0.789λ and 1.98λ respectively.

5.7 Conclusion

In this chapter, we have explored the concept of apodization and its effect in the PSF among high-numerical-aperture (NA) optical system. Each optical system has a different apodization function. We analytically derived the apodization functions of the aplanatic lens, the parabolic mirror, and the Cartesian Oval.

To rigorously compare these apodizations, we employed the Richards-Wolf integrals under radially polarized illumination. This comparative framework illustrate the differences between optical system with different apodizations. Being the parabolic mirror the system that achieves the smallest PSF.

In the next chapter we are going to study the stigmatic lens. The stigmatic lens in general is free of spherical aberration, but not necessary to follow the Abbe since condition. Getting the general expression of the stigmatic lens and its apodization function is not trivial. The purpose of the following chapter is to show the rigorous computation of the diffraction pattern of the stigmatic lens.

5.8 Problems

Problem 1: Energy conservation and apodization derivation

The fundamental relation for energy conservation between the incident plane and the reference sphere is given by

$$|\mathbf{E}_{\text{inc}}|^2\, 2\pi h\, dh = |\mathbf{E}_{\infty}|^2\, 2\pi f^2 \sin\theta\, \mathrm{d}\theta.$$

(a) Show that this leads to the general expression for the apodization factor:

$$q(\theta) = \frac{1}{f}\sqrt{\frac{h(\theta)\, h'(\theta)}{\sin\theta}},$$

where $h(\theta) = f\, g(\theta)$ and $h'(\theta) = \frac{dh}{\mathrm{d}\theta}$.

(b) Explain physically why $q(\theta) \to 1$ when $\theta \to 0$ (small-angles/paraxial-case).

Problem 2: Aplanatic lens apodization
The aplanatic lens satisfies the Abbe sine condition: $h = f \sin\theta$.

(a) Starting from $h = f\sin\theta$, derive step-by-step that the apodization function is

$$q(\theta) = \sqrt{\cos\theta}.$$

(b) What physical property of the aplanatic system is preserved thanks to this particular apodization function?

Problem 3: Parabolic mirror apodization
For a parabolic mirror focusing a collimated beam ($z_o \to -\infty$), the surface is described by $z_a = \frac{r_a^2}{4f}$ and it was shown that

$$h = 2f\tan(\theta/2).$$

(a) Verify that the apodization function for the parabolic mirror is

$$q(\theta) = \sec^2(\theta/2).$$

(b) Explain why a parabolic mirror *does not* satisfy the Abbe sine condition and mention one important aberration that is typically strong in such systems.

Problem 4: Cartesian oval with object at infinity
For a stigmatic Cartesian oval with object at infinity and refractive indices n_o (incident) and n_i (image space), the radial height is

$$r_a = \frac{(n_i - n_o)z_i \sin\theta}{n_i - n_o\cos\theta}.$$

(a) Derive the corresponding apodization function and show that it can be written as

$$q(\theta) = \sqrt{\frac{(n_i - n_o)^2(n_i\cos\theta - n_o)}{(n_i - n_o\cos\theta)^3}}.$$

(b) Evaluate $q(\theta)$ at $\theta = 0$ and comment on the normalization.

(c) Determine the maximum focusing angle $\theta_{\max}$ and explain its physical meaning.

Problem 5: Comparison of apodization behaviors
The following systems are compared in the chapter:

- Aplanatic lens: $q(\theta) = \sqrt{\cos\theta}$
- Parabolic mirror: $q(\theta) = \sec^2(\theta/2)$

- Cartesian oval (object at infinity)

(a) Qualitatively sketch or describe how $q(\theta)$ behaves from $\theta = 0$ to $\theta_{\max}$ for each system.

(b) Which system has the strongest amplitude weighting (compression) at large angles?

(c) Which system has the weakest amplitude weighting (closest to uniform) at large angles?

Problem 6: Influence on the point spread function (conceptual)
The Richards–Wolf integrals for radially polarized light contain the apodization term $q(\theta)$:

$$\begin{bmatrix} E_r \\ E_z \end{bmatrix} \propto \int_0^{\theta_{\max}} q(\theta) \sin\theta \begin{bmatrix} \mathrm{i}\cos\theta\, \mathrm{J}_1(kr\sin\theta) \\ \sin\theta\, \mathrm{J}_0(kr\sin\theta) \end{bmatrix} e^{-\mathrm{i}kz\cos\theta}\, \mathrm{d}\theta.$$

(a) Explain physically why a stronger decay of $q(\theta)$ at large θ usually leads to a broader focal spot (larger FWHM).

(b) According to the chapter comparison, which system gives the *smallest* focal volume when using uniform radial illumination: aplanatic lens, parabolic mirror, or Cartesian oval (positive index)?

(c) Why is the parabolic mirror particularly effective for obtaining a very small spot size?

Problem 7: Numerical / qualitative estimation
Assume a maximum focusing angle $\theta_{\max} = 60°$ and uniform radial illumination.

(a) Compute the value of $q(60°)$ for:
  - aplanatic lens
  - parabolic mirror
  - Cartesian oval with $n_o = 1$, $n_i = 2$

(b) Rank the three systems according to the strength of amplitude apodization (weighting) at the edge of the pupil (strongest → weakest).

(c) Which of these three systems is expected to have the *largest* longitudinal field component E_z on axis (qualitative reasoning)?

Problem 8: Open conceptual question
The chapter concludes that the parabolic mirror gives the smallest PSF compared to the aplanatic lens and the Cartesian oval under the same maximum angle and uniform radial illumination.

(a) Explain why this result is somewhat counter-intuitive (considering that aplanatic systems are usually considered *ideal* in classical optics).

(b) Discuss one important practical disadvantage of the parabolic mirror compared to an aplanatic lens when used in real high-NA imaging systems.

6

Stigmatic lenses and stigmatic vector diffraction

This chapter presents a rigorous derivation for designing stigmatic singlet lenses, which perfectly eliminate spherical aberration for a given on-axis object–image conjugate pair without necessarily satisfying the Abbe sine condition (not aplanatic). Starting from Fermat's principle and the vector form of Snell's law, closed-form expressions are derived. The ray mapping naturally yields the exact pupil apodization function $q(\theta)$. Coupling these results with Richards–Wolf vectorial diffraction integrals enables accurate computation of non-paraxial diffraction patterns of stigmatic lenses. Comparisons between the non-paraxial diffraction patterns of the stigmatic lens and aplanatic lens are discussed.

Stigmatic optical systems are conceived with the fundamental goal of imaging a single object point exactly onto a single image point, thereby completely eliminating spherical aberration for the chosen conjugate pair. In the last chapter we studied some of the most popular stigmatic optical systems, the aplanatic lens, the parabolic mirror and the Cartesian Oval.

The aplanatic lens is a special case of the stigmatic lens, a stigmatic lens that full fills the Abbe sine condition. The general equation of the stigmatic lens has been found recently [43, 23, 85, 95, 97]. This equation consider all the possible stigmatic lenses, including the ones that fulfill the Abbe sine condition (aplanatic lenses).

For many decades, obtaining the exact surface profiles of stigmatic lenses was considered an extremely difficult, if not practically intractable, problem once one moved beyond paraxial approximations or purely numerical optimization techniques. It is only in recent years that a general, exact analytical solution has been derived for the design of stigmatic singlets, valid for arbitrary object-image conjugates and free of any small-angle assumptions [43, 23, 85]. We are interested in the stigmatic lenses because from them we can get the apodization function that characterize them. Once we get the apodization

DOI: 10.1201/9781003726494-9

function, we can use the Richards-Wolf integral to compute the diffraction pattern produced by a stigmatic lens [48]. Therefore the present chapter pursues three main objectives:

1. To provide a self-contained, rigorous derivation of the general closed-form analytical expression that defines the stigmatic singlet lens, starting from first principles (Fermat's principle combined with the vector formulation of Snell's law).
2. To extend the formulation to the particularly important practical case of the *stigmatic collimator*, and to demonstrate how the resulting ray mapping naturally yields the pupil apodization function $q(\theta)$.
3. Compute the diffraction pattern produced by a stigmatic lens via the Richards-Wolf integrals and the apodization function obtained for the stigmatic lens $q(\theta)$.

This chapter therefore aims to bridge the gap between exact analytical lens design and modern stigmatic vector diffraction.

6.1 The stigmatic lens

The theory of *stigmatic optics* begins with Fermat's principle and Snell's law, without numerical nor paraxial approximations. Its central problem is the general equation of the stigmatic lens [43, 22, 98] with any of the mentioned approximations. Once we get this close-form expression we can get a close-form expression for the apodization function of the stigmatic lens. With the a apodization function of the stigmatic lens $q(\theta)$, finally we can compute the diffraction patterns via the Richards-Wolf integrals [48].

We consider two cases: finite object–image distance, and *stigmatic collimator*. The first case is a preliminary step to obtain the second case, which indeed is the one the apodization function can be extracted. The design problem is: given the first surface, object and image positions, refractive indices of object, image, and glass, and the central thickness, how should the second surface be shaped to eliminate spherical aberration?

6.1.1 Finite object–image distance

The solution requires determining the second surface (z_b, r_b) from the known first surface (z_a, r_a). Here r_a is the only independent variable, while z_b, r_b, and z_a are functions of r_a. The sagitta of the second surface is z_b with radius r_b; the first surface sagitta is z_a with radius r_a. The coordinate origin is set at the vertex of z_a (see Fig. 6.1).

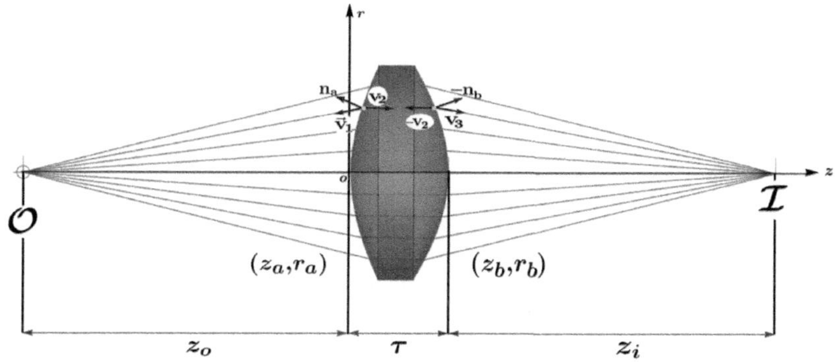

FIGURE 6.1
Illustration of an on-axis stigmatic singlet. The first and second surfaces are denoted by (z_a, r_a) and (z_b, r_b), respectively. The axial separation between the first surface and the object is z_o, while the axial diameter of the lens is τ, and the axial distance between the second surface and the image is z_i. The refractive index in the object space is n_o, within the lens it is n, and in the image space, it is n_i.

For a stigmatic design, the optical path length must be consistent for all rays originating from a point object on the axis and converging onto a point image on the axis. Consequently, the optical path length remains constant for all rays.

According to Fermat's principle, light propagates along the path of least time. Combining this with the preceding conditions, we can compare the optical path length (OPL) of the axial ray with that of a non-axial ray.

We start with the axial ray. Emerging from $-z_o$, it passes through the lens over a distance τ, and then, from the second surface, continues a distance z_i to the image point (see Fig. 6.1). The OPL of the axial ray is therefore given by

$$-n_o z_o + n\tau + n_i z_i = \text{constant}. \tag{6.1}$$

The optical path length of a non-axial ray is more complex, as it varies depending on whether the object/image is real or virtual. Let's consider the more straightforward scenario first, where both the object and the image are real, denoted by $z_o < 0$ and $z_i > 0$.

$$\begin{aligned} -n_o z_o + n\tau + n_i z_i &= n_o\sqrt{r_a^2 + (z_a - z_o)^2} \\ &+ n\sqrt{(r_b - r_a)^2 + (z_b - z_a)^2} \\ &+ n_i\sqrt{r_b^2 + (z_b - \tau - z_i)^2}, \end{aligned} \tag{6.2}$$

When both $z_o > 0$ and $z_i > 0$, the object is virtual while the image remains real.

$$\begin{aligned} -n_o z_o + n\tau + n_i z_i &= -n_o\sqrt{r_a^2 + (z_a - z_o)^2} \\ &+ n\sqrt{(r_b - r_a)^2 + (z_b - z_a)^2} \\ &+ n_i\sqrt{r_b^2 + (z_b - \tau - z_i)^2}. \end{aligned} \tag{6.3}$$

In the case of a real object and a virtual image, the coordinates satisfy $z_o < 0$ and $z_i < 0$. This means that the object is physically located on the optical axis before the lens, while the image does not form on the opposite side but instead appears to diverge from a point on the same side as the object. Such a configuration is characteristic of lenses that do not produce a real convergence of rays but rather extend them backward to form a virtual image.

$$\begin{aligned} -n_o z_o + n\tau + n_i z_i &= n_o\sqrt{r_a^2 + (z_a - z_o)^2} \\ &+ n\sqrt{(r_b - r_a)^2 + (z_b - z_a)^2} \\ &- n_i\sqrt{r_b^2 + (z_b - \tau - z_i)^2}. \end{aligned} \tag{6.4}$$

Finally, when both the object and the image are virtual, the coordinates satisfy $z_o > 0$ and $z_i < 0$. In this situation, the object appears to be located on the side opposite to the incoming light, so it does not physically exist in the object space. Similarly, the image also does not form as a real convergence of rays but appears to diverge from a point in the image space. Such a configuration arises in optical systems where the lens produces virtual images for virtual objects, and the rays must be traced accordingly to determine the optical path lengths and focusing behavior.

$$\begin{aligned} -n_o z_o + n\tau + n_i z_i &= -n_o\sqrt{r_a^2 + (z_a - z_o)^2} \\ &+ n\sqrt{(r_b - r_a)^2 + (z_b - z_a)^2} \\ &- n_i\sqrt{r_b^2 + (z_b - \tau - z_i)^2}, \end{aligned} \tag{6.5}$$

The four cases described above—real object to real image, real object to virtual image, virtual object to real image, and virtual object to virtual image—can all be expressed in a unified form. By carefully accounting for the signs of z_o and z_i, as well as the corresponding ray directions, these separate scenarios can be incorporated into a single general equation that describes the optical path length for any combination of real or virtual objects and images. This unified expression can be conveniently written as

$$\begin{aligned} -n_o z_o + n\tau + n_i z_i = \; & -n_o \mathrm{sgn}(z_o)\sqrt{r_a^2 + (z_a - z_o)^2} \\ & + n\sqrt{(r_b - r_a)^2 + (z_b - z_a)^2} \\ & + n_i \mathrm{sgn}(z_i)\sqrt{r_b^2 + (z_b - \tau - z_i)^2}. \end{aligned} \tag{6.6}$$

Please note that sgn(·) denotes the sign of its argument, taking values +1 or −1 depending on whether the input is positive or negative, and is undefined when the argument is zero. This function plays a crucial role in correctly handling the direction of rays for the various object–image configurations.

It is also important to recognize that the two unknowns in our problem are (z_b, r_b), representing the sagitta and radius of the lens's second surface. At this stage, however, we have only a single equation relating these variables. Consequently, the system is underdetermined, and we need a second independent equation to fully determine the lens shape. This additional constraint will be derived from Snell's law, which governs the refraction of light at the lens surfaces and ensures that the resulting design satisfies the conditions for a stigmatic, spherical aberration-free lens.

Let us now turn our attention to Snell's law, which governs the refraction of light as it passes from one medium to another, as we saw in chapter 2. This fundamental principle dictates how the direction of a light ray changes depending on the refractive indices of the two media. In the context of lens design, Snell's law provides the necessary relationship between the incident and refracted rays at each surface, ensuring proper focusing.

At the first surface of the lens, the vector form of Snell's law can be expressed as,

$$\mathbf{v_2} = \frac{n_o}{-\text{sgn}(z_o)n}[\mathbf{v_1} - (\mathbf{n_a} \cdot \mathbf{v_1})\mathbf{n_a}] - \mathbf{n_a}\sqrt{1 - \frac{n_o^2}{n^2}(\mathbf{n_a} \times \mathbf{v_1})^2} \tag{6.7}$$

In this expression, $\mathbf{v}_1$ denotes the unit vector along the incident ray, while $\mathbf{v}_2$ represents the unit vector of the refracted ray after passing through the first surface. The vector $\mathbf{n}_a$ corresponds to the normal of the first surface at the point of incidence, which determines the direction in which the ray bends according to Snell's law. The factor $-\text{sgn}(z_o)$ is included to account for whether the object is real or virtual, ensuring that the sign convention correctly reflects the orientation of the incoming ray relative to the optical axis. For a visual reference, see Fig. 6.1, which illustrates the geometry of the incident and refracted rays at the first lens surface.

The unit vector of the incident ray, $\mathbf{v}_1$, can be expressed in terms of the object position and the coordinates of the first surface. It defines the direction along which light travels from the object to the point of incidence on the lens. By normalizing the displacement vector from the object to the surface point, we obtain

$$\mathbf{v_1} = \frac{r_a\hat{\mathbf{x}} + (z_a - z_o)\hat{\mathbf{y}}}{\sqrt{r_a^2 + (z_a - z_o)^2}}. \tag{6.8}$$

The unit vector of the refracted ray, $\mathbf{v}_2$, is determined by the geometry of the first surface and the refractive indices of the media on either side of the interface. Its explicit expression can be written as

$$\mathbf{v_2} = \frac{(r_b - r_a)\hat{\mathbf{x}} + (z_b - z_a)\hat{\mathbf{y}}}{\sqrt{(r_b - r_a)^2 + (z_b - z_a)^2}}, \tag{6.9}$$

Finally, we consider the normal vector of the first surface, $\mathbf{n}_a$, which is perpendicular to the tangent plane at any point on the surface. This vector is essential for applying Snell's law, as it determines the orientation relative to the incident ray $\mathbf{v}_1$ and the resulting refracted ray $\mathbf{v}_2$. Accurate knowledge of $\mathbf{n}_a$ allows us to compute the direction of light within the lens and ultimately ensures that the optical path lengths are consistent for stigmatic imaging. The normal vector can be expressed as

$$\mathbf{n_a} = \frac{z_a'\hat{\mathbf{x}} - \hat{\mathbf{y}}}{\sqrt{1 + z_a'^2}}, \tag{6.10}$$

Here, $\hat{\mathbf{x}}$ denotes the unit vector in the radial (r) direction, while $\hat{\mathbf{y}}$ represents the unit vector along the axial (z) direction. These basis vectors provide a convenient coordinate system for describing the geometry of the lens surfaces and the directions of incident and refracted rays. Expressing vectors such as $\mathbf{v}_1$, $\mathbf{v}_2$, and $\mathbf{n}_a$ in terms of $\hat{\mathbf{x}}$ and $\hat{\mathbf{y}}$ allows us to decompose the problem into radial and axial components, simplifying the computation of optical path lengths and ensuring precise modeling of the lens for stigmatic imaging.

We begin by focusing on the evaluation of $(\mathbf{n}_a \times \mathbf{v}_1)^2$, which plays a key role in the substitution process within Eq. (6.7).

$$(\mathbf{n_a} \times \mathbf{v_1}) = \frac{[r_a + (z_a - z_o)z_a']}{\sqrt{r_a^2 + (z_a - z_o)^2}\sqrt{1 + z_a'^2}}(\hat{\mathbf{x}} \times \hat{\mathbf{y}}), \tag{6.11}$$

where $(\hat{\mathbf{x}} \times \hat{\mathbf{y}})^2 = \mathbf{e_3}^2 = 1$,

$$(\mathbf{n_a} \times \mathbf{v_1})^2 = \frac{[r_a + (z_a - z_o)z_a']^2}{[r_a^2 + (z_a - z_o)^2](1 + z_a'^2)}. \tag{6.12}$$

Next, we turn our attention to the square root term in Eq. (6.7).

$$-\mathbf{n_a}\sqrt{1 + \frac{1}{n^2}(\mathbf{n_a} \times \mathbf{v_1})^2} = -\frac{(z_a'\hat{\mathbf{x}} - \hat{\mathbf{y}})}{\sqrt{1 + z_a'^2}}\sqrt{1 - \frac{n_o^2[r_a + (z_a - z_o)z_a']^2}{n^2[r_a^2 + (z_a - z_o)^2](1 + z_a'^2)}}. \tag{6.13}$$

Expression (6.7) has $\mathbf{v_1} - (\mathbf{n_a} \cdot \mathbf{v_1})\mathbf{n_a}$,

$$\begin{aligned}\mathbf{v_1} - (\mathbf{n_a} \cdot \mathbf{v_1})\mathbf{n_a} = &\frac{r_a\hat{\mathbf{x}} + (z_a - z_o)\hat{\mathbf{y}}}{\sqrt{r_a^2 + (z_a - z_o)^2}} \\ &- \left[\frac{r_a z_a' - (z_a - z_o)}{\sqrt{r_a^2 + (z_a - z_o)^2}\sqrt{1 + z_a'^2}}\right]\frac{(z_a'\hat{\mathbf{x}} - \hat{\mathbf{y}})}{\sqrt{1 + z_a'^2}}.\end{aligned} \tag{6.14}$$

After simplification and multiplication by $\frac{n_o}{(-\mathrm{sgn}(z_o))n}$ above's expression,

$$\frac{n_o}{-\mathrm{sgn}(z_o)n}[\mathbf{v_1}-(\mathbf{n_a}\cdot\mathbf{v_1})\mathbf{n_a}] = \frac{n_o[(z_a-z_o)z_a'+r_a]}{-\mathrm{sgn}(z_o)n\sqrt{r_a^2+(z_o-z_a)^2}\,(1+z_a'^2)}\hat{\mathbf{x}} + \frac{n_o\left[r_a+(z_a-z_o)z_a'\right]z_a'}{-\mathrm{sgn}(z_o)n\sqrt{r_a^2+(z_o-z_a)^2}\,(1+z_a'^2)}\hat{\mathbf{y}}. \quad (6.15)$$

By combining Eqs. (6.13) and (6.15), we can express the refracted ray $\mathbf{v}_2$ as the solution of Eq. (6.7).

$$\mathbf{v}_2 = \frac{n_o[(z_a-z_o)z_a'+r_a]}{-\mathrm{sgn}(z_o)n\sqrt{r_a^2+(z_o-z_a)^2}\,(1+z_a'^2)}\hat{\mathbf{x}} + \frac{n_o\left[r_a+(z_a-z_o)z_a'\right]z_a'}{-\mathrm{sgn}(z_o)n\sqrt{r_a^2+(z_o-z_a)^2}\,(1+z_a'^2)}\hat{\mathbf{y}} - \frac{(z_a'\hat{\mathbf{x}}-\hat{\mathbf{y}})}{\sqrt{1+z_a'^2}}\sqrt{1-\frac{n_o^2[r_a+(z_a-z_o)z_a']^2}{n^2[r_a^2+(z_a-z_o)^2](1+z_a'^2)}}. \quad (6.16)$$

By separating the terms along the basis vectors $\hat{\mathbf{x}}$ and $\hat{\mathbf{y}}$,

$$\frac{r_b-r_a}{\sqrt{(z_b-z_a)^2+(r_b-r_a)^2}} = \frac{n_o[(z_a-z_o)z_a'+r_a]}{-\mathrm{sgn}(z_o)n\sqrt{r_a^2+(z_o-z_a)^2}\,(1+z_a'^2)} - z_a'\frac{\sqrt{1-\frac{n_o^2\left[r_a+(z_a-z_o)z_a'\right]^2}{n^2\left[r_a^2+(z_o-z_a)^2\right](1+z_a'^2)}}}{\sqrt{1+z_a'^2}}, \quad (6.17)$$

and,

$$\frac{z_b-z_a}{\sqrt{(z_b-z_a)^2+(r_b-r_a)^2}} = \frac{n_o\left[r_a+(z_a-z_o)z_a'\right]z_a'}{-\mathrm{sgn}(z_o)n\sqrt{r_a^2+(z_o-z_a)^2}\,(1+z_a'^2)} + \frac{\sqrt{1-\frac{n_o^2\left[r_a+(z_a-z_o)z_a'\right]^2}{n^2\left[r_a^2+(z_o-z_a)^2\right](1+z_a'^2)}}}{\sqrt{1+z_a'^2}}, \quad (6.18)$$

The unknowns z_b and r_b appear only on the left-hand side of Eqs. (6.17) and (6.18). It is important to note that the right-hand sides of these equations correspond to the direction cosines of the refracted ray within the lens. By definition, these direction cosines satisfy the following normalization condition

$\wp_r^2 + \wp_z^2 = 1$. $\wp_r$ is the component along the radial direction $\hat{\mathbf{x}}$, and $\wp_z$ is the component along the axial direction $\hat{\mathbf{y}}$.

By substituting $\wp_r$ and $\wp_z$ into the right-hand sides of Eqs. (6.17)) and (6.18)), we obtain a complete set of expressions that relate the unknown surface coordinates (r_b, z_b) to the lens geometry and ray directions. These expressions form the basis for computing the second surface shape required to achieve a stigmatic, spherical aberration-free lens.

$$\wp_r = \frac{n_o[(z_a - z_o)z_a' + r_a]}{-\text{sgn}(z_o) n \sqrt{r_a^2 + (z_o - z_a)^2}\,(1 + z_a'^2)} - z_a' \frac{\sqrt{1 - \dfrac{n_o^2 \left[r_a + (z_a - z_o) z_a'\right]^2}{n^2 \left[r_a^2 + (z_o - z_a)^2\right](1 + z_a'^2)}}}{\sqrt{1 + z_a'^2}}, \tag{6.19}$$

and,

$$\wp_z = \frac{n_o \left[r_a + (z_a - z_o) z_a'\right] z_a'}{-\text{sgn}(z_o) n \sqrt{r_a^2 + (z_o - z_a)^2}\,(1 + z_a'^2)} + \frac{\sqrt{1 - \dfrac{n_o^2 \left[r_a + (z_a - z_o) z_a'\right]^2}{n^2 \left[r_a^2 + (z_o - z_a)^2\right](1 + z_a'^2)}}}{\sqrt{1 + z_a'^2}}. \tag{6.20}$$

The separation between the first and second lens surfaces is embedded implicitly in Eqs. (6.17) and (6.18) through the following parameter,

$$\vartheta \equiv \sqrt{(z_b - z_a)^2 + (r_b - r_a)^2}. \tag{6.21}$$

By substituting ϑ into Eqs. (6.17) and (6.18), the resulting expressions become,

$$\frac{z_b - z_a}{\vartheta} = \wp_z, \qquad \frac{r_b - r_a}{\vartheta} = \wp_r. \tag{6.22}$$

From the above expressions, the unknowns z_b and r_b can be directly determined, yielding,

$$z_b = z_a + \vartheta \wp_z, \qquad r_b = r_a + \vartheta \wp_r. \tag{6.23}$$

The expressions obtained above provide the solution we are seeking, specifying point by point the shape of the second surface required to achieve stigmatic performance. The only remaining unknown is the parameter ϑ, which represents the distance along the ray between the first and second surfaces. This ambiguity can be resolved by invoking the Fermat principle. Next, we will show how to use this principle to determine ϑ and complete the lens design.

Thus far, we have determined z_b and r_b in terms of ϑ, but an additional equation is needed to fully specify the solution. This extra constraint comes from the Fermat principle, which ensures that all rays connecting the object and image points have equal optical path lengths. It is important to note that while z_b and r_b are the only unknowns, the system initially appears to contain three equations: two from Snell's law at the first surface and one from the Fermat principle. The apparent redundancy occurs because the two equations from Snell's law are not independent. To proceed, we return to Eq. (6.6) as the starting point for applying Fermat's principle,

$$\begin{aligned} -n_o z_o + n\tau + n_i z_i = \quad & -n_o \text{sgn}(z_o)\sqrt{r_a^2 + (z_a - z_o)^2} \\ & +n\sqrt{(r_b - r_a)^2 + (z_b - z_a)^2} \\ & +n_i \text{sgn}(z_i)\sqrt{r_b^2 + (z_b - \tau - z_i)^2}. \end{aligned} \tag{6.6}$$

To streamline the previous expression, we introduce the following parameter,

$$z_f = -n_o z_o + n\tau + n_i z_i + \text{sgn}(z_o) n_0 \sqrt{r_a^2 + (z_o - z_a)^2} \tag{6.24}$$

By combining Eq. (6.24) with Eq. (6.6),

$$z_f = n\sqrt{(r_b - r_a)^2 + (z_b - z_a)^2} + n_i\sqrt{r_b^2 + (-\tau + z_b - z_i)^2} \tag{6.25}$$

By substituting Eq. (6.23) into Eq. (6.25),

$$z_f = n\vartheta + n_i\sqrt{(r_a + \vartheta\wp_r^2 + (-\tau + z_a - z_i + \vartheta\wp_z)^2}. \tag{6.26}$$

Furthermore, we introduce an additional parameter to simplify the form of Eq. (6.26),

$$z_\tau = -\tau + z_a - z_i. \tag{6.27}$$

Next, by substituting Eq. (6.27) into Eq. (6.26),

$$z_f = n\vartheta + n_i\sqrt{(r_a + \vartheta\wp_r)^2 + (z_\tau + \vartheta\wp_z)^2}. \tag{6.28}$$

By squaring both sides of the preceding equation and carrying out the necessary algebraic manipulations, we obtain a simplified expression that isolates the desired terms. This procedure allows us to eliminate the square root and combine like terms, making it easier to solve for the unknown parameter ϑ. The resulting equation can be written as

$$(z_f - n\vartheta)^2 = n_i^2\left[(r_a + \vartheta\wp_r)^2 + (z_\tau + \vartheta\wp_z)^2\right] \tag{6.29}$$

expanding,

$$z_f^2 - 2z_f n\vartheta + n^2\vartheta^2 = n_i^2\left[r_a^2 + 2r_a\vartheta\wp_r + \vartheta^2\wp_r^2 + z_\tau^2 + 2z_\tau\vartheta\wp_z + \vartheta^2\wp_z^2\right] \tag{6.30}$$

simplify with $\vartheta^2 \wp_r^2 + \vartheta^2 \wp_z^2 = \vartheta^2$,

$$z_f^2 - 2z_f n\vartheta + n^2\vartheta^2 = n_i^2\left[r_a^2 + 2r_a\vartheta\wp_r + \vartheta^2 + z_\tau^2 + 2z_\tau\vartheta\wp_z\right] \tag{6.31}$$

collecting terms with ϑ,

$$\vartheta^2\left(n_i^2 - n^2\right) + \vartheta\left(2z_f n + 2n_i^2 r_a\wp_r + 2n_i^2 z_\tau\wp_z\right) + n_i^2(r_a^2 + z_\tau^2) - z_f^2 = 0 \tag{6.32}$$

multiplying equation (6.32) by $4(n_i^2 - n^2)$,

$$\begin{aligned}4(n_i^2 - n^2)^2\vartheta^2 &+ 4(n_i^2 - n^2)(2z_f n + 2n_i^2 r_a\wp_r + 2n_i^2 z_\tau\wp_z)\vartheta \\ &= -4(n_i^2 - n^2)[n_i^2(r_a^2 + z_\tau^2) - z_f^2]\end{aligned} \tag{6.33}$$

summing $(2z_f n + 2n_i^2 r_a\wp_r + 2n_i^2 z_\tau\wp_z)^2$ in both side of Eq. (6.33),

$$\begin{aligned}4(n_i^2 - n^2)^2\vartheta^2 &+ 4(n_i^2 - n^2)(2z_f n + 2n_i^2 r_a\wp_r + 2n_i^2 z_\tau\wp_z)\vartheta \\ &+ (2z_f n + 2n_i^2 r_a\wp_r + 2n_i^2 z_\tau\wp_z)^2 \\ &= (2z_f n + 2n_i^2 r_a\wp_r + 2n_i^2 z_\tau\wp_z)^2 - 4(n_i^2 - n^2)[n_i^2(r_a^2 + z_\tau^2) - z_f^2]\end{aligned} \tag{6.34}$$

simplifying,

$$\begin{aligned}&[2(n_i^2 - n^2)\vartheta + (2z_f n + 2n_i^2 r_a\wp_r + 2n_i^2 z_\tau\wp_z)]^2 \\ &= (2z_f n + 2n_i^2 r_a\wp_r + 2n_i^2 z_\tau\wp_z)^2 - 4(n_i^2 - n^2)[n_i^2(r_a^2 + z_\tau^2) - z_f^2]\end{aligned} \tag{6.35}$$

applying the square root on both sides,

$$\begin{aligned}&2(n_i^2 - n^2)\vartheta + (2z_f n + 2n_i^2 r_a\wp_r + 2n_i^2 z_\tau\wp_z) \\ &= \pm\sqrt{(2z_f n + 2n_i^2 r_a\wp_r + 2n_i^2 z_\tau\wp_z)^2 - 4(n_i^2 - n^2)[n_i^2(r_a^2 + z_\tau^2) - z_f^2]}.\end{aligned} \tag{6.36}$$

Finally, ϑ si expressed as,

$$\vartheta = \frac{-[z_f n + n_i^2(r_a \wp_r + z_T \wp_z)]}{n_i^2 - n^2} - \frac{\sqrt{4\left(n_i^2 - n^2\right)\left[n_i^2\left(r_a^2 + z_T^2\right) - z_f^2\right] + 4\left[z_f n + n_i^2(r_a \wp_r + z_T \wp_z)\right]^2}}{2n_i^2 - 2n^2} \tag{6.37}$$

eliminating the number 2,

$$\vartheta = \frac{-[z_f n + n_i^2(r_a \wp_r + z_T \wp_z)]}{n_i^2 - n^2} - \frac{\sqrt{\left(n_i^2 - n^2\right)\left[n_i^2\left(r_a^2 + z_T^2\right) - z_f^2\right] + \left[z_f n + n_i^2(r_a \wp_r + z_T \wp_z)\right]^2}}{n_i^2 - n^2}. \tag{6.38}$$

Using Eqs. (6.23), and (6.38), the second surface turns to,

$$z_b = z_a + \vartheta \wp_z, \tag{6.39}$$

and

$$r_b = r_a + \vartheta \wp_r. \tag{6.40}$$

Eqs. (6.39) and (6.40) represent the most pivotal results of this chapter. They provide a point-by-point prescription for shaping the second surface of the lens to achieve stigmatic performance. It is important to emphasize that these equations apply specifically to singlet lenses, where rays within the lens do not intersect. In other words, for each point on the first surface, there exists a unique corresponding point on the second surface that ensures stigmatism.

It is also noteworthy that the expression for ϑ Eq. (6.38) does not include $\text{sgn}(t_b)$, indicating that the second surface adapts its shape according to the image location, regardless of whether the image is real or virtual.

Furthermore, Eq. (6.38) contains a $\pm$ sign in front of the square root, yielding two potential solutions. The correct choice depends on the sign of the refractive index n and whether the object and image are real or virtual. In the following sections, we will illustrate this with several examples.

Before presenting those examples, let us provide an expanded version of Eqs. (6.39) and (6.40) to make the derivation and their application clearer.

$$
\begin{aligned}
z_b(r_a) &= z_a(r_a) \\
&+ \frac{1}{n^2-n_i^2}\left(\frac{n_o\left[z_a'(r_a)\left(r_a+(z_a(r_a)-z_o)z_a'(r_a)\right)\right]}{n\sqrt{r_a+(z_o-z_a(r_a))^2}\left(r_a^2+z_a'(r_a)^2\right)}+\frac{\sqrt{1-\frac{n_o^2\left[r_a+(z_a(r_a)-z_o)z_a'(r_a)\right]^2}{n^2\left[r_a+(z_o-z_a(r_a))^2\right]\left(r_a^2+z_a'(r_a)^2\right)}}}{\sqrt{r_a^2+z_a'(r_a)^2}}\right)\times \\
&\Bigg(n\left(n_iz_i-n_oz_o+n\tau+n_o\,\mathrm{Sign}[z_o]\sqrt{r_a+(z_o-z_a(r_a))^2}\right) \\
&\quad + n_i^2\Bigg\{\Bigg[r_a\times\left(\frac{n_o\left[z_a'(r_a)\left(r_a+(z_a(r_a)-z_o)z_a'(r_a)\right)\right]}{n\sqrt{r_a+(z_o-z_a(r_a))^2}\left(r_a^2+z_a'(r_a)^2\right)}-\frac{z_a'(r_a)\sqrt{1-\frac{n_o^2\left[r_a+(z_a(r_a)-z_o)z_a'(r_a)\right]^2}{n^2\left[r_a+(z_o-z_a(r_a))^2\right]\left(r_a^2+z_a'(r_a)^2\right)}}}{\sqrt{r_a^2+z_a'(r_a)^2}}\right)\Bigg] \\
&\qquad + \Bigg[(z_a(r_a)-z_i-\tau)\times\left(\frac{n_o\left[z_a'(r_a)\left(r_a+(z_a(r_a)-z_o)z_a'(r_a)\right)\right]}{n\sqrt{r_a+(z_o-z_a(r_a))^2}\left(r_a^2+z_a'(r_a)^2\right)}+\frac{\sqrt{1-\frac{n_o^2\left[r_a+(z_a(r_a)-z_o)z_a'(r_a)\right]^2}{n^2\left[r_a+(z_o-z_a(r_a))^2\right]\left(r_a^2+z_a'(r_a)^2\right)}}}{\sqrt{r_a^2+z_a'(r_a)^2}}\right)\Bigg]\Bigg\} \\
&\quad - n_i\,\mathrm{Sign}[n_1]\times \\
&\quad \Bigg(\left(n_iz_i-n_oz_o+n\tau+n_o\,\mathrm{Sign}[z_o]\sqrt{r_a+(z_o-z_a(r_a))^2}\right)^2+n^2\left(r_a+(z_a(r_a)-z_i-\tau)^2\right) \\
&\quad - n_i^2\Bigg\{\Bigg[(z_a(r_a)-z_i-\tau)\times\left(\frac{n_o\left[z_a'(r_a)\left(r_a+(z_a(r_a)-z_o)z_a'(r_a)\right)\right]}{n\sqrt{r_a+(z_o-z_a(r_a))^2}\left(r_a^2+z_a'(r_a)^2\right)}-\frac{z_a'(r_a)\sqrt{1-\frac{n_o^2\left[r_a+(z_a(r_a)-z_o)z_a'(r_a)\right]^2}{n^2\left[r_a+(z_o-z_a(r_a))^2\right]\left(r_a^2+z_a'(r_a)^2\right)}}}{\sqrt{r_a^2+z_a'(r_a)^2}}\right)\Bigg] \\
&\qquad + \Bigg[-r_a\times\left(\frac{n_o\left[z_a'(r_a)\left(r_a+(z_a(r_a)-z_o)z_a'(r_a)\right)\right]}{n\sqrt{r_a+(z_o-z_a(r_a))^2}\left(r_a^2+z_a'(r_a)^2\right)}+\frac{\sqrt{1-\frac{n_o^2\left[r_a+(z_a(r_a)-z_o)z_a'(r_a)\right]^2}{n^2\left[r_a+(z_o-z_a(r_a))^2\right]\left(r_a^2+z_a'(r_a)^2\right)}}}{\sqrt{r_a^2+z_a'(r_a)^2}}\right)\Bigg]\Bigg\}^2 \\
&\quad + 2n\left(n_iz_i-n_oz_o+n\tau+n_o\,\mathrm{Sign}[z_o]\sqrt{r_a+(z_o-z_a(r_a))^2}\right) \\
&\quad \times\Bigg\{\Bigg[r_a\times\left(\frac{n_o\left[z_a'(r_a)\left(r_a+(z_a(r_a)-z_o)z_a'(r_a)\right)\right]}{n\sqrt{r_a+(z_o-z_a(r_a))^2}\left(r_a^2+z_a'(r_a)^2\right)}-\frac{z_a'(r_a)\sqrt{1-\frac{n_o^2\left[r_a+(z_a(r_a)-z_o)z_a'(r_a)\right]^2}{n^2\left[r_a+(z_o-z_a(r_a))^2\right]\left(r_a^2+z_a'(r_a)^2\right)}}}{\sqrt{r_a^2+z_a'(r_a)^2}}\right)\Bigg] \\
&\qquad + \Bigg[(z_a(r_a)-z_i-\tau)\times\left(\frac{n_o\left[z_a'(r_a)\left(r_a+(z_a(r_a)-z_o)z_a'(r_a)\right)\right]}{n\sqrt{r_a+(z_o-z_a(r_a))^2}\left(r_a^2+z_a'(r_a)^2\right)}+\frac{\sqrt{1-\frac{n_o^2\left[r_a+(z_a(r_a)-z_o)z_a'(r_a)\right]^2}{n^2\left[r_a+(z_o-z_a(r_a))^2\right]\left(r_a^2+z_a'(r_a)^2\right)}}}{\sqrt{r_a^2+z_a'(r_a)^2}}\right)\Bigg]\Bigg\}\Bigg)^{1/2}\Bigg)
\end{aligned}
$$

$$
\begin{aligned}
r_b(r_a) = {} & r_a \\
& + \frac{1}{n^2 - n_i^2}\left(\frac{n_o\left[z'_a(r_a)\left(r_a+(z_a(r_a)-z_o)z'_a(r_a)\right)\right]}{n\sqrt{r_a+(z_o-z_a(r_a))^2}\left(r_a^2+z'_a(r_a)^2\right)} - \frac{z'_a(r_a)\sqrt{1-\frac{n_o^2\left[r_a+(z_a(r_a)-z_o)z'_a(r_a)\right]^2}{n^2\left[r_a+(z_o-z_a(r_a))^2\right]\left(r_a^2+z'_a(r_a)^2\right)}}}{\sqrt{r_a^2+z'_a(r_a)^2}}\right) \times \\
& \Bigg(n\left(n_i z_i - n_o z_o + n\tau + n_o \operatorname{Sign}[z_o]\sqrt{r_a+(z_o-z_a(r_a))^2}\right) \\
& + n_i^2 \left\{ \left[r_a \times \left(\frac{n_o\left[z'_a(r_a)\left(r_a+(z_a(r_a)-z_o)z'_a(r_a)\right)\right]}{n\sqrt{r_a+(z_o-z_a(r_a))^2}\left(r_a^2+z'_a(r_a)^2\right)} - \frac{z'_a(r_a)\sqrt{1-\frac{n_o^2\left[r_a+(z_a(r_a)-z_o)z'_a(r_a)\right]^2}{n^2\left[r_a+(z_o-z_a(r_a))^2\right]\left(r_a^2+z'_a(r_a)^2\right)}}}{\sqrt{r_a^2+z'_a(r_a)^2}}\right)\right] \right. \\
& \left. + \left[(z_a(r_a)-z_i-\tau) \times \left(\frac{n_o\left[z'_a(r_a)\left(r_a+(z_a(r_a)-z_o)z'_a(r_a)\right)\right]}{n\sqrt{r_a+(z_o-z_a(r_a))^2}\left(r_a^2+z'_a(r_a)^2\right)} + \frac{\sqrt{1-\frac{n_o^2\left[r_a+(z_a(r_a)-z_o)z'_a(r_a)\right]^2}{n^2\left[r_a+(z_o-z_a(r_a))^2\right]\left(r_a^2+z'_a(r_a)^2\right)}}}{\sqrt{r_a^2+z'_a(r_a)^2}}\right)\right] \right\} \\
& - n_i \operatorname{Sign}[n_1] \times \\
& \Bigg(\left(n_i z_i - n_o z_o + n\tau + n_o \operatorname{Sign}[z_o]\sqrt{r_a+(z_o-z_a(r_a))^2}\right)^2 + n^2\left(r_a+(z_a(r_a)-z_i-\tau)^2\right) \\
& - n_i^2 \left\{ \left[(z_a(r_a)-z_i-\tau) \times \left(\frac{n_o\left[z'_a(r_a)\left(r_a+(z_a(r_a)-z_o)z'_a(r_a)\right)\right]}{n\sqrt{r_a+(z_o-z_a(r_a))^2}\left(r_a^2+z'_a(r_a)^2\right)} - \frac{z'_a(r_a)\sqrt{1-\frac{n_o^2\left[r_a+(z_a(r_a)-z_o)z'_a(r_a)\right]^2}{n^2\left[r_a+(z_o-z_a(r_a))^2\right]\left(r_a^2+z'_a(r_a)^2\right)}}}{\sqrt{r_a^2+z'_a(r_a)^2}}\right)\right] \right. \\
& \left. + \left[-r_a \times \left(\frac{n_o\left[z'_a(r_a)\left(r_a+(z_a(r_a)-z_o)z'_a(r_a)\right)\right]}{n\sqrt{r_a+(z_o-z_a(r_a))^2}\left(r_a^2+z'_a(r_a)^2\right)} + \frac{\sqrt{1-\frac{n_o^2\left[r_a+(z_a(r_a)-z_o)z'_a(r_a)\right]^2}{n^2\left[r_a+(z_o-z_a(r_a))^2\right]\left(r_a^2+z'_a(r_a)^2\right)}}}{\sqrt{r_a^2+z'_a(r_a)^2}}\right)\right] \right\}^2 \\
& + 2n\left(n_i z_i - n_o z_o + n\tau + n_o \operatorname{Sign}[z_o]\sqrt{r_a+(z_o-z_a(r_a))^2}\right) \\
& \times \left\{ \left[r_a \times \left(\frac{n_o\left[z'_a(r_a)\left(r_a+(z_a(r_a)-z_o)z'_a(r_a)\right)\right]}{n\sqrt{r_a+(z_o-z_a(r_a))^2}\left(r_a^2+z'_a(r_a)^2\right)} - \frac{z'_a(r_a)\sqrt{1-\frac{n_o^2\left[r_a+(z_a(r_a)-z_o)z'_a(r_a)\right]^2}{n^2\left[r_a+(z_o-z_a(r_a))^2\right]\left(r_a^2+z'_a(r_a)^2\right)}}}{\sqrt{r_a^2+z'_a(r_a)^2}}\right)\right] \right. \\
& \left. + \left[(z_a(r_a)-z_i-\tau) \times \left(\frac{n_o\left[z'_a(r_a)\left(r_a+(z_a(r_a)-z_o)z'_a(r_a)\right)\right]}{n\sqrt{r_a+(z_o-z_a(r_a))^2}\left(r_a^2+z'_a(r_a)^2\right)} + \frac{\sqrt{1-\frac{n_o^2\left[r_a+(z_a(r_a)-z_o)z'_a(r_a)\right]^2}{n^2\left[r_a+(z_o-z_a(r_a))^2\right]\left(r_a^2+z'_a(r_a)^2\right)}}}{\sqrt{r_a^2+z'_a(r_a)^2}}\right)\right] \right\} \Bigg)^{1/2} \Bigg)
\end{aligned}
$$

While the derivation of Eq. (6.39) may appear straightforward when reading sequentially, the equation itself is complex, spanning two pages when fully expanded. Historically, many attempts to find a general closed-form solution failed, largely because traditional Snell's law in terms of angles does not directly relate to optical paths. By using Snell's law in vector form, the connection between optical paths and direction cosines becomes clear, enabling the solution. Importantly, not only is a closed-form solution difficult, but it is also unique.

6.1.2 The stigmatic collimator

Now that we have introduced the stigmatic lens, we can focus of the Stigmatic aspheric collimator, which is a stigmatic lens that has an object or image at infinity.

Fig. 6.2 illustrates the geometry of the stigmatic lens under study. For convenience in obtaining a closed-form solution and the apodization function, the order of the lens surfaces has been reversed. This adjustment arises naturally from the design equations when the Abbe sine condition—which relates the ray height at the first interface to the focusing angle—is not applied. In this configuration, the first surface is defined by (r_b, z_b) and the second by (r_a, z_a), with r_a remaining the only independent variable [99]. The coordinate system is centered at the vertex of (r_a, z_a). The object lies at a finite distance z_o from the vertex, the lens has a central thickness τ, and the image is assumed at infinity ($z_i \to -\infty$). The refractive indices in the object, lens, and image spaces are n_o, n, and n_i, respectively.

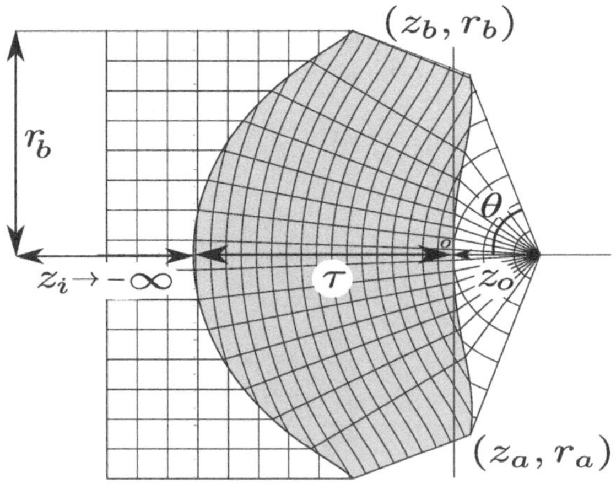

FIGURE 6.2
Diagram of a stigmatic singlet for a distant image. The first surface is (z_b, r_b) and the second surface is defined by (z_a, r_a). A ray with height r_b has a focusing angle θ.

The idea is to get the relations for (r_b, z_b) in such a way that the rays before the lens are perfectly collimated along the axis and then express these functions in terms of θ instead of r_a. This will lead ultimately to compute the apodization factor. According to the geometry, the unit vector definition for the stigmatic lens should be

$$\mathbf{v}_1 = \frac{r_a\hat{\mathbf{x}} + (z_a - z_o)\hat{\mathbf{y}}}{\sqrt{r_a^2 + (z_a - z_o)^2}}, \qquad \mathbf{n}_a = \frac{z'_a\hat{\mathbf{x}} - \hat{\mathbf{y}}}{\sqrt{1 + z'^2_a}},$$

$$\mathbf{v}_2 = [\wp_r, \wp_z] = \frac{(r_b - r_a)\hat{\mathbf{x}} + (z_b - z_a)\hat{\mathbf{y}}}{\sqrt{(r_b - r_a)^2 + (z_b - z_a)^2}}. \tag{6.41}$$

Using these with Eq. (6.7), one can compute the cosine directors $\wp_r$ and $\wp_z$ for a stigmatic lens.

$$\wp_r = \frac{n_o[(z_a - z_o)z'_a + r_a]}{n\sqrt{r_a^2 + (z_o - z_a)^2}\,(1 + z'^2_a)} - z'_a\frac{\sqrt{1 - \dfrac{n_o^2\,[r_a + (z_a - z_o)z'_a]^2}{n^2\,[r_a^2 + (z_o - z_a)^2]\,(1 + z'^2_a)}}}{\sqrt{1 + z'^2_a}}, \tag{6.42}$$

$$\wp_z = \frac{n_o\,[r_a + (z_a - z_o)z'_a]\,z'_a}{n\sqrt{r_a^2 + (z_o - z_a)^2}\,(1 + z'^2_a)} + \frac{\sqrt{1 - \dfrac{n_o^2\,[r_a + (z_a - z_o)z'_a]^2}{n^2\,[r_a^2 + (z_o - z_a)^2]\,(1 + z'^2_a)}}}{\sqrt{1 + z'^2_a}}, \tag{6.43}$$

Eqs. (6.39) and (6.40) are unchanged with the inversion of the model. Therefore, we only need to express ϑ with dependency on known parameters (i.e. z_a and r_a). Recalling the Fermat principle, the constant optical path length (OPL) of every ray should be,

$$-n_o z_o + n\tau + n_i z_i = n_o\sqrt{r_a^2 + (z_a - z_o)^2} + n\sqrt{(r_b - r_a)^2 + (z_b - z_a)^2} + n_i\sqrt{r_b^2 + (z_b - \tau - z_i)^2}. \tag{6.44}$$

The left side of Eq. (6.44) is the OPL of the axial ray and the right side is the OPL of a non-axial ray. To recover the parallelism of the rays we need to compute the limit when $z_i \to -\infty$. We obtain a relation for the OPL and an avenue to solve ϑ without dependency upon z_b or r_b,

$$(-n_i + n)\tau - n_o z_o + n_i z_b - n \left[\lim_{z_i \to \infty} (\vartheta) \right] = n_o \sqrt{r_a^2 + (z_a - z_o)^2}, \qquad (6.45)$$

Solving for $[\lim_{z_i \to \infty}(\vartheta)]$,

$$\left[\lim_{z_i \to \infty} (\vartheta) \right] = \frac{(-n_i + n)\tau - n_o z_o + n_i z_a - n_o \sqrt{r_a^2 + (z_a - z_o)^2}}{n - n_i \wp_z}. \qquad (6.46)$$

Thus, we can compute the sag definition of a general stigmatic lens with the object at infinity. Therefore, Eqs. (6.39) and (6.40) and (6.46) can be used to get the apodization factor of a stigmatic lens with a given second surface z_a.

$$\begin{cases} \left[\lim_{z_i \to \infty} (z_b) \right] = z_a + \left[\lim_{z_i \to -\infty} (\vartheta) \right] \wp_z, \\ \left[\lim_{z_i \to \infty} (r_b) \right] = r_a + \left[\lim_{z_i \to -\infty} (\vartheta) \right] \wp_r. \end{cases} \qquad (6.47)$$

The last equation express show much be the second surface of a given lens such that singlet is a *stigmatic collimator*.

6.2 Apodization function

For an optical system with collimated rays input, it relates the angle θ with the height r_b of these collimated rays. See Fig. 6.2. Because of the inversion of the lens surfaces, the apodization factor $q(\theta)$ can be obtain from last Chapter [60, 96, 47],

$$\frac{1}{q(\theta)} = \frac{1}{f} \sqrt{\frac{r_b(\theta) r_b'(\theta)}{\sin\theta}}, \qquad (6.48)$$

where r_b' is the derivative of r_b in respect to θ. However, the relation between r_b and θ is not trivial. Therefore, the apodization function should be in term of the independent variable of the design r_a. To do so, we define a composite function $\bar{r}_b$ between r_b and r_a such

$$\bar{r}_b(\theta) = (r_b \circ r_a)(\theta) = r_b(r_a(\theta)). \qquad (6.49)$$

Then, the composite function is useful as it give an avenue to express the function with the variable θ since r_a and θ are related by

$$\tan\theta = \frac{r_a}{z_o - z_a(r_a)}. \tag{6.50}$$

The focal length f of the system works as a normalization factor. The focal distance for a stigmatic lens with a distant object is given by the multiplication of the distance between the image and the second surface by the radial derivative of the first surface evaluated in its vertex [23],

$$f = -z_o r_b'(0). \tag{6.51}$$

Consequently, the focal properties of the stigmatic lens vary with different surface definitions $z_a(r_a)$, as expected. The sagitta $z_a(r_a)$ directly influences the apodization factor $q(\theta)$, which determines how the amplitude distribution varies across the exit pupil. In the following subsections, we analyze two important cases: first, when the second surface is planar, and second, when it is a spherical interface. These examples illustrate how surface geometry affects both focusing behavior and diffraction characteristics of the lens.

We can define a maximum focusing angle, $\theta_{\max}$, for the stigmatic lens, which characterizes the largest angle between the optical axis and the marginal rays that the lens can focus. This angle depends on the lens geometry, the refractive indices, and the surface sagitta $z_a(r_a)$. Mathematically, it can be expressed as,

$$\theta_{\max} = \tan^{-1}\left[\frac{r_a}{z_o - z_a(r_a)}\right]\Bigg|_{r_a = r_{a_{\max}}}. \tag{6.52}$$

6.2.1 The apodization factor of a aspheric-plane stigmatic lens

To obtain the apodization factor of a aspheric-plane stigmatic lens with a distance object, the following relation r_a should be respected

$$r_a = z_o \tan\theta. \tag{6.53}$$

while solving (6.50) with $z_a = 0$. Then, one can obtain the relation $\bar{r}_b$ in terms of θ by using (6.53) in (6.49)following

$$\bar{r}_b(\theta) = z_o \tan\theta - \frac{2n_o \sin\theta[(n-1)\tau + z_o(\sec\theta - 1)]}{n\left(2n - \sqrt{\frac{2n_o^2[\cos(2\theta)-1]}{n^2} + 4}\right)}, \tag{6.54}$$

With (6.54) and $f = -z_o + \tau/n$, we can directly compute the apodization factor of a aspheric-plane stigmatic lens following (6.48).

6.2.2 The apodization factor of a aspheric-spherical stigmatic lens

Similarly, the sag definition z_a when considering that the second interface is spherical with radius r_a should follow,

$$z_a(r_a) = r_a - \sqrt{r_a^2 - r_a^2}, \tag{6.55}$$

where r_a is the radius of the spheric surface. Using (6.55) in (6.50) and solving for r_a leads to

$$r_a = \cos\theta \left[\sin\theta(r_a - z_o) - \cos\theta\sqrt{r_a^2 \tan^2(\theta) + z_o \tan^4(\theta)(2r_a - z_o)} \right]. \tag{6.56}$$

Then, one can obtain $r_b(\theta)$ from (6.49),

$$\bar{r}_b(\theta) = r_b\left(\cos\theta \left[\sin\theta(r_a - z_o) - \cos\theta\sqrt{r_a^2 \tan^2(\theta) + z_o \tan^4(\theta)(2r_a - z_o)} \right] \right) \tag{6.57}$$

The explicit expression for $\bar{r}_b(\theta)$ has not been included, as it is very large and cumbersome for a general aspheric–spherical lens, making it impractical to present in full within the manuscript. However, the apodization function $q(\theta)$ can still be determined using Eq. (6.48). In the particular case where the second surface is spherical, the normalized focal length f is given by an expression that can be obtained using symbolic algebra software, allowing the computation of the lens properties without manually handling the lengthy formulas,

$$f = \frac{\tau[nz_o + n_o(r_a - z_o)]}{nr_a} - z_o. \tag{6.58}$$

6.3 Diffraction of a stigmatic lens

From this point onward we adopt the Richards–Wolf vectorial diffraction theory [79], studied in Chapter 4, to describe focusing in high-NA systems. As mentioned before, this formalism correctly incorporates the vector nature of light and remains valid well beyond the paraxial regime.

We are going to use the case of a radially polarized incident beam focused by an axisymmetric optical system, the transverse and longitudinal electric

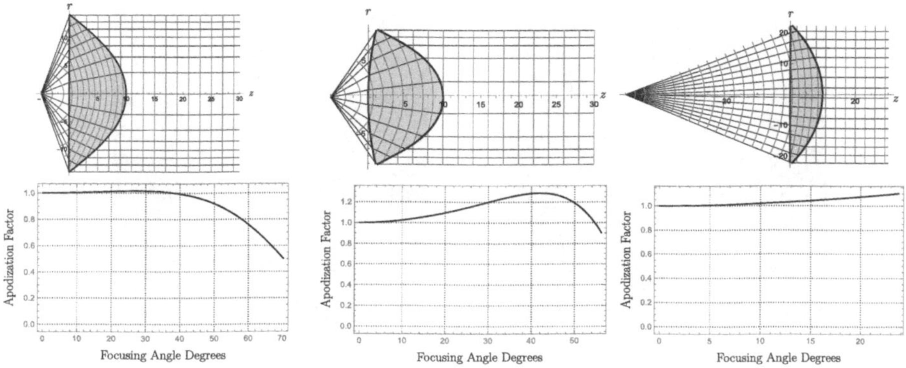

FIGURE 6.3
Plot of the apodization factor of several optical systems. The design specifications are: *Left:* $z_a = 0$, $z_o = 5\text{mm}$ $n_o = n_i = 1$, $n = 2$ and $\tau = 10\text{mm}$. *Centre:* $z_a = 40 - \sqrt{40^2 - r_a^2}$, $z_o = -50\text{mm}$ $n_o = n_i = 1$, $n = 1.5$ and $\tau = 10\text{mm}$. *Right:* $z_a = 0$, $z_o = -50\text{mm}$ $n_o = n_i = 1$, $n = 1.5$ and $\tau = 10\text{mm}$.

field components in the focal region are expressed as [100, 61],

$$\begin{bmatrix} E_r \\ E_z \end{bmatrix} = E_0 \int_0^{\theta_{max}} q(\theta) l_0(\theta) \sin\theta \begin{bmatrix} \mathrm{i}\cos\theta \mathrm{J}_1(kr\sin\theta) \\ \sin\theta \mathrm{J}_0(kr\sin\theta) \end{bmatrix} \exp[-\mathrm{i}kz\cos\theta]\mathrm{d}\theta, \tag{6.59}$$

where $l_0(\theta)$ is the illumination pattern. We will consider that the properties of the lens are not affected by the polarization (transmission coefficient, etc.) but the integrals vary with the considered polarization.

The design of multiple stigmatic lenses following the aspheric-plane or the aspheric-spherical relations have been implemented. Imposing various focal length, height and sag definition z_a, the designed lenses are perfectly stigmatic. However, as one can observe, the apodization function is clearly different as shown in Fig. 6.3. The focusing properties of those lens will be affected and we present an avenue to compute it.

The left panel of Fig. 6.4 shows the normalized intensity distribution produced by the stigmatic aspheric-plane lens whose apodization appears in the *center* panel of Fig. 6.3. The right panel displays, for comparison, the focal field of a high-NA aplanatic lens designed to have identical focal length and maximum focusing angle $\theta_{\max}$.

The two patterns are strikingly close, with relative differences typically not exceeding 2 % under uniform illumination. This small discrepancy results in a radial FWHM that is ~0.8 % smaller and an axial FWHM that is ~3 % larger for the stigmatic design. As illustrated in Fig. 6.3, the apodization function $q(\theta)$ can be dramatically altered by appropriate choice of the sag profile of the first optical surface, but not necessary the diffraction pattern. The reader

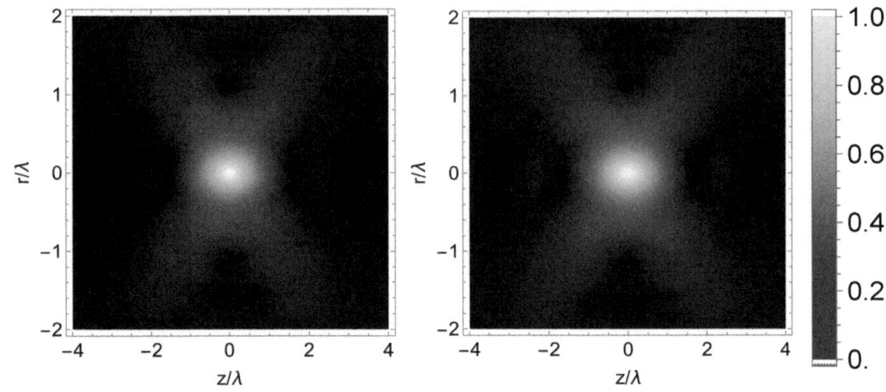

FIGURE 6.4
Normalized intensity of the diffraction for a high NA stigmatic lens (aspheric-plane) (*left*) and an aplanatic lens (*right*) for a radially polarized. See characterizations of the first design in the caption of Fig. 6.3.

is invited to compute the comparison of the aplanatic lens with the two other lenses of Fig. 6.3.

6.4 Conclusion

In this chapter we have shown that the long-standing problem of finding an exact analytical expression for a stigmatic singlet lens, free from spherical aberration but not necessarily aplanatic, admits a general closed-form solution when formulated in terms of optical path length and vector Snell's law.

The derived expressions allow systematic design of stigmatic lenses for arbitrary combinations of real/virtual object and image locations, including the practically very important case of collimating lenses (object at finite distance, image at infinity). The solutions naturally provide the exact ray mapping and therefore the pupil apodization function $q(\theta)$, which turns out to be highly sensitive to the chosen first surface shape.

Coupling these analytically exact pupil functions with the Richards–Wolf formalism enables, quantitatively accurate vectorial computation of the focal field produced by physically realizable stigmatic singlets of high numerical aperture.

Comparative calculations reveal that the on-axis focusing performance of well-designed stigmatic lenses can approach, and in some cases rival, that of classical aplanatic objectives, despite the fundamental difference in aberration correction, spherical aberration correction versus sine condition. This result opens interesting perspectives for future high-NA optical systems in which controlled deviation from aplanatism is deliberately exploited to tailor the intensity distribution, polarization state, or depth of focus in the focal region.

Problems

Problem 1: Basic conceptual difference

Explain in your own words the most important conceptual difference between

(a) a *stigmatic* lens (as presented in the chapter),

(b) a classical *aplanatic* lens.

Which aberration is perfectly corrected in a stigmatic lens? Which additional condition is satisfied by aplanatic lenses (but usually *not* by stigmatic ones)?

Problem 2: Fermat's principle—sign convention

A stigmatic singlet must produce a constant optical path length (OPL) for all rays connecting the object point to the image point.

(a) Write the **general expression** of this constant OPL using the sign convention of the chapter (including the terms $\mathrm{sgn}(z_o)$ and $\mathrm{sgn}(z_i)$).

(b) Explain physically why the term $-n_o z_o$ changes sign depending on whether the object is real or virtual.

Problem 3: Direction cosines after refraction

After refraction at the first surface, the direction cosines of the ray inside the lens are denoted $\wp_r$ and $\wp_z$.

(a) Write the normalization condition these two quantities must satisfy.

(b) Explain physically why this condition *must* be fulfilled.

(c) In which situation could this condition fail numerically during computation? (Hint: think about total internal reflection.)

Problem 4: Number of mathematical solutions for ϑ

The final expression for the optical path length parameter ϑ contains a $\pm$ sign in front of the square root.

(a) How many *mathematical* solutions for ϑ can exist for a given ray height r_a?

(b) Which physical considerations usually allow us to discard one of them?

(c) In which (rather unusual) optical configurations might *both* solutions be physically meaningful?

Problem 5: Stigmatic collimator (image at infinity)
Consider the practically very important case of a stigmatic lens with the image at infinity (collimator).

(a) Write the simplified limit expression for the parameter ϑ when $z_i \to -\infty$.

(b) Explain why the term $n_i z_i$ disappears from the final expression even though $z_i \to \infty$.

(c) Why do many authors reverse the order of the surfaces (making the exit surface the first one in the equations) when treating this collimating case?

Problem 6: Qualitative behavior of the apodization function
The chapter shows that different choices of the first surface $z_a(r_a)$ produce very different apodization functions $q(\theta)$.

(a) Qualitatively describe what happens to $q(\theta)$ when the first surface becomes flatter (tends toward a plane).

(b) What kind of focal spot intensity distribution would you expect when $q(\theta)$ decreases very rapidly with increasing θ?

(c) What happens to the diffraction pattern when $q(\theta)$ is almost constant (similar to the classical aplanatic case)?

Problem 7: Comparison stigmatic vs. aplanatic lens
Consider two high-NA lenses (NA ≈ 0.9) that have the same focal length and the same maximum focusing angle θ_{max}:

(a) Lens A: classical *aplanatic* singlet,

(b) Lens B: *stigmatic aspheric-plane* singlet (first surface plane).

According to the results and figures shown in the chapter:

(a) Which lens will most probably have *stronger sidelobes* in the focal spot for radially polarized light?

(b) Which lens will most probably produce a *slightly longer depth of focus*?

(c) In which practical situation would you deliberately prefer the stigmatic design over the aplanatic one, even if the difference in performance is only a few percent?

Problem 8: Richards-Wolf integral—longitudinal field

The Richards-Wolf integral for a radially polarized incident beam contains terms with Bessel functions J_0 and J_1.

(a) Which field component (E_r or E_z) is responsible for creating the very strong *longitudinal* electric field near focus in high-NA systems?

(b) Explain physically why this longitudinal component is usually much weaker (or even negligible) when using low-NA objectives.

(c) How does the apodization function $q(\theta)$ influence the relative strength of the longitudinal field E_z compared to the transverse component?

Bonus (open discussion question):

If you could freely choose the apodization function $q(\theta)$ of a stigmatic lens, would you try to imitate the classical aplanatic case exactly, or would you deliberately design a different apodization to obtain some special focal-field property (e.g. flatter depth of focus, stronger longitudinal field, smaller spot size, etc.)? Briefly justify your choice.

Part IV

Freeform vector diffraction theory

7

Generalization of the Richards-Wolf equations to freeform wavefronts

This chapter presents a rigorous generalization of the well-known Richards–Wolf vector diffraction integrals to arbitrary freeform wavefronts. Starting from the angular spectrum representation and incorporating energy conservation, the far-field approximation, and the Malus–Dupin theorem, a unified vector diffraction integral is derived in Cartesian pupil coordinates. The formulation explicitly accounts for the wavefront shape $w(x, y)$, spatially varying polarization via the angle $\alpha(x, y)$, and arbitrary pupil geometries through the mask function $P(x, y)$. The resulting double integrals naturally reduce to the classical Richards–Wolf expressions for spherical wavefronts and support radial, azimuthal, linear, and complex vector beams, offering a versatile tool for high-NA focusing, aberrations, and advanced beam-shaping applications in modern optics. The generalized integral can successfully compute diffraction of freeform wavefronts.

7.1 Introduction

Rayleigh–Sommerfeld and scalar Kirchhoff wavefront propagation models are convenient ways to describe the propagation of light in paraxial conditions. Nevertheless, these wavefront propagation methods cannot be used for high-numerical-aperture optical systems. This limitation arises from the scalar nature of the models, which neglect the state of polarization of light and therefore fail to capture essential vectorial effects that become prominent under strong focusing. To accurately describe such phenomena, the scalar theory must be extended to a full vectorial treatment of Maxwell's equations.

DOI: 10.1201/9781003726494-11

The first formalism that considered the vector character of the fields can be traced back to Ignatowski's pioneering studies of diffraction based directly on Maxwell's equations [101, 102]. Subsequently, Kottler developed a model reformulating the problem as an integral equation using Green's function techniques [103]. This framework was later generalized by Stratton and Chu, who introduced a rigorous vector diffraction theory grounded in boundary conditions and field equivalence principles [104]. A different but related approach was introduced by Richards and Wolf, who derived a diffraction integral tailored to the focusing of light in high-numerical-aperture systems [105, 56, 57, 2]. The foundations of this integral are energy conservation and the far-field approximation, making it particularly well-suited for describing the formation of focal fields in stigmatic, non-aberrated systems.

It is important to remark that all these formalisms were originally conceived for monochromatic light, although they have later been extended or applied in contexts involving broadband or pulsed fields. The models of Stratton–Chu and Richards–Wolf have been extensively used to study tight focusing and related vectorial effects in modern optics, including applications in microscopy, optical trapping, beam engineering, and laser–matter interaction [106, 107, 62, 65, 63, 64, 66, 68, 69, 108, 109, 100, 61, 67, 58, 48]. In particular, while the Richards–Wolf model has become the standard tool in many high-resolution optical designs, it remains limited to ideal stigmatic and non-aberrated systems, highlighting the need for more general formulations to account for aberrations and freeform optical elements. We deeply studied Richards–Wolf integrals in Chapter 4 and show its application in stigmatic optical systems in Chapters 5 and 6. But in general optical systems are not stigmatic and therefore a generalized version of Richards–Wolf integrals is need it. Since aberrated wavefronts have mostly an arbitrary shape, this wavefront must be consider as freeform $w(x, y)$.

This chapter introduces a vector diffraction integral, capable of computing the diffraction patterns for aspheric and freeform wavefronts under arbitrary pupil shapes and polarization states. This formulation extends beyond the limitations of classical models by incorporating the full vectorial character of the electromagnetic field, thus enabling accurate predictions in high-numerical-aperture and non-stigmatic systems. The proposed integral is derived from fundamental principles, namely energy conservation, the far-field approximation, and the angular spectrum representation of optical fields, ensuring both physical consistency and mathematical rigor.

An important advantage of this approach is its generality: it accommodates wavefronts defined by complex surface geometries, including non-rotationally symmetric. Furthermore, it provides a unified framework that can handle arbitrary input polarization distributions—ranging from simple linear or circular states to space-variant vector beams such as radial, azimuthal, or hybrid polarizations. This versatility makes the integral particularly suitable for studying the diffraction phenomena with full rigor in optical systems where the polarization state of the wave aberrations cannot be ignored.

7.2 From angular spectrum representation to freeform wavefronts

The model presented here is based on the angular spectrum representation of the optical fields. This representation provides a rigorous mathematical framework for describing optical propagation in homogeneous media using the concepts of Fourier analysis. As we saw in Chapter 4, in this approach, an arbitrary field distribution is decomposed into a continuous superposition of plane waves and evanescent components, both of which are exact solutions of Maxwell's equations. Each plane-wave component carries a well-defined spatial frequency and propagates independently, while the evanescent waves account for the high spatial frequency content that decays exponentially with distance from the aperture.

Physically, the direction of propagation of each plane wave corresponds to the direction of the associated geometric optical ray, thereby establishing a natural link between wave optics and ray optics. This connection makes the angular spectrum method not only rigorous but also intuitively meaningful, as it enables the analysis of complex wavefronts in terms of their angular content. Moreover, the approach inherently incorporates the vectorial nature of the field, allowing the polarization state to be preserved and accurately propagated through the system.

Acknowledging the aforementioned considerations and the content studied in Chapter 4, the electric field $\mathbf{E}(r, \varphi, z)$, where (r, φ, z) are cylindrical coordinates, at the focus can be expressed with,

$$\mathbf{E}(r, \varphi, z) = \int_{-\infty}^{\infty} \int_{-\infty}^{\infty} \check{\mathbf{E}}(k_x, k_y, 0) \exp(\mathrm{i}k_x r \cos\varphi + \mathrm{i}k_y r \sin\varphi \pm \mathrm{i}k_z z)\mathrm{d}k_x \mathrm{d}k_y \tag{7.1}$$

where $\check{\mathbf{E}}(k_x, k_y, 0)$ denotes the Fourier transform of $\mathbf{E}(r, \varphi, z)$ with respect to k_x and k_y evaluated at $z = 0$. Here, k_x, k_y, and k_z are the Cartesian spatial frequency components, with k_x and k_y describing the transverse spatial frequencies and k_z determined by the dispersion relation $k_z = \sqrt{k^2 - k_x^2 - k_y^2}$. This representation emphasizes that the propagation of the field can be fully characterized in terms of its angular spectrum.

To simplify the formulation, one can apply the far-field approximation together with the energy conservation theorem [55]. The far-field approximation ensures that only propagating components of the angular spectrum are retained, while evanescent contributions—which decay exponentially with distance—are neglected. In parallel, the energy conservation condition guarantees that the field amplitude is properly normalized when expressed in angular coordinates. Under these assumptions, Eq. (7.1) simplifies to a tractable expression that directly relates the field distribution at the aperture plane to the field at the focal region,

$$\mathbf{E}(r, \varphi, z) = \frac{E_0}{2\pi} \int_{-\infty}^{\infty} \int_{-\infty}^{\infty} \frac{\mathbf{E}_\infty(k_x, k_y)}{k_z} \exp(-\mathrm{i}\mathbf{k} \cdot \mathbf{r}) \mathrm{d}k_x \mathrm{d}k_y \tag{7.2}$$

E_0 is a constant amplitude that ensures consistency with the far-field approximation and the energy conservation theorem. The vector $\mathbf{k} \equiv [k_x, k_y, k_z]$ represents the spatial frequency components of the wave, while $\mathbf{r} \equiv [r\cos\varphi, r\sin\varphi, z]$ is the position vector expressed in cylindrical coordinates at the focal region. The term $\mathbf{E}_\infty(k_x, k_y) \equiv q(k_x, k_y) l_0(k_x, k_y) \hat{\mathbf{a}}(k_x, k_y)$ encapsulates the spectral weighting of the plane-wave components that form the angular spectrum.

Here, $q(k_x, k_y)$ is the apodization factor, derived from the energy conservation theorem, as we saw in Chapter 5. It ensures that the mapping between the pupil plane and the angular spectrum preserves the total optical power. The function $l_0(k_x, k_y)$ corresponds to the amplitude distribution of the input wavefront $w(x, y)$, and thus carries the information about the illumination and the geometry of the pupil. Finally, $\hat{\mathbf{a}}(k_x, k_y)$ is a unit vector describing the local polarization orientation of each plane-wave component in the angular spectrum. This term guarantees that the vectorial nature of the electric field is preserved after propagation and correctly represented in the focal region.

In this way, the field at focus is expressed as a coherent superposition of angular components, each one weighted by amplitude, apodization, and polarization. This formalism provides a powerful and general description of how arbitrary wavefronts and input polarization states contribute to the focal field distribution, thereby extending the applicability of diffraction integrals to freeform and non-stigmatic systems.

Eq. (7.2) seems to be very abstract and cumbersome since the integration is over the frequency domain defined by k_x, and k_y. It will be more understandable if Eq. (7.2) will be in terms of the spatial domain since the wavefronts are represented in the spatial domain. In other words, a wavefront is typically represented by a function $w(x, y)$ that depends in two independent variables x and y instead of k_x, and k_y. The main goal of this chapter is to express Eq. (7.2) in a more intuitive way in the spatial domain for any wavefront $w(x, y)$. Richards-Wolf did the same but when $w(x, y)$ is a spherical surface, here we want to consider a board general wavefront, a freeform wavefront $w(x, y)$. In a manner that makes more sense in the context of the spatial domain, enhancing its interpretability.

7.2.1 The propagation vector

In Fig. 7.1, it is evident that Eq. (7.2) only considers plane waves. This implies that the wavevector components $[k_x, k_y, k_z]$ are real-valued. Recognizing this,

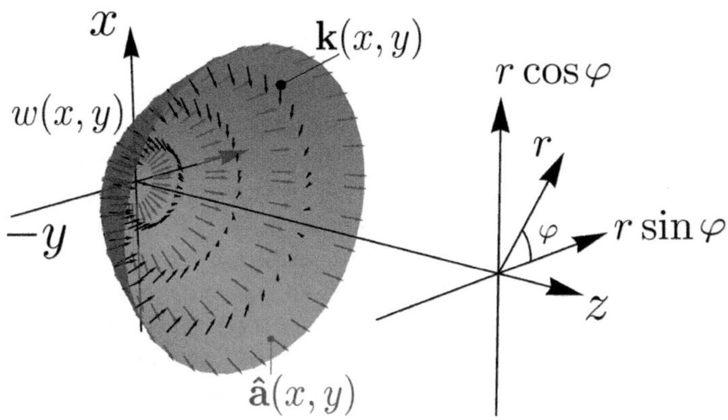

FIGURE 7.1
Diagram of the diffraction phenomena. The wavefront $w(x, y)$ is placed in the coordinate system (x, y). The vector $\mathbf{k}$ is normal to $w(x, y)$ and $\hat{\mathbf{a}}$ is normal to $\mathbf{k}$. The diffraction pattern is in the coordinate system $(r\cos\varphi, r\sin\varphi)$ and z is the axis of propagation.

and drawing from the Malus-Dupin theorem [23], we can characterize the propagation vector of a wavefront $w(x, y)$ as the normal vector to the surface defined by the wavefront. This provides a direct link between the frequency-domain formulation and the spatial domain representation of the wavefront.

By incorporating this geometric interpretation, the propagation of wavefronts becomes clearer, allowing for a more tangible understanding of how the wavefront evolves across space. The aim here is to bridge the gap between the abstract nature of the frequency domain and the more accessible, spatial domain, which directly relates to physical measurements and observations. So let $\mathbf{k}$ the propagation vector be equal to the normal vector of a wavefront $w(x, y)$, using Eq. (1.81),

$$\begin{aligned}\mathbf{k} = [k_x, k_y, k_z] = k\frac{-\partial_x w}{\sqrt{\partial_x w^2 + \partial_y w^2 + 1}}\hat{\mathbf{x}} \\ + k\frac{-\partial_y w}{\sqrt{\partial_x w^2 + \partial_y w^2 + 1}}\hat{\mathbf{y}} \\ + k\frac{1}{\sqrt{\partial_x w^2 + \partial_y w^2 + 1}}\hat{\mathbf{z}}. \quad (7.3)\end{aligned}$$

In this context, $\partial_x w$ represents partial derivative of the function w with respect to the x-coordinate, while $\partial_y w$ denotes partial derivative of w with respect to the y-coordinate. These quantities are the partial derivatives, which quantify how the wavefront $w(x, y)$ varies along a specific direction while holding the other variables constant.

Additionally, k_x, k_y, and k_z represent the directional cosines of the wave vector $\mathbf{k}$ along the x-, y-, and z-axes, respectively. These directional cosines provide a measure of the orientation of the wave vector in three-dimensional space, indicating the relative contribution of the wave's propagation in each spatial direction. Specifically, the components k_x, k_y, and k_z describe the alignment of the wave vector with the coordinate axes and are essential in understanding the propagation characteristics of the wave. Notice that from Eq. (7.3) we had described successfully described the propagation vector in terms of the spacial independent variables (x, y) that maps wavefront $w(x, y)$. Our next task is to do the same but for the polarization state vector.

7.2.2 The polarization vector

Let us focus on $\mathbf{a}$, a polarization state vector associated with a given wavefront $w(x, y)$. The polarization vector $\mathbf{a}$ defines the orientation of the electric field relative to the direction propagation $\mathbf{k}$. The only requirements that must be satisfied by $\mathbf{a}$ are that it must be orthogonal to the wave vector $\mathbf{k}$ and that its magnitude must be equal to one, ensuring that it represents a unit vector.

The condition of orthogonality to $\mathbf{k}$ is essential, as the polarization direction lies in a plane perpendicular to the direction of wave propagation. In other words, the polarization vector must not have any component along the direction of propagation of the wave. We know this fact from Eq. (4.11) and Eq. (4.12).

Additionally, the condition that $\mathbf{a}$ is a unit vector, i.e., $|\mathbf{a}| = 1$, ensures that the polarization state is normalized. This is a necessary constraint, as it guarantees that the physical properties of the wave, such as its energy density or amplitude, are well-defined and normalized. With this in mind lets define a polarization vector $\mathbf{a}$ as,

$$\begin{aligned}\mathbf{a}(x, y) = &\frac{\partial_x w}{\sqrt{\partial_x w^2 + \partial_y w^2 + 1}\sqrt{\partial_x w^2 + \partial_y w^2}}\hat{\mathbf{x}} \\ &+ \frac{\partial_y w}{\sqrt{\partial_x w^2 + \partial_y w^2 + 1}\sqrt{\partial_x w^2 + \partial_y w^2}}\hat{\mathbf{y}} \\ &+ \sqrt{1 - \frac{1}{\partial_x w^2 + \partial_y w^2 + 1}}\hat{\mathbf{z}}. \end{aligned} \tag{7.4}$$

This vector is a particular polarization state of the wavefront $w(x, y)$ at point (x, y). It can easily proven with the equations presented in Chapter 1, at the section of tangent and normal vectors. But which polarization does Eq. (7.4) represents? It is still not clear but from it we can span all the possible polarization state for the $w(x, y)$ at point (x, y).

According to Rodrigues' rotation formula, Eq. (1.14), we can rotate **a** with a desire angle $\alpha(x, y)$. Therefore, all the possible vectors orthogonal to **k** can be represented with Rodrigues' rotation formula,

$$\hat{\mathbf{a}} = \mathbf{a}\cos\alpha + (\mathbf{k}\times\mathbf{a})\sin\alpha + \mathbf{k}(\mathbf{k}\cdot\mathbf{a})(1-\cos\alpha). \tag{7.5}$$

where $\hat{\mathbf{a}}$ is unit vector which represent a polarization state, rotated $\alpha(x, y)$ along the perpendicular plane to **k**. Thus, this vector must be perpendicular to the propagation vector **k**, $\mathbf{k}\perp\mathbf{a}$. Replacing Eqs. (7.3) and (7.4) in Eq. (7.5), we have,

$$
\begin{aligned}
\hat{\mathbf{a}}(x,y) = \\
&\left[\frac{\cos\alpha\partial_x w}{\sqrt{\partial_x w^2+\partial_y w^2}\sqrt{\partial_x w^2+\partial_y w^2+1}}\right.\\
&\sin\alpha\left(-\frac{\partial_y w\sqrt{1-\frac{1}{\partial_x w^2+\partial_y w^2+1}}}{\sqrt{\partial_x w^2+\partial_y w^2+1}}+\frac{-\partial_y w+\cos\alpha\partial_x w}{\sqrt{\partial_x w^2+\partial_y w^2}\left(\partial_x w^2+\partial_y w^2+1\right)}\right)\\
&\left.-\frac{(1-\cos\alpha)\partial_x w\left(\frac{-\partial_x w^2-\partial_y w^2}{\sqrt{\partial_x w^2+\partial_y w^2}(\partial_x w^2+\partial_y w^2+1)}+\frac{\sqrt{1-\frac{1}{\partial_x w^2+\partial_y w^2+1}}}{\sqrt{\partial_x w^2+\partial_y w^2+1}}\right)}{\sqrt{\partial_x w^2+\partial_y w^2+1}}\right]\hat{\mathbf{x}}\\
&+\left[\frac{\cos\alpha\partial_x w}{\sqrt{\partial_x w^2+\partial_y w^2}\sqrt{\partial_x w^2+\partial_y w^2+1}}\right.\\
&\sin\alpha\left(\frac{\partial_x w\sqrt{1-\frac{1}{\partial_x w^2+\partial_y w^2+1}}}{\sqrt{\partial_x w^2+\partial_y w^2+1}}+\frac{\partial_x w+\cos\alpha\partial_y w}{\sqrt{\partial_x w^2+\partial_y w^2}\left(\partial_x w^2+\partial_y w^2+1\right)}\right)\\
&\left.-\frac{(1-\cos\alpha)\partial_y w\left(\frac{-\partial_x w^2-\partial_y w^2}{\sqrt{\partial_x w^2+\partial_y w^2}(\partial_x w^2+\partial_y w^2+1)}+\frac{\sqrt{1-\frac{1}{\partial_x w^2+\partial_y w^2+1}}}{\sqrt{\partial_x w^2+\partial_y w^2+1}}\right)}{\sqrt{\partial_x w^2+\partial_y w^2+1}}\right]\hat{\mathbf{y}}\\
&+\left[\frac{(1-\cos\alpha)\left(\frac{-\partial_x w^2-\partial_y w^2}{\sqrt{\partial_x w^2+\partial_y w^2}(\partial_x w^2+\partial_y w^2+1)}+\frac{\sqrt{1-\frac{1}{\partial_x w^2+\partial_y w^2+1}}}{\sqrt{\partial_x w^2+\partial_y w^2+1}}\right)}{\sqrt{\partial_x w^2+\partial_y w^2+1}}\right.\\
&\left.+\cos\alpha\sqrt{1-\frac{1}{\partial_x w^2+\partial_y w^2+1}}\right]\hat{\mathbf{z}}.
\end{aligned}
\tag{7.6}
$$

Let's simplify the last expression, which is quite large. Notice that the following term $\mathbf{k}(\mathbf{k}\cdot\mathbf{a})(1-\cos\alpha)$ is actually zero, since **k** is perpendicular to $\mathbf{a}$, $\mathbf{k} \perp \mathbf{a}$,

$$\frac{(1-\cos\alpha)\left(\frac{-\partial_x w^2-\partial_y w^2}{\sqrt{\partial_x w^2+\partial_y w^2(\partial_x w^2+\partial_y w^2+1)}}+\frac{\sqrt{1-\frac{1}{\partial_x w^2+\partial_y w^2+1}}}{\sqrt{\partial_x w^2+\partial_y w^2+1}}\right)}{\sqrt{\partial_x w^2+\partial_y w^2+1}}=0. \tag{7.7}$$

So, eliminating the mentioned terms. The vector $\hat{\mathbf{a}}(x, y)$ can be reduced to a much simpler expression given by Eq. (7.8),

$$\begin{aligned}\hat{\mathbf{a}}(x,y) = & \\ & \left[\frac{\cos\alpha\partial_x w}{\sqrt{\partial_x w^2+\partial_y w^2}\sqrt{\partial_x w^2+\partial_y w^2+1}}\right. \\ & \left.+\sin\alpha\left(-\frac{\partial_y w\sqrt{1-\frac{1}{\partial_x w^2+\partial_y w^2+1}}}{\sqrt{\partial_x w^2+\partial_y w^2+1}}+\frac{-\partial_y w+\cos\alpha\partial_x w}{\sqrt{\partial_x w^2+\partial_y w^2}\left(\partial_x w^2+\partial_y w^2+1\right)}\right)\right]\hat{\mathbf{x}} \\ & +\left[\frac{\cos\alpha\partial_x w}{\sqrt{\partial_x w^2+\partial_y w^2}\sqrt{\partial_x w^2+\partial_y w^2+1}}\right. \\ & \left.+\sin\alpha\left(\frac{\partial_x w\sqrt{1-\frac{1}{\partial_x w^2+\partial_y w^2+1}}}{\sqrt{\partial_x w^2+\partial_y w^2+1}}+\frac{\partial_x w+\cos\alpha\partial_x w}{\sqrt{\partial_x w^2+\partial_y w^2}\left(\partial_x w^2+\partial_y w^2+1\right)}\right)\right]\hat{\mathbf{y}} \\ & +\cos\alpha\sqrt{1-\frac{1}{\partial_x w^2+\partial_y w^2+1}}\hat{\mathbf{z}}. \end{aligned} \tag{7.8}$$

After a long manipulation Eq. (7.8) can be expressed as,

$$\begin{aligned}\hat{\mathbf{a}}(x,y) = & \frac{\frac{\cos\alpha\partial_x w}{\sqrt{\partial_x w^2+\partial_y w^2+1}}-\sin\alpha\partial_y w}{\sqrt{\partial_x w^2+\partial_y w^2}}\hat{\mathbf{x}} \\ & +\frac{\frac{\cos\alpha\partial_y w}{\sqrt{\partial_x w^2+\partial_y w^2+1}}+\sin\alpha\partial_x w}{\sqrt{\partial_x w^2+\partial_y w^2}}\hat{\mathbf{y}} \\ & +\cos\alpha\sqrt{1-\frac{1}{\partial_x w^2+\partial_y w^2+1}}\hat{\mathbf{z}}, \end{aligned} \tag{7.9}$$

Depending on the value of α, the polarization state of the wave can vary significantly. The parameter α governs the orientation and configuration of the

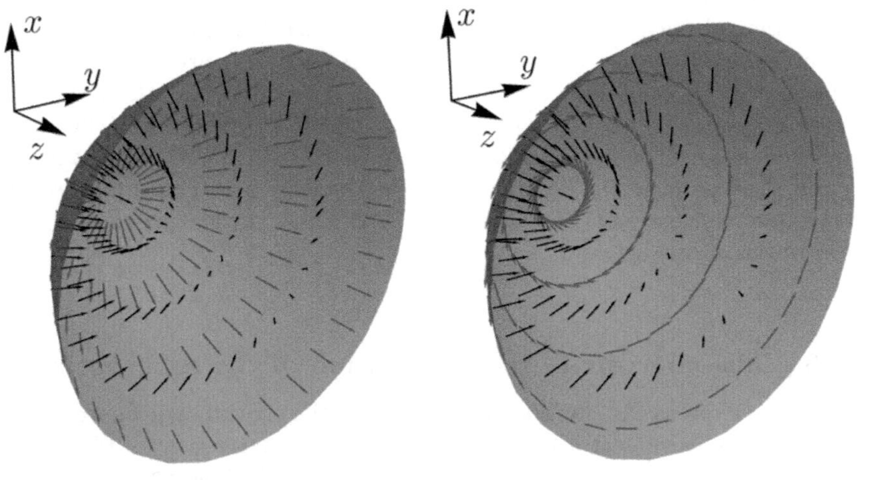

FIGURE 7.2
The polarization state depends of $\alpha(x, y)$. For a radially symmetric wavefront $w(x, y)$, *left* if $\alpha = 0$, we get a radial polarization, *right* $\alpha = \pi/2$ we have an azimuthal polarization.

polarization vector, thus directly influencing the nature of the wave's polarization at any given point. For example, in the case of a radially symmetric wavefront $w(x, y)$, if $\alpha = 0$, the wave exhibits radial polarization, where the electric field vectors point outward from the center in a radial pattern. On the other hand, if $\alpha = \pi/2$, the wave shows azimuthal polarization, where the electric field vectors are oriented tangentially around the axis of propagation, following an azimuthal direction. This behavior is illustrated in Fig. 7.2, which depicts the two distinct polarization states in a visual manner.

More generally, $\alpha(x, y)$ can be treated as a position-dependent function, meaning that the polarization state can vary spatially across the wavefront $w(x, y)$ profile. In this case, $\alpha(x, y)$ represents the polarization aberration function, which describes how the polarization evolves across the wavefront $w(x, y)$'s cross-section. This variation is crucial in many optical systems, such as those involving wavefront $w(x, y)$ shaping, laser optics, or systems with optical aberrations. The spatial dependence of α reflects how the polarization might be manipulated or distorted due to factors like lensing effects, material properties, or external fields.

It is important to recall that $\hat{\mathbf{a}}$, as defined in Eq. (7.9), is a unit vector. This means that the magnitude of the polarization vector is normalized to 1, ensuring that it only represents the direction of polarization without affecting the overall amplitude of the wave. This normalization is critical for maintaining a consistent physical interpretation of the polarization state.

Furthermore, notice that both Eqs. (7.3) and (7.9) are orthogonal. This orthogonality is a fundamental property, as it ensures that the polarization vector $\hat{\mathbf{a}}$ remains perpendicular to the wave vector $\mathbf{k}$, maintaining the physical requirement that the polarization lies in a plane orthogonal to the wave's direction of propagation. This orthogonality condition is essential for describing the transverse nature of the polarization, which is a key feature of electromagnetic waves in free space.

7.2.3 The coordinate transformation of the differentials

In order to express the integration in terms of the spatial coordinates x and y instead of the wave vector components k_x and k_y, it is essential to perform a coordinate transformation. This transformation will allow us to switch from the wave vector space (Fourier space) to the spatial domain, which is often required in physical applications like diffraction analysis or imaging. The wave vectors k_x and k_y correspond to the components of the wave vector $\mathbf{k}$ in the x- and y-directions, respectively, and are related to the spatial coordinates via Fourier transforms.

Let's focus now on the differentials $\mathrm{d}k_x$ and $\mathrm{d}k_y$, which describe small changes in the wave vector components. In order to express the integrals in terms of the spatial variables, we need to understand how the differential elements in the wave vector space relate to those in the spatial domain. This transformation is accomplished using the Jacobian matrix, which describes the rate of change of the new variables (here, x and y) with respect to the old variables (the components of the wave vector, k_x and k_y).

The Jacobian matrix, as discussed in Chapter 1, provides the necessary information for transforming the differentials from the wave vector space to the spatial domain. Specifically, it helps us account for the changes in the x- and y-coordinates due to small variations in the wave vector components.

$$\mathrm{d}k_x \mathrm{d}k_y = \mathbf{Jcb}(x, y) \mathrm{d}x \mathrm{d}y, \tag{7.10}$$

By applying the Jacobian, we obtain the following relations for the differentials,

$$\mathbf{Jcb}(x, y) = \begin{bmatrix} \dfrac{\partial k_x}{\partial x} & \dfrac{\partial k_x}{\partial y} \\ \dfrac{\partial k_y}{\partial x} & \dfrac{\partial k_y}{\partial y} \end{bmatrix}. \tag{7.11}$$

These derivatives encapsulate how the wave vector changes as the position in space varies, and they allow us to properly transform the integration variables. Computing the determinant of the Jacobian matrix,

$$\begin{aligned}\mathbf{Jcb}(x,y) = k^2 &\left(\frac{\partial_y w\left(2\partial_x w \partial_{xy} w + 2\partial_y w \partial_y^2 w\right)}{2\left(\partial_x w^2 + \partial_y w^2 + 1\right)^{3/2}} - \frac{\partial_y^2 w}{\sqrt{\partial_x w^2 + \partial_y w^2 + 1}}\right)\\ &\times \left(\frac{\partial_x w\left(2\partial_x w \partial_x^2 w + 2\partial_y w \partial_{xy} w\right)}{2\left(\partial_x w^2 + \partial_y w^2 + 1\right)^{3/2}} - \frac{\partial_x^2 w}{\sqrt{\partial_x w^2 + \partial_y w^2 + 1}}\right)\\ &- k^2 \left(\frac{\partial_x w\left(2\partial_x w \partial_{xy} w + 2\partial_y w \partial_y^2 w\right)}{2\left(\partial_x w^2 + \partial_y w^2 + 1\right)^{3/2}} - \frac{\partial_{xy} w}{\sqrt{\partial_x w^2 + \partial_y w^2 + 1}}\right)\\ &\times \left(\frac{\partial_y w\left(2\partial_x w \partial_x^2 w + 2\partial_y w \partial_{xy} w\right)}{2\left(\partial_x w^2 + \partial_y w^2 + 1\right)^{3/2}} - \frac{\partial_{xy} w}{\sqrt{\partial_x w^2 + \partial_y w^2 + 1}}\right),\end{aligned} \tag{7.12}$$

where $\partial_{xy} w$ is the derivative of w respect to x and y. After a long manipulation, the last equation can be expressed as

$$\mathbf{Jcb}(x,y) = k^2 \frac{\partial_x^2 w \partial_y^2 w - \left(\partial_{xy} w\right)^2}{\left(\partial_x w^2 + \partial_y w^2 + 1\right)^2}. \tag{7.13}$$

By applying this transformation, we were able to rewrite the integral over the wave vector space into an integral over the spatial coordinates, effectively mapping the problem from Fourier space back to real space. This was a crucial step, where it was necessary to express quantities like the diffraction pattern or the field distribution in real space, rather than in the abstract wave vector space.

Notice that in Eq. (7.2), the integrant is mutiplied by 1/kz. So lets divide the Jacobean by $\mathbf{Jcb}(x,y)$ by k_z,

$$\frac{\mathbf{Jcb}(x,y)}{k_z} = k \frac{\partial_x^2 w \partial_y^2 w - \left(\partial_{xy} w\right)^2}{\left(\partial_x w^2 + \partial_y w^2 + 1\right)^{3/2}} \tag{7.14}$$

This last expression does not have a particularly physical meaning but it will more initiative for the substitution on Eq. (7.2), a procedure we will perform in the next section.

7.3 The general diffraction integral

If we replace Eqs. (7.3), (7.9), and (7.14) into Eq. (7.2), we obtain the general vector diffraction integral that has been proposed. This step involves substituting the expressions for the wave vector components, the Jacobian terms, and the polarization vector into the original diffraction formula, which allows us to reformulate the diffraction problem in terms of the spatial coordinates.

By performing these substitutions, we obtain a more comprehensive expression for the diffraction integral, which accounts for both the wave propagation and the polarization state of the wave at each point. This general form is crucial because it enables the incorporation of various physical factors, such as polarization aberrations and wavevector variations, that influence the diffraction process.

The resulting vector diffraction integral serves as a powerful tool in computational optics, providing a framework for simulating and analyzing diffraction patterns in systems where the polarization state of the wave cannot be ignored. It also offers insight into how different factors, including the geometry of the system and the properties of the wavefront, affect the resulting diffraction pattern. The general vector diffraction integral thus extends the classical scalar diffraction theory, providing a more accurate and complete description of wave propagation in optical systems, The main equation is given by,

$$\begin{aligned}
\mathbf{E}(r,\varphi,z) &= \frac{E_0 k}{2\pi}\int\int_{\Omega} P(x,y)q(x,y)l_0(x,y)\left[\frac{\partial_x^2 w \partial_y^2 w - (\partial_{xy} w)^2}{(\partial_x w^2 + \partial_y w^2 + 1)^{3/2}}\right] \\
&\times \left[\frac{\frac{\cos\alpha\partial_x w}{\sqrt{\partial_x w^2+\partial_y w^2+1}} - \sin\alpha\partial_y w}{\sqrt{\partial_x w^2 + \partial_y w^2}}\hat{\mathbf{x}} + \frac{\frac{\cos\alpha\partial_y w}{\sqrt{\partial_x w^2+\partial_y w^2+1}} + \sin\alpha\partial_x w}{\sqrt{\partial_x w^2 + \partial_y w^2}}\hat{\mathbf{y}}\right. \\
&\left. + \cos\alpha\sqrt{1 - \frac{1}{\partial_x w^2 + \partial_y w^2 + 1}}\hat{\mathbf{z}}\right] \exp\left[\frac{-izk}{\sqrt{\partial_x w^2 + \partial_y w^2 + 1}}\right] \\
&\times \exp\left[-\frac{ir\cos\varphi k\partial_y w}{\sqrt{\partial_x w^2 + \partial_y w^2 + 1}} - \frac{ir\sin\varphi k\partial_x w}{\sqrt{\partial_x w^2 + \partial_y w^2 + 1}}\right] dx\, dy. \qquad (7.15)
\end{aligned}$$

Eq. (7.15) is the most important equation of this treatise, as it provides a comprehensive description of the diffraction pattern of a polarized freeform wavefront. This equation encapsulates the interaction between the spatial structure of the wavefront, its polarization state, and any optical elements that may alter the wave's characteristics, such as masks or apertures. The fundamental assumption is that the wavefront $w(x, y)$ is a continuous and differentiable function that governs the shape of the wave as it propagates. For practical purposes, $w(x, y)$ is often expressed in terms of Zernike polynomials, or any other suitable smooth, continuous function that can represent arbitrary phase and amplitude variations of the wavefront.

The polarization of the wavefront is explicitly dependent on the function $\alpha(x, y)$, which describes the spatial variation of the polarization state across the wavefront. This function $\alpha(x, y)$ can be adjusted to account for various polarization configurations, such as linear, circular, or elliptical polarization, at different points on the wavefront. The spatial dependence of $\alpha(x, y)$ allows for the modeling of complex polarization aberrations, which are often encountered in real optical systems, such as in imaging or beam-shaping applications.

The limits of the integral given by Eq. (7.15) are given by the aperture at the exist pupil Ω. The aperture Ω is limited by $x \in (x_{\min}, x_{\max})$, $y \in (y_{\min}, y_{\max})$ in Eq. (7.15). Notice in Chapter 4, Ω was related to the solid angle, because the original Richards-Wolf integrals the aperture is in terms of angles. We will use the symbol Ω as the limits of the apertures (diffraction integrals).

In the generalized form of Richards-Wolf integrals, Eq. (7.15), we also introduce the mask function $P(x, y)$, which represents the shape of any optical mask or aperture that might be applied to the wavefront. The mask function $P(x, y)$ can encode information about the mask's geometry, including its spatial structure and any transmissive or reflective properties. The inclusion of $P(x, y)$ makes the diffraction integral more flexible, enabling it to capture the effects of complex optical systems.

By default, in the absence of any mask or optical elements, Eq. (7.15) assumes a rectangular pupil shape, since the integration is performed over the spatial coordinates x and y. This rectangular shape is typical in many optical setups, particularly in free-space propagation or standard imaging systems. However, the flexibility of the mask term $P(x, y)$ allows for the consideration of more complex pupil shapes, including circular, annular, or even irregularly shaped apertures. As such, the equation remains valid in a wide range of scenarios, including those involving aberrated or non-ideal optical systems.

Thus, Eq. (7.15) provides a unified framework for understanding and calculating the diffraction patterns of polarized freeform wavefronts in the presence of arbitrary optical elements. It is a powerful tool that can be applied in a variety of settings, including wavefront engineering, holography, and adaptive optics, where the precise control of both the spatial and polarization characteristics of light is essential.

Since we know that the magnetic field is always perpendicular to the direction of propagation, as defined by the wave vector $\mathbf{k}$, and that the electric field is perpendicular to both the magnetic field and the propagation direction, we compute the magnetic field $\mathbf{B}(r, \varphi, z)$ as the cross product of $\mathbf{k}$ with $\mathbf{E}(r, \varphi, z)$

divided by c,

$$\mathbf{B}(r,\varphi,z) = \frac{E_0 k}{2\pi c}\int\int_{\Omega} P(x,y)q(x,y)l_0(x,y)\left[\frac{\partial_x^2 w \partial_y^2 w - (\partial_{xy}w)^2}{(\partial_x w^2 + \partial_y w^2 + 1)^{3/2}}\right]$$
$$\times\left[\left(-\frac{\cos\alpha\partial_y w\sqrt{1-\frac{1}{\partial_x w^2+\partial_y w^2+1}}}{\sqrt{\partial_x w^2 + \partial_y w^2 + 1}} - \frac{\cos\alpha\partial_y w - \sin\alpha\partial_x w}{\sqrt{\partial_x w^2 + \partial_y w^2}\,(\partial_x w^2 + \partial_y w^2 + 1)}\right)\hat{\mathbf{x}}\right.$$
$$+\left(\frac{\cos\alpha\partial_x w\sqrt{1-\frac{1}{\partial_x w^2+\partial_y w^2+1}}}{\sqrt{\partial_x w^2 + \partial_y w^2 + 1}} + \frac{\cos\alpha\partial_x w - \sin\alpha\partial_y w}{\sqrt{\partial_x w^2 + \partial_y w^2}\,(\partial_x w^2 + \partial_y w^2 + 1)}\right)\hat{\mathbf{y}}$$
$$\left. - \frac{\sin\alpha\partial_x w^2 + \sin\alpha\partial_y w^2}{\sqrt{\partial_x w^2 + \partial_y w^2}\sqrt{\partial_x w^2 + \partial_y w^2 + 1}}\hat{\mathbf{z}}\right]$$
$$\times \exp\left[\frac{-izk}{\sqrt{\partial_x w^2 + \partial_y w^2 + 1}} - \frac{ir\cos\varphi k\partial_y w}{\sqrt{\partial_x w^2 + \partial_y w^2 + 1}} - \frac{ir\sin\varphi k\partial_y w}{\sqrt{\partial_x w^2 + \partial_y w^2 + 1}}\right] dxdy. \tag{7.16}$$

After carrying out a considerably long and detailed mathematical derivation involving several intermediate steps of algebraic manipulation and rearrangement, the above expression can be reduced to the following much simpler form,

$$\mathbf{B}(r,\varphi,z) = \frac{E_0 k}{2\pi c}\int\int_{\Omega} P(x,y)q(x,y)l_0(x,y)\left[\frac{\partial_x^2 w \partial_y^2 w - (\partial_{xy}w)^2}{(\partial_x w^2 + \partial_y w^2 + 1)^{3/2}}\right]$$
$$\times\left[-\frac{\sin\alpha\partial_x w\sqrt{\partial_x w^2 + \partial_y w^2 + 1} + \cos\alpha\left(\partial_x w^2 + 1\right)\partial_y w + \cos\alpha\partial_y w^3}{\sqrt{\partial_x w^2 + \partial_y w^2}\,(\partial_x w^2 + \partial_y w^2 + 1)}\hat{\mathbf{x}}\right.$$
$$\frac{-\sin\alpha\partial_y w\sqrt{\partial_x w^2 + \partial_y w^2 + 1} + \cos\alpha\partial_x w\left(\partial_y w^2 + 1\right) + \cos\alpha\partial_x w^3}{\sqrt{\partial_x w^2 + \partial_y w^2}\,(\partial_x w^2 + \partial_y w^2 + 1)}\hat{\mathbf{y}}$$
$$\left. - \sin\alpha\sqrt{\frac{\partial_x w^2 + \partial_y w^2}{\partial_x w^2 + \partial_y w^2 + 1}}\hat{\mathbf{z}}\right]\exp\left[\frac{-izk}{\sqrt{\partial_x w^2 + \partial_y w^2 + 1}}\right]$$
$$\times \exp\left[-\frac{ir\cos\varphi k\partial_y w}{\sqrt{\partial_x w^2 + \partial_y w^2 + 1}} - \frac{ir\sin\varphi k\partial_y w}{\sqrt{\partial_x w^2 + \partial_y w^2 + 1}}\right] dxdy. \tag{7.17}$$

Although along the book we will consider Eq. (7.15) for the examples and illustrative scenarios. It is important to remark that Eq. (7.17) gives the description of the distribution of the magnetic field $\mathbf{B}(r,\varphi,z)$ in the image plane of a polarized freeform wavefront.

7.3.1 Aligning an input normalized polarization vector

In order to align an input normalized polarization vector $\mathbf{E}_{\text{inc}} = [e_x, e_y, 0]^T$ with the local polarization vector $\hat{\mathbf{a}}(x, y)$, the polarization angle $\alpha(x, y)$ must be carefully chosen. This angle α defines the orientation of the polarization vector in the transverse plane of the propagation vector $\mathbf{k}$ and determines how the electric field is distributed between the x- and y-components at each point (x, y) on the wavefront $w(x, y)$.

To achieve alignment between the input polarization vector and the desired local polarization vector $\hat{\mathbf{a}}(x, y)$, we must ensure that the angle α corresponds to the direction of the input polarization. In other words, the projection of the input vector onto the local polarization basis must be maximized, or equivalently, the angle α should be selected such that the two vectors are parallel.

The angle α is obtained by computing the dot product between the input normalized polarization vector $\mathbf{E}_{\text{inc}} = [e_x, e_y, 0]^T$ and the local polarization vector $[a_x, a_y, a_z]^T$, and then setting this dot product equal to zero. This condition ensures that the two vectors are orthogonal, which is necessary for aligning the input polarization with the desired local polarization state.

In other words, α is found by solving the equation,

$$[e_x, e_y, 0]^T \cdot [a_x, a_y, a_z]^T = 0, \tag{7.18}$$

which explicitly relates the components of the input vector and the polarization vector. By substituting the expressions for a_x, a_y, and a_z (which depend on α) into the dot product, one obtains an equation in terms of α. Replacing Eq. (7.9) in the last expression,

$$\frac{\frac{\cos\alpha\partial_x w}{\sqrt{\partial_x w^2+\partial_y w^2+1}} - \sin\alpha\partial_y w}{\sqrt{\partial_x w^2 + \partial_y w^2}} e_x + \frac{\frac{\cos\alpha\partial_y w}{\sqrt{\partial_x w^2+\partial_y w^2+1}} + \sin\alpha\partial_x w}{\sqrt{\partial_x w^2 + \partial_y w^2}} e_y = 0. \tag{7.19}$$

Solving this equation for α allows us to determine the precise polarization angle that aligns the input polarization vector with the local polarization vector $\hat{\mathbf{a}}(x, y)$,

$$\alpha = \arctan\left[\frac{e_y\partial_x w - e_x\partial_y w}{\sqrt{\partial_x w^2 + \partial_y w^2 + 1}(e_x\partial_x w + e_y\partial_y w)}\right], \tag{7.20}$$

This process is critical because it provides a way to map an arbitrary input polarization state onto a spatially varying polarization distribution described by $\alpha(x, y)$. This angle is essential in practical applications involving vector beam shaping, polarization control, or designing optical elements. By controlling $\alpha(x, y)$, one can impose a spatially varying polarization pattern onto the beam. Thus, replacing Eq. (7.20) in Eq. (7.9), the polarization turns to, $\hat{\mathbf{a}}(x,y)$

$$\hat{\mathbf{a}}(x,y) = \frac{1}{\sqrt{\partial_x w^2+\partial_y w^2}}\Bigg[\times \frac{\partial_x w}{\sqrt{\partial_x w^2+\partial_y w^2+1}\sqrt{\frac{(e_y\partial_x w-e_x\partial_y w)^2}{(\partial_x w^2+\partial_y w^2+1)(e_x\partial_x w+e_y\partial_y w)^2}+1}} - \frac{\partial_y w(e_y\partial_x w-e_x\partial_y w)}{\sqrt{\partial_x w^2+\partial_y w^2+1}(e_x\partial_x w+e_y\partial_y w)\sqrt{\frac{(e_y\partial_x w-e_x\partial_y w)^2}{(\partial_x w^2+\partial_y w^2+1)(e_x\partial_x w+e_y\partial_y w)^2}+1}}\Bigg]\hat{\mathbf{x}}$$
$$+\frac{1}{\sqrt{\partial_x w^2+\partial_y w^2}}\Bigg[\frac{\partial_y w}{\sqrt{\partial_x w^2+\partial_y w^2+1}\sqrt{\frac{(e_y\partial_x w-e_x\partial_y w)^2}{(\partial_x w^2+\partial_y w^2+1)(e_x\partial_x w+e_y\partial_y w)^2}+1}} - \frac{\partial_x w(e_y\partial_x w-e_x\partial_y w)}{\sqrt{\partial_x w^2+\partial_y w^2+1}(e_x\partial_x w+e_y\partial_y w)\sqrt{\frac{(e_y\partial_x w-e_x\partial_y w)^2}{(\partial_x w^2+\partial_y w^2+1)(e_x\partial_x w+e_y\partial_y w)^2}+1}}\Bigg]\hat{\mathbf{y}}$$
$$+\frac{\sqrt{1-\frac{1}{\partial_x w^2+\partial_y w^2+1}}}{\sqrt{\frac{(e_y\partial_x w-e_x\partial_y w)^2}{\left(\partial_x w^2+\partial_y w^2+1\right)\left(e_x\partial_x w+e_y\partial_y w\right)^2}+1}}\hat{\mathbf{z}}. \tag{7.21}$$

Therefore, Replacing last expression in Eq. (7.15), $\mathbf{E}(r,\varphi,z)$ becomes,

$$\mathbf{E}(r,\varphi,z) = \frac{E_0 k}{2\pi}\int\int_\Omega P(x,y)q(x,y)l_0(x,y)\left[\frac{\partial_x^2 w\partial_y^2 w-(\partial_{xy}w)^2}{(\partial_x w^2+\partial_y w^2+1)^{3/2}}\right]$$
$$\times\Bigg[\frac{1}{\sqrt{\partial_x w^2+\partial_y w^2}}\Bigg(\frac{\partial_x w}{\sqrt{\partial_x w^2+\partial_y w^2+1}\sqrt{\frac{(e_y\partial_x w-e_x\partial_y w)^2}{(\partial_x w^2+\partial_y w^2+1)(e_x\partial_x w+e_y\partial_y w)^2}+1}} - \frac{\partial_y w(e_y\partial_x w-e_x\partial_y w)}{\sqrt{\partial_x w^2+\partial_y w^2+1}(e_x\partial_x w+e_y\partial_y w)\sqrt{\frac{(e_y\partial_x w-e_x\partial_y w)^2}{(\partial_x w^2+\partial_y w^2+1)(e_x\partial_x w+e_y\partial_y w)^2}+1}}\Bigg)\hat{\mathbf{x}}$$
$$+\frac{1}{\sqrt{\partial_x w^2+\partial_y w^2}}\Bigg(\frac{\partial_y w}{\sqrt{\partial_x w^2+\partial_y w^2+1}\sqrt{\frac{(e_y\partial_x w-e_x\partial_y w)^2}{(\partial_x w^2+\partial_y w^2+1)(e_x\partial_x w+e_y\partial_y w)^2}+1}} - \frac{\partial_x w(e_y\partial_x w-e_x\partial_y w)}{\sqrt{\partial_x w^2+\partial_y w^2+1}(e_x\partial_x w+e_y\partial_y w)\sqrt{\frac{(e_y\partial_x w-e_x\partial_y w)^2}{(\partial_x w^2+\partial_y w^2+1)(e_x\partial_x w+e_y\partial_y w)^2}+1}}\Bigg)\hat{\mathbf{y}}$$
$$+\frac{\sqrt{1-\frac{1}{\partial_x w^2+\partial_y w^2+1}}}{\sqrt{\frac{(e_y\partial_x w-e_x\partial_y w)^2}{\left(\partial_x w^2+\partial_y w^2+1\right)\left(e_x\partial_x w+e_y\partial_y w\right)^2}+1}}\hat{\mathbf{z}}\Bigg]$$
$$\times\exp\left[\frac{-izk}{\sqrt{\partial_x w^2+\partial_y w^2+1}}-\frac{ir\cos\varphi k\partial_y w}{\sqrt{\partial_x w^2+\partial_y w^2+1}}-\frac{ir\sin\varphi k\partial_y w}{\sqrt{\partial_x w^2+\partial_y w^2+1}}\right]dxdy. \tag{7.22}$$

From above's expression it is quite clear that $\mathbf{E}(r, \varphi, z)$ reduces to,

$$\mathbf{E}(r,\varphi,z) = \frac{E_0 k}{2\pi} \int\int_{\Omega} P(x,y) q(x,y) l_0(x,y) \left[\frac{\partial_x^2 w \partial_y^2 w - (\partial_{xy} w)^2}{(\partial_x w^2 + \partial_y w^2 + 1)^{3/2}} \right] \\ \times \left[\frac{\sqrt{1 - \frac{1}{\partial_x w^2 + \partial_y w^2 + 1}} (e_x \hat{\mathbf{x}} + e_y \hat{\mathbf{y}})}{(e_x \partial_x w + e_y \partial_y w) \sqrt{\frac{(e_y \partial_x w - e_x \partial_y w)^2}{(\partial_x w^2 + \partial_y w^2 + 1)(e_x \partial_x w + e_y \partial_y w)^2} + 1}} + \right. \\ \left. \frac{\sqrt{1 - \frac{1}{\partial_x w^2 + \partial_y w^2 + 1}} \hat{\mathbf{z}}}{\sqrt{\frac{(e_y \partial_x w - e_x \partial_y w)^2}{(\partial_x w^2 + \partial_y w^2 + 1)(e_x \partial_x w + e_y \partial_y w)^2} + 1}} \right] \exp\left[\frac{-izk}{\sqrt{\partial_x w^2 + \partial_y w^2 + 1}} \right] \\ \times \exp\left[-\frac{ir \cos\varphi k \partial_y w}{\sqrt{\partial_x w^2 + \partial_y w^2 + 1}} - \frac{ir \sin\varphi k \partial_y w}{\sqrt{\partial_x w^2 + \partial_y w^2 + 1}} \right] dx dy. \quad (7.23)$$

Eq. (7.23) is similar to Eq. (7.15), but it is more accessible for the input polarization normalized vector $\mathbf{E}_{\text{inc}} = [e_x, e_y, 0]^T$. This representation is especially convenient when dealing with standard linear polarizations, such as horizontal or vertical polarization, or when the polarization state is defined point by point across the input aperture, i.e., $\mathbf{E}_{\text{inc}} = [e_x, e_y, 0]^T$. In contrast, Eq. (7.15) is more practical when treating cylindrically symmetric cases such as radial and azimuthal polarizations, where the polarization vectors naturally adapt to the geometry of the system. Thus, both formulations are complementary: Eq. (7.23) emphasizes general Cartesian-based input fields, while Eq. (7.15) is tailored for fields with intrinsic cylindrical symmetry.

But it is an important remark to mention here that the polarization mapping done in with expression Eq. (7.23) is not the dipole coordinates or the double pole coordinates [53].

The most popular coordinate system for the surface of a sphere is the latitude-longitude coordinate system, here referred to as dipole coordinates. Dipole coordinates are defined relative to a polar axis, specified by an axis vector. Each unit propagation vector $\mathbf{k}$ is expressed in terms of its latitude and longitude. The latitude and longitude form a local coordinate system. Observe Fig.7.3, the local longitude-coordinates are presented at the *left* of the figure. The latitude-coordinates are presented on the *center* of the figure. Notice that both coordinates system are similar, they are just rotated one from the other. The *right*-side of the figure shows a side view of a longitude/latitude-coordinates. The longitude-coordinates resemble the linear polarization in $\hat{\mathbf{x}}$, while the latitude-coordinates resemble the linear polarization in $\hat{\mathbf{y}}$. A proper rotation of the sphere can easily turn the longitude/latitude-coordinates to resemble the azimuthal and radial polarizations. Now, observe, the problem with this mapping is that dipole local latitude-coordinates and dipole local

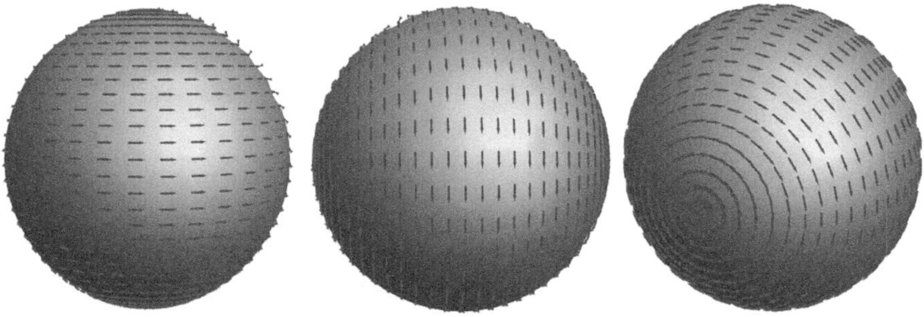

FIGURE 7.3
The dipole coordinates are a latitude/longitude coordinate system. In the *left* of the figure are longitude-coordinates. At the *center* latitude-coordinates are presented. Finally at *right* of the figure is a rotation that show the two singularities provided by this mapping.

longitude-coordinates change direction rapidly near the poles for small changes in direction, naturally giving rise to two singularities, one at each pole [53].

The double-pole coordinate system is another system to map the polarization over a sphere. It is less popular than the dipole coordinates. But according to R. Chipman [53], the double pole coordinates have the two main advantages (1) Double pole coordinates match the form of polarization that naturally occurs with lens and mirror systems. (2) Double pole coordinates scale (stretch) in such a way as to match magnification changes and changes of NA. The mathematical method for the mapping is well described in Chapter 11 of [53]. The double pole coordinates are presented in Fig.7.4. The front sphere is in

FIGURE 7.4
The double pole coordinates. The front sphere is in the *left* of the figure, and on the *right* is the rare side.

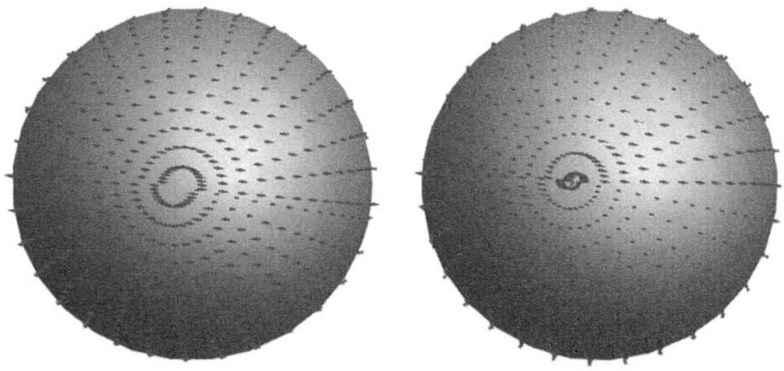

FIGURE 7.5
The map proposed by Eq. (7.23). The sign of the vector changes with the quadrant along the sphere. The front sphere is in the *left* of the figure, and on the *right* is the rare side.

the *left* of the figure, and on the *right* is the rare side. The problem is that the rear side of the sphere for the double pole coordinates shows an interesting double singularity at the pole where the basis vector rotates by 4π for a single circuit around the double singularity. With the proper rotation the front side of the sphere the double-pole coordinate system can resemble linear polarization in $\hat{\mathbf{x}}$ and $\hat{\mathbf{y}}$.

The polarization mapping presented in Eq. (7.23) is not limited to a unit sphere, for it applies to a freeform wavefront $w(x, y)$. If $w(x, y)$ is reduced to a sphere, the mapping is different from the dipole coordinates or the double pole coordinates. Fig. 7.5 shows the polarization mapping of Eq. (7.23) over a unit sphere; the front part of the sphere is shown at the *left* and the rear side is at the *right*. Notice that the direction of the vector changes with the quadrant along the sphere and the distribution of rays looks more similar to the one of the double pole coordinates, but the singularities are placed for the poles, like the dipole coordinates map. Just like the double pole coordinates, this mapping shows an interesting pattern at the rare side, which can be neglected in a focusing system. No focusing system focuses the whole solid angle.

With the polarization mapping explained in Eq. (7.23), we can now focus on the magnetic field $\mathbf{B}(r, \varphi, z)$. In that case, we know that it is perpedicular to the electric field, $\mathbf{B}(r,\varphi,z) = (1/c)\mathbf{k}\times\mathbf{E}(r,\varphi,z)$ turns to

$$\begin{aligned}
\mathbf{B}(r,\varphi,z) &= \frac{E_0 k}{2\pi c}\int\int_{\Omega} P(x,y)q(x,y)l_0(x,y)\left[\frac{\partial_x^2 w\partial_y^2 w - (\partial_{xy}w)^2}{(\partial_x w^2+\partial_y w^2+1)^{3/2}}\right] \\
&\times\Bigg[\Bigg(-\frac{\partial_y w\sqrt{1-\frac{1}{\partial_x w^2+\partial_y w^2+1}}}{\sqrt{\partial_x w^2+\partial_y w^2+1}\sqrt{\frac{(e_y\partial_x w-e_x\partial_y w)^2}{(\partial_x w^2+\partial_y w^2+1)(e_x\partial_x w+e_y\partial_y w)^2}+1}} \\
&- \frac{e_y\sqrt{1-\frac{1}{\partial_x w^2+\partial_y w^2+1}}}{\sqrt{\partial_x w^2+\partial_y w^2+1}(e_x\partial_x w+e_y\partial_y w)\sqrt{\frac{(e_y\partial_x w-e_x\partial_y w)^2}{(\partial_x w^2+\partial_y w^2+1)(e_x\partial_x w+e_y\partial_y w)^2}+1}}\Bigg)\hat{\mathbf{x}} \\
&+\Bigg(\frac{\partial_x w\sqrt{1-\frac{1}{\partial_x w^2+\partial_y w^2+1}}}{\sqrt{\partial_x w^2+\partial_y w^2+1}\sqrt{\frac{(e_y\partial_x w-e_x\partial_y w)^2}{(\partial_x w^2+\partial_y w^2+1)(e_x\partial_x w+e_y\partial_y w)^2}+1}} \\
&+ \frac{e_x\sqrt{1-\frac{1}{\partial_x w^2+\partial_y w^2+1}}}{\sqrt{\partial_x w^2+\partial_y w^2+1}(e_x\partial_x w+e_y\partial_y w)\sqrt{\frac{(e_y\partial_x w-e_x\partial_y w)^2}{(\partial_x w^2+\partial_y w^2+1)(e_x\partial_x w+e_y\partial_y w)^2}+1}}\Bigg)\hat{\mathbf{y}} \\
&+\Bigg(\frac{e_x\partial_y w\sqrt{1-\frac{1}{\partial_x w^2+\partial_y w^2+1}}}{\sqrt{\partial_x w^2+\partial_y w^2+1}(e_x\partial_x w+e_y\partial_y w)\sqrt{\frac{(e_y\partial_x w-e_x\partial_y w)^2}{(\partial_x w^2+\partial_y w^2+1)(e_x\partial_x w+e_y\partial_y w)^2}+1}} \\
&- \frac{e_y\partial_x w\sqrt{1-\frac{1}{\partial_x w^2+\partial_y w^2+1}}}{\sqrt{\partial_x w^2+\partial_y w^2+1}(e_x\partial_x w+e_y\partial_y w)\sqrt{\frac{(e_y\partial_x w-e_x\partial_y w)^2}{(\partial_x w^2+\partial_y w^2+1)(e_x\partial_x w+e_y\partial_y w)^2}+1}}\Bigg)\hat{\mathbf{z}}\Bigg] \\
&\times\exp\left[\frac{-izk}{\sqrt{\partial_x w^2+\partial_y w^2+1}} - \frac{ir\cos\varphi k\partial_y w}{\sqrt{\partial_x w^2+\partial_y w^2+1}} - \frac{ir\sin\varphi k\partial_y w}{\sqrt{\partial_x w^2+\partial_y w^2+1}}\right]dxdy.
\end{aligned} \tag{7.24}$$

Simplifying,

$$\begin{aligned}
\mathbf{B}(r,\varphi,z) &= \frac{E_0 k}{2\pi c}\int\int_{\Omega} P(x,y)q(x,y)l_0(x,y)\left[\frac{\partial_x^2 w\partial_y^2 w - (\partial_{xy}w)^2}{(\partial_x w^2+\partial_y w^2+1)^{3/2}}\right] \\
&\times\Bigg[-\frac{\sqrt{1-\frac{1}{\partial_x w^2+\partial_y w^2+1}}\left(e_x\partial_x w\partial_y w+e_y\partial_y w^2+e_y\right)}{\sqrt{\partial_x w^2+\partial_y w^2+1}(e_x\partial_x w+e_y\partial_y w)\sqrt{\frac{(\partial_x w^2+\partial_y w^2)\left((e_x\partial_x w+e_y\partial_y w)^2+e_x^2+e_y^2\right)}{(\partial_x w^2+\partial_y w^2+1)(e_x\partial_x w+e_y\partial_y w)^2}}}\hat{\mathbf{x}} \\
&+\frac{\sqrt{1-\frac{1}{\partial_x w^2+\partial_y w^2+1}}\left(e_x\partial_x w^2+e_y\partial_x w\partial_y w+e_x\right)}{\sqrt{\partial_x w^2+\partial_y w^2+1}(e_x\partial_x w+e_y\partial_y w)\sqrt{\frac{(\partial_x w^2+\partial_y w^2)\left((e_x\partial_x w+e_y\partial_y w)^2+e_x^2+e_y^2\right)}{(\partial_x w^2+\partial_y w^2+1)(e_x\partial_x w+e_y\partial_y w)^2}}}\hat{\mathbf{y}} \\
&+\frac{\sqrt{1-\frac{1}{\partial_x w^2+\partial_y w^2+1}}(e_x\partial_y w-e_y\partial_x w)}{\sqrt{\partial_x w^2+\partial_y w^2+1}(e_x\partial_x w+e_y\partial_y w)\sqrt{\frac{(\partial_x w^2+\partial_y w^2)\left((e_x\partial_x w+e_y\partial_y w)^2+e_x^2+e_y^2\right)}{(\partial_x w^2+\partial_y w^2+1)(e_x\partial_x w+e_y\partial_y w)^2}}}\hat{\mathbf{z}}\Bigg] \\
&\times\exp\left[\frac{-izk}{\sqrt{\partial_x w^2+\partial_y w^2+1}} - \frac{ir\cos\varphi k\partial_y w}{\sqrt{\partial_x w^2+\partial_y w^2+1}} - \frac{ir\sin\varphi k\partial_y w}{\sqrt{\partial_x w^2+\partial_y w^2+1}}\right]dxdy.
\end{aligned} \tag{7.25}$$

Eq. (7.25) is analogous to Eq. (7.15), but it is more convenient when working with input polarization vectors of the form $\mathbf{E}_{\text{inc}} = [e_x, e_y, 0]^T$. This includes, for example, uniform linear polarizations or spatially varying polarizations specified point by point as $\mathbf{E}_{\text{inc}}(x, y) = [e_x(x, y), e_y(x, y), 0]^T$. In contrast, Eq. (7.17) is particularly well-suited for describing radial and azimuthal polarizations.

7.4 Diffraction pattern of a plane wave

Eq. (7.15) constitutes a remarkably robust expression, as it is capable of computing the diffraction pattern generated by freeform wavefronts under a wide variety of conditions, including different polarization states and arbitrary pupil shapes. In this sense, it can be legitimately regarded as the general diffraction integral for freeform wavefronts, encompassing polarization diversity and aperture flexibility. Nevertheless, it is essential to emphasize a key mathematical requirement for its validity: Eq. (7.15) holds if and only if the Jacobian of the phase function $w(x, y)$ does not vanish. As an illustrative counterexample, consider the case of a single plane wave. For such a wave, the Jacobian identically vanishes, which invalidates the use of Eq. (7.15). This observation is consistent with physical intuition, since a single plane wave cannot form a focus and thus lies outside the applicability of the formalism.

7.5 Reduction to Richards-Wolf integrals

Before exploring the diffraction patterns generated by Eq. (7.15), it is important to recall its origin from Eq. (7.2). In fact, Eq. (7.2) can be regarded as the root expression of the Richards–Wolf integrals. The well-known Richards–Wolf formulation arises as the special case of Eq. (7.2) when the input wavefront is spherical. Similarly, Eq. (7.15) represents the generalization of Eq. (7.2) for an arbitrary freeform wavefront $w(x, y)$. Consequently, when $w(x, y)$ corresponds to a spherical wavefront, Eq. (7.15) must naturally reduce to Eq. (7.2). In the following discussion, we will explore this scenario to verify the consistency of the formulation and highlight the connection between the classical Richards–Wolf integrals and their freeform generalization.

So let's assume that the wavefront $w(x, y)$,

$$w(x, y) = R - \sqrt{R^2 - x^2 - y^2} \tag{7.26}$$

where R is the radius of the spherical wavefront. We will see later that the value of this radius does not affect at all in our computations, The partial derivatives are given by,

$$\partial_x w = \frac{x}{\sqrt{R^2 - x^2 - y^2}}, \quad \partial_y w = \frac{y}{\sqrt{R^2 - x^2 - y^2}} \tag{7.27}$$

Since we are going to reduce Eq. (7.15) to the the classical Richards–Wolf integrals, Eq. (7.2), let set our independent variables x and y as,

$$x = R\sin\theta\cos\phi, \quad y = R\sin\theta\sin\phi \tag{7.28}$$

where θ and ϕ are the polar and azimuthal angles of spherical coordinates. They are the same angles proposed by B. Richards and E. Wolf. Then replacing Eq. (7.28) in Eq. (7.27), we get the partial derivatives,

$$\partial_x w = \tan\theta\cos\phi, \quad \partial_y w = \tan\theta\sin\phi \tag{7.29}$$

With the partial derivatives, last expression, and Eq. (7.9), we can express the polarization vector in terms of θ and ϕ, so from $\hat{\mathbf{a}}(x, y)$ to $\hat{\mathbf{a}}(\theta, \phi)$. Replacing Eq. (7.29) in Eq. (7.9),

$$\hat{\mathbf{a}}(\theta,\phi) = \left[\frac{\frac{\cos\alpha\tan\theta\cos\phi}{\sqrt{\tan^2\theta\sin^2\phi+\tan^2\theta\cos^2\phi+1}} - \sin\alpha\tan\theta\sin\phi}{\sqrt{\tan^2\theta\sin^2\phi+\tan^2\theta\cos^2\phi}}\hat{\mathbf{x}} \right.$$
$$\frac{\frac{\cos\alpha\tan\theta\sin\phi}{\sqrt{\tan^2\theta\sin^2\phi+\tan^2\theta\cos^2\phi+1}} + \sin\alpha\tan\theta\cos\phi}{\sqrt{\tan^2\theta\sin^2\phi+\tan^2\theta\cos^2\phi}}\hat{\mathbf{y}}$$
$$\left. \cos\alpha\sqrt{1 - \frac{1}{\tan^2\theta\sin^2\phi+\tan^2\theta\cos^2\phi+1}}\hat{\mathbf{z}} \right] \tag{7.30}$$

Simplifying the last expression, we can reduce $\hat{\mathbf{a}}(\theta, \phi)$ to,

$$\hat{\mathbf{a}}(\theta,\phi) = [\cos\alpha\cos\theta\cos\phi - \sin\alpha\sin\phi]\hat{\mathbf{x}}$$
$$+ [\cos\alpha\cos\theta\sin\phi + \sin\alpha\cos\phi]\hat{\mathbf{y}} + \cos\alpha\sin\theta\hat{\mathbf{z}} \tag{7.31}$$

If our generalization is correct, then for some values of α the polarization vector should be reduced to the same polarization vectors of the classical Richards–Wolf integrals. We are going to find value of α for radial, azimuthal and linear polarizations.

7.5.1 Radial polarization

The radial polarization vector for a spherical wavefront can be achieved from Eq. (7.31) by setting α to be,

$$\alpha = 0. \tag{7.32}$$

By replacing $\alpha = 0$, in Eq. (7.31),

$$\hat{\mathbf{a}}(\theta, \phi) = \cos\theta\cos\phi\hat{\mathbf{x}} + \cos\theta\sin\phi\hat{\mathbf{y}} + \sin\theta\hat{\mathbf{z}} \tag{7.33}$$

The last expression its the same expression studied previously used in Chapter 4. The general diffraction integral, Eq. (7.15), smoothly converges to the Richards–Wolf integrals.

7.5.2 Azimuthal polarization

The next study case is the azimuthal polarization, for it we set α to be

$$\alpha = \frac{\pi}{2}. \tag{7.34}$$

By setting $\alpha = \pi/2$ in Eq. (7.31), we have,

$$\hat{\mathbf{a}}(\theta, \phi) = -\sin\phi\hat{\mathbf{x}}, + \cos\phi\hat{\mathbf{y}} \tag{7.35}$$

The last expression is the same as the one used for Richards–Wolf integrals, for an azimuthal polarization. Richards–Wolf integrals are a special case of the main integral of this book, Eq. (7.15).

7.5.3 Linear polarization

The finally example of this section is the linear polarization, for that α should negatively change along ϕ. Thus, in a more mathematical way α should be,

$$\alpha = -\phi. \tag{7.36}$$

Let us set $\alpha = -\phi$ in Eq. (7.31), $\hat{\mathbf{a}}(\theta, \phi)$ turns to be,

$$\hat{\mathbf{a}}(\theta, \phi) = [\cos\theta\cos^2\phi + \sin^2\phi]\hat{\mathbf{x}} + [(\cos\theta - 1)\sin\phi\cos\phi]\hat{\mathbf{y}} + \sin\theta\cos\phi\hat{\mathbf{z}} \tag{7.37}$$

As expected $\hat{\mathbf{a}}(\theta, \phi)$ turns to be the same as Eq. (4.82) from Chapter 4. Observe that we have part from Eq. (7.2), which is the same root of the classical Richards-Wolf integrals. We have not added any new physical concept, and we have just applied well-known proved theorems from analytical geometry and vector calculus. Therefore, it should be no surprise that the general integral, Eq. (7.15), can reduce the classical Richards-Wolf integrals for a spherical wavefront.

Observe that all these values of α are strictly valid only in the case of a spherical wavefront. If the wavefront instead corresponds to a more general freeform shape, no guarantee setting $\alpha = -\phi$ results in a linearly polarized field. Similarly, assigning $\alpha = 0$ or $\alpha = \pi/2$ does not necessarily imply radial or azimuthal polarization, respectively, when the wavefront deviates from spherical symmetry. These interpretations of α hold true only under the assumption of spherical wavefront geometry, and should not be generalized to arbitrary wavefronts without additional justification.

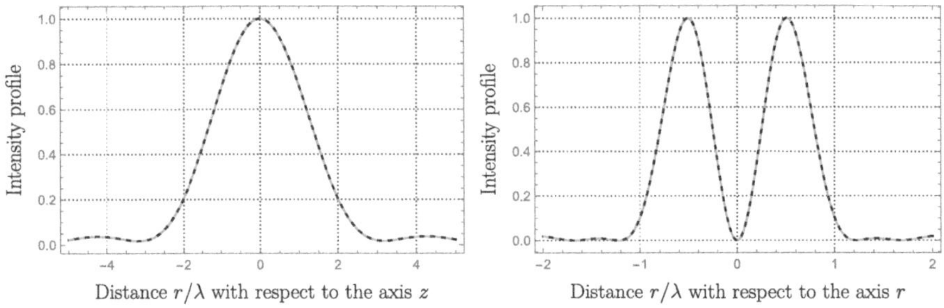

FIGURE 7.6
Normalized intensity of the diffraction produced by a spherical wavefront, in dashed black and gray is PSF computed with Eq. (7.15) and Ref. [2], respectively. *Left* is intensity profile with respect to the z axis. *Right* is the intensity profile with respect to the r axis.

7.5.4 The diffraction pattern of a spherical wavefront

In order to validate Eq. (7.15), we present in Fig. 7.6 two examples of different radially symmetric wavefronts $w(x, y)$. Specifically, the dashed black curve and the gray curve represent the normalized intensity of the diffraction patterns of spherical wavefronts, Eq. (4.62), as predicted by Ref. [2] and Eq. (7.15), respectively. The profiles are shown in Fig. 7.6, which correspond to the intensity distributions with respect to the axial distance z and the radial distance r, respectively. The caption of the figure indicates the input values of the example being considered. For this example we consider $q(x, y) = 1$ and $l_0(x, y) = 1$.

It is evident from these plots that the dashed black and dashed gray curves closely overlap, indicating that the prediction made by Eq. (7.15) matches the results of Ref. [2] for spherical wavefronts. This agreement validates the applicability of Eq. (7.15) to describe the diffraction pattern of spherical waves. All wavefronts analyzed in this example are radially polarized, meaning that the polarization is aligned along the radial direction from the optical axis. Moreover, it is important to note that Eq. (7.15) is versatile and can be used to derive other classical results. For instance, when the field is linearly polarized, Eq. (7.23) can be employed to predict the corresponding diffraction patterns, further demonstrating the utility and flexibility of the framework presented by Eq. (7.15).

7.6 Rectangular aperture

Fig. 6 presents an example inspired by the work in [110]. In this figure, the diffraction pattern produced by an off-axis parabolic mirror with a rectangular aperture is illustrated. Since the mirror is off-axis and the aperture is not circular, the resulting wavefront is spherical but truncated by a rectangular boundary. For this simulation, we consider a radially polarized incident beam, which introduces unique features in the resulting intensity distribution. So let be the wavefront for this example $w(x, y)$ be spherical,

$$w(x, y) = R - \sqrt{R^2 - x^2 - y^2} \tag{7.38}$$

where R is the radius of the spherical wavefront. The partial derivatives are,

$$\partial_x w = \frac{x}{\sqrt{R^2 - x^2 - y^2}}, \quad \partial_y w = \frac{y}{\sqrt{R^2 - x^2 - y^2}} \tag{7.39}$$

The propagation vector $\mathbf{k}(x, y)$ is,

$$\mathbf{k}(x, y) = -\frac{xk}{R}\hat{\mathbf{x}} - \frac{yk}{R}\hat{\mathbf{y}} + \frac{k}{R\sqrt{\frac{1}{R^2 - x^2 - y^2}}}\hat{\mathbf{z}}. \tag{7.40}$$

The polarization is radial; then the angle α turns to $\alpha = 0$ along the wavefront $w(x, y)$. Considering this and the spherical shape of the wavefront, the polarization vector is $\hat{\mathbf{a}}(x, y)$,

$$\hat{\mathbf{a}}(x, y) = \frac{x\sqrt{R^2 - x^2 - y^2}}{R\sqrt{x^2 + y^2}}\hat{\mathbf{x}} + \frac{y\sqrt{R^2 - x^2 - y^2}}{R\sqrt{x^2 + y^2}}\hat{\mathbf{y}} + \frac{\sqrt{x^2 + y^2}}{R}\hat{\mathbf{z}} \tag{7.41}$$

From Eq. (7.14), we know that the Jacobian divided by the k_z is given by

$$\mathbf{Jcb}(x, y)/k_z = \frac{\partial_x^2 w \partial_y^2 w - (\partial_{xy} w)^2}{(\partial_x w^2 + \partial_y w^2 + 1)^{3/2}}. \tag{7.42}$$

Considering that the wavefront is spherical, the $\mathbf{Jcb}(x, y)$ turns to,

$$\frac{\partial_x^2 w \partial_y^2 w - (\partial_{xy} w)^2}{(\partial_x w^2 + \partial_y w^2 + 1)^{3/2}} = \frac{1}{R\sqrt{R^2 - x^2 - y^2}} \tag{7.43}$$

Replacing Eqs. (7.38), (7.39), (7.40), (7.41), (7.43) in Eq.(7.15), we have that $\mathbf{E}(r, \varphi, z)$ is given by,

$$\begin{aligned}\mathbf{E}(r, \varphi, z) = {} & \frac{E_0 k}{2\pi} \int\int_\Omega P(x, y) q(x, y) l_0(x, y) \left[\frac{1}{R\sqrt{R^2 - x^2 - y^2}}\right] \\ & \times \left[\frac{x\sqrt{R^2 - x^2 - y^2}}{R\sqrt{x^2 + y^2}}\hat{\mathbf{x}} + \frac{y\sqrt{R^2 - x^2 - y^2}}{R\sqrt{x^2 + y^2}}\hat{\mathbf{y}} + \frac{\sqrt{x^2 + y^2}}{R}\hat{\mathbf{z}}\right] \\ & \times \exp\left[\frac{-izk\sqrt{R^2 - x^2 - y^2}}{R} - \frac{ir\cos\varphi kx}{R} - \frac{ir\sin\varphi ky}{R}\right] dxdy.\end{aligned} \tag{7.44}$$

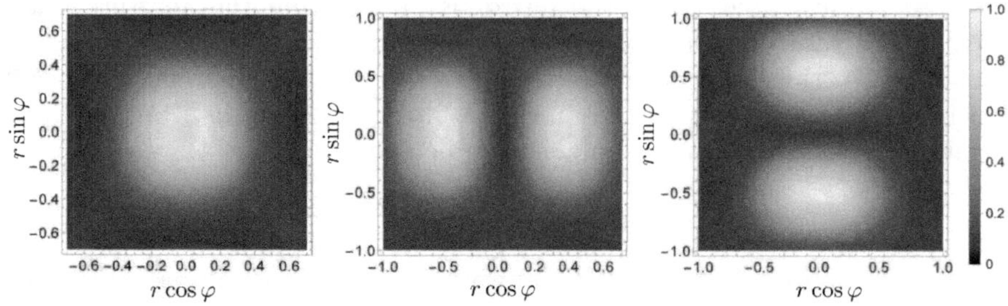

FIGURE 7.7
Normalized intensity of the diffraction pattern produced by spherical wavefront radial polarization $\alpha = 0$. At *right*, *center*, and *left* are the intensity profiles with respect to the x, y, and z axes, respectively, in the axis of the image plane. The mask for all the examples in this figure is $P(x, y) = 1$, and $\alpha(x, y) = 0$, and $\alpha(x, y) = x^2 + y^2$. The domain of the wavefront is $x \in \{-0.7, 0.7\}$ and $y \in \{-0.7, 0.7\}$.

We use the last expression directly to compute the diffraction patterns presented in Fig. 7.7. Due to the rectangular shape of the aperture, the diffracted pattern does not exhibit cylindrical or spherical symmetry. This asymmetry is evident in the intensity distribution along the x, y, and z directions in the image plane. Specifically, the intensity profile along the optical axis (z) reveals modulation patterns that differ from those expected from a circular aperture. Similarly, the lateral intensity profiles along the x and y directions reflect the rectangular geometry of the aperture.

The non-spherical form of the pattern and the distinct side-lobe structure are consistent with the behavior reported in [110], where the influence of aperture shape on the focal field distribution was analyzed. These results highlight the strong dependence of the diffraction characteristics on the aperture geometry, particularly under radial polarization, which enhances the field confinement and sharpens the features of the intensity distribution.

7.7 Example of polarization function distribution

In Fig. 7.8, we present the normalized intensity of the diffraction pattern resulting from a paraboloid wavefront with a spatially varying polarization defined by $\alpha(x, y) = \sqrt{x^2 + y^2}$. The aperture in these examples is considered to be circular. The paraboloid wavefront is given by the following expression

$$w(x, y) = ax^2 + by^2, \tag{7.45}$$

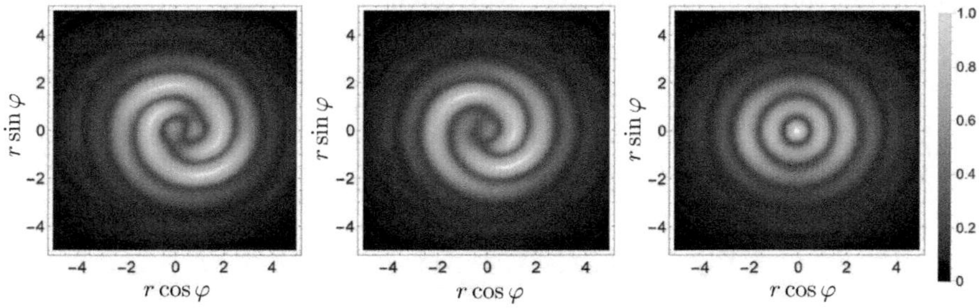

FIGURE 7.8
Normalized intensity of the diffraction produced by $w(x,y) = cx^2 + (c/2)y^2$, with $c = 0.05$. At *right*, *center*, and *left* are the intensity profiles with respect to the x, y, and z axes, respectively, in the image plane. The mask for all the examples in this figure is $P(x,y) = \{x^2+y^2 < 8 \implies 1, \text{otherwise} \implies 0\}$ and $\alpha(x,y) = 0$ and $\alpha(x,y) = x^2+y^2$. The domain of the wavefront is $x \in \{0,5\}$ and $y \in \{0,5\}$.

where a and b are constants. The partial derivatives are,

$$\partial_x w = 2ax, \quad \partial_y w = 2by. \tag{7.46}$$

In order to compute the diffraction pattern presented in Fig. 4 over a circular aperture, we need to made coordinate transformation,

$$x = \rho \cos \varsigma, y = \rho \sin \varsigma. \tag{7.47}$$

This leads to expressing the partial derivatives as,

$$\partial_x w = 2a\rho \cos \varsigma, \quad \partial_y w = 2b\rho \sin \varsigma. \tag{7.48}$$

The angle $\alpha(x,y)$ under this coordinate transformation can be expressed as,

$$\alpha = \sqrt{(\rho \cos \varsigma)^2 + (\rho \sin \varsigma)^2} = \rho. \tag{7.49}$$

Replacing Eq. (7.48) in Eq. (7.9),

$$\begin{aligned}\mathbf{k}(\rho,\varsigma) = &-\frac{2a\rho k \cos \varsigma}{\sqrt{4\rho^2\left(a^2\cos^2\varsigma + b^2\sin^2\varsigma\right)+1}}\hat{\mathbf{x}} \\ &-\frac{2b\rho k \sin \varsigma}{\sqrt{4\rho^2\left(a^2\cos^2\varsigma + b^2\sin^2\varsigma\right)+1}}\hat{\mathbf{y}} \\ &+\frac{k}{\sqrt{4\rho^2\left(a^2\cos^2\varsigma + b^2\sin^2\varsigma\right)+1}}\hat{\mathbf{z}}.\end{aligned} \tag{7.50}$$

Considering that $\alpha = \rho$, in the last expression,

$$\hat{\mathbf{a}}(\rho,\varsigma) = \frac{\frac{2a\rho\cos(\rho)\cos\varsigma}{\sqrt{4\rho^2(a^2\cos^2\varsigma+b^2\sin^2\varsigma)+1}} - 2b\rho\sin(\rho)\sin\varsigma}{2\sqrt{\rho^2\left(a^2\cos^2\varsigma+b^2\sin^2\varsigma\right)}}\hat{\mathbf{x}} + \frac{\rho\left(\frac{b\cos(\rho)\sin\varsigma}{\sqrt{4\rho^2(a^2\cos^2\varsigma+b^2\sin^2\varsigma)+1}} + a\sin(\rho)\cos\varsigma\right)}{\sqrt{\rho^2\left(a^2\cos^2\varsigma+b^2\sin^2\varsigma\right)}}\hat{\mathbf{y}} + \cos(\rho)\sqrt{\frac{1}{-4\rho^2\left(a^2\cos^2\varsigma+b^2\sin^2\varsigma\right)-1}+1}\,\hat{\mathbf{z}} \quad (7.51)$$

Considering that the wavefront is a paraboloid, we have the Jacobian divided by the k_z turns to,

$$\frac{\partial_x^2 w \partial_y^2 w - (\partial_{xy} w)^2}{(\partial_x w^2 + \partial_y w^2 + 1)^{3/2}} = \frac{4ab}{(4a^2(\rho\cos\varsigma)^2 + 4b^2(\rho\sin\varsigma)^2 + 1)^{3/2}} \quad (7.52)$$

Substituting Eqs. (7.50), (7.51), (7.52) in $\mathbf{E}(r,\varphi,z)$, Eq. (7.15),

$$\mathbf{E}(r,\varphi,z) = \frac{E_0 k}{2\pi}\int_0^{2\pi}\int_0^{\rho_{\max}} \frac{P(\rho,\varsigma)q(\rho,\varsigma)l_0(\rho,\varsigma)4ab}{(4a^2(\rho\cos\varsigma)^2 + 4b^2(\rho\sin\varsigma)^2 + 1)^{3/2}} \times\left[-\frac{2a\rho\cos\varsigma}{\sqrt{4\rho^2\left(a^2\cos^2\varsigma+b^2\sin^2\varsigma\right)+1}}\hat{\mathbf{x}} - \frac{2b\rho\sin\varsigma}{\sqrt{4\rho^2\left(a^2\cos^2\varsigma+b^2\sin^2\varsigma\right)+1}}\hat{\mathbf{y}} + \frac{1}{\sqrt{4\rho^2\left(a^2\cos^2\varsigma+b^2\sin^2\varsigma\right)+1}}\hat{\mathbf{z}}\right]\exp\left[\frac{-izk}{\sqrt{4\rho^2\left(a^2\cos^2\varsigma+b^2\sin^2\varsigma\right)+1}} - \frac{2a\rho\cos\varsigma ir\cos\varphi k}{\sqrt{4\rho^2\left(a^2\cos^2\varsigma+b^2\sin^2\varsigma\right)+1}} - \frac{2b\rho\sin\varsigma ir\sin\varphi k}{\sqrt{4\rho^2\left(a^2\cos^2\varsigma+b^2\sin^2\varsigma\right)+1}}\right]\mathrm{d}\rho\mathrm{d}\varsigma. \quad (7.53)$$

This polarization function, $\alpha = \rho$, introduces a non-uniform distribution of the electric field orientation across the wavefront, leading to complex diffraction behavior. The resulting diffraction patterns are analyzed along the $r\cos\varphi$ and $r\sin\varphi$ axes, with the intensity components along the x-, y-, and z-directions shown in the *right*, *center*, and *left* panels, respectively. These intensity components are obtained by decomposing the diffracted electric field into its Cartesian vector components, allowing a detailed visualization of the energy distribution in space.

Due to the aberrations introduced by the freeform nature of the wavefront, the diffraction patterns along the x and y directions exhibit noticeable asymmetries, as expected. These asymmetries are particularly evident in the differing side-lobe structures and intensity distributions across the lateral directions. Additionally, at the center of the diffraction field, the electric field vanishes for both the x and y components. This null at the center arises from the vectorial interference caused by the specific polarization distribution, which effectively cancels the transverse field components at the origin.

The z-component of the electric field, on the other hand, shows a distinct distribution, highlighting the impact of polarization-induced longitudinal fields, which become significant in tightly focused or strongly aberrated systems. Overall, the results illustrate how the combination of wavefront geometry and spatially varying polarization strongly influences the structure and symmetry of the diffracted field.

7.8 Example of an aberrated linearly polarized wavefront

Next is the diffraction pattern of a linearly polarized wavefront using Eq. (7.23). In other words $\mathbf{E}_{\text{inc}} = [e_x, e_y, 0]^T = [1, 0, 0]^T$. The linearly polarized wavefront can be found in Fig. 7.9. At *right*, *center*, and *left* are the intensity profiles respect to the x, y, and z axes, respectively. In this example, the

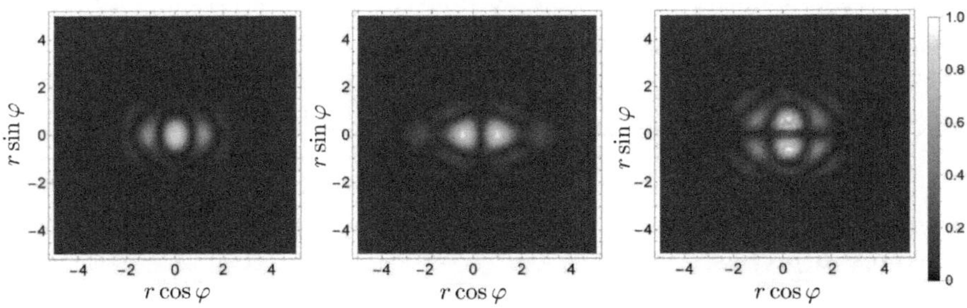

FIGURE 7.9
Normalized intensity of the diffraction pattern produced by $w(x, y) = cx^2 + (c/2)y^2$, with $c = 0.05$. *Top left* is the shape of the wavefront. At *right*, *center*, and *left* are the intensity profiles respect to the x, y, and z axes, respectively, in the axis of the image plane. The mask for all the examples in this figure is $P(x, y) = \{x^2 + y^2 < 8 \implies 1, \text{otherwise} \implies 0\}$, and the wavefront is linearly polarized.

diffraction pattern is computed with Eq. (7.23) rather than Eq. (7.15), since it is straightforward to express a linearly polarized wavefront with Eq. (7.23).

Recalling paraboloid wavefront,

$$w(x, y) = ax^2 + by^2, \tag{7.54}$$

and its partial derivatives,

$$\partial_x w = 2ax, \quad \partial_y w = 2by. \tag{7.55}$$

Recalling the coordinate transformation,

$$x = \rho \cos \varsigma, y = \rho \sin \varsigma. \tag{7.56}$$

Calling the partial derivatives under the coordinates transformation,

$$\partial_x w = 2a\rho \cos \varsigma, \quad \partial_y w = 2b\rho \sin \varsigma. \tag{7.57}$$

Notice that the propagation vector is also the same as in Eq. (7.9), we have the same wavefront but different polarization angle α. So for this example $\mathbf{k}(\rho, \varsigma)$ is also given by,

$$\begin{aligned}\mathbf{k}(\rho, \varsigma) = &-\frac{2a\rho k \cos \varsigma}{\sqrt{4\rho^2 \left(a^2 \cos^2 \varsigma + b^2 \sin^2 \varsigma\right) + 1}}\hat{\mathbf{x}} \\ &-\frac{2b\rho k \sin \varsigma}{\sqrt{4\rho^2 \left(a^2 \cos^2 \varsigma + b^2 \sin^2 \varsigma\right) + 1}}\hat{\mathbf{y}} \\ &+\frac{k}{\sqrt{4\rho^2 \left(a^2 \cos^2 \varsigma + b^2 \sin^2 \varsigma\right) + 1}}\hat{\mathbf{z}}\end{aligned} \tag{7.58}$$

The angle $\alpha(\rho, \varsigma)$ this coordinate transformation can be expressed as,

$$\alpha(\rho, \varsigma) = \tan^{-1}\left(\frac{2ae_y\rho \cos \varsigma - 2be_x\rho \sin \varsigma}{\sqrt{4a^2\rho^2 \cos^2 \varsigma + 4b^2\rho^2 \sin^2 \varsigma + 1}(2ae_x\rho \cos \varsigma + 2be_y\rho \sin \varsigma)}\right) \tag{7.59}$$

By replacing Eq. (7.59) in Eq. (7.21), and after a very long simplification, we can express $\hat{\mathbf{a}}(\rho, \varsigma)$ as,

$$\begin{aligned}\hat{\mathbf{a}}(\rho, \varsigma) = &\\ &\frac{\sec \varsigma \sqrt{\rho^2 \left(a^2 \cos^2 \varsigma + b^2 \sin^2 \varsigma\right)}\hat{\mathbf{x}}}{a\rho\sqrt{4\rho^2 \left(a^2 \cos^2 \varsigma + b^2 \sin^2 \varsigma\right) + 1}\sqrt{\frac{b^2 \tan^2 \varsigma}{a^2(4\rho^2(a^2 \cos^2 \varsigma + b^2 \sin^2 \varsigma)+1)} + 1}} \\ &+\frac{\sqrt{\frac{1}{-4\rho^2(a^2 \cos^2 \varsigma + b^2 \sin^2 \varsigma)-1} + 1}}{\sqrt{\frac{b^2 \tan^2 \varsigma}{a^2(4\rho^2(a^2 \cos^2 \varsigma + b^2 \sin^2 \varsigma)+1)} + 1}}\hat{\mathbf{z}}.\end{aligned} \tag{7.60}$$

Therefore the $\mathbf{E}(r, \varphi, z)$ can be obtained by replacing Eq. (7.55), (7.58), (7.60), in Eq. (7.23),

$$\mathbf{E}(r,\varphi,z) = \frac{E_0 k}{2\pi}\int_0^{2\pi}\int_0^{\rho_{\max}} \frac{P(\rho,\varsigma)q(\rho,\varsigma)l_0(\rho,\varsigma)4ab}{\left(4a^2(\rho\cos\varsigma)^2+4b^2(\rho\sin\varsigma)^2+1\right)^{3/2}}$$
$$\times\left[\frac{\sec\varsigma\sqrt{\rho^2\left(a^2\cos^2\varsigma+b^2\sin^2\varsigma\right)}\hat{\mathbf{x}}}{a\rho\sqrt{4\rho^2\left(a^2\cos^2\varsigma+b^2\sin^2\varsigma\right)+1}\sqrt{\frac{b^2\tan^2\varsigma}{a^2(4\rho^2(a^2\cos^2\varsigma+b^2\sin^2\varsigma)+1)}+1}}\right.$$
$$\left.+\frac{\sqrt{\frac{1}{-4\rho^2(a^2\cos^2\varsigma+b^2\sin^2\varsigma)-1}+1}}{\sqrt{\frac{b^2\tan^2\varsigma}{a^2(4\rho^2(a^2\cos^2\varsigma+b^2\sin^2\varsigma)+1)}+1}}\hat{\mathbf{z}}\right]\exp\left[\frac{-izk}{\sqrt{4\rho^2\left(a^2\cos^2\varsigma+b^2\sin^2\varsigma\right)+1}}\right.$$
$$\left.-\frac{2a\rho\cos\varsigma ir\cos\varphi k}{\sqrt{4\rho^2\left(a^2\cos^2\varsigma+b^2\sin^2\varsigma\right)+1}}-\frac{2b\rho\sin\varsigma ir\sin\varphi k}{\sqrt{4\rho^2\left(a^2\cos^2\varsigma+b^2\sin^2\varsigma\right)+1}}\right]\mathrm{d}\rho\mathrm{d}\varsigma. \tag{7.61}$$

7.9 Other generalizations of Richards-Wolf integrals

One of the most appealing aspects of Eq. (7.15) is that it can be interpreted as a conceptual bridge between two fundamental paradigms of optics: geometrical optics and diffraction theory. Indeed, the integral is derived from the angular spectrum representation of optical fields, while the propagation vector $\mathbf{k}$ naturally aligns with the geometrical ray direction. Furthermore, the formalism guarantees that the vectors $\mathbf{k}$ and $\hat{\mathbf{a}}$ remain orthogonal for all admissible values of $\alpha(x, y)$, a property that is crucial for physical consistency but frequently neglected in previously proposed diffraction integrals [96, 47, 109, 111, 108, 112].

The most popular generalization of Richards-Wolf integrals are made by researcher Rishi Kant. His works on this topic is presented in the following three papers,

- Rishi Kant, *An analytical solution of vector diffraction for focusing optical systems*, Journal of Modern Optics, 40:11, 2293-2310 [108].
- Rishi Kant, *An Analytical Solution of Vector Diffraction for Focusing Optical Systems with Seidel Aberrations*, Journal of Modern Optics, 40:2, 337-347 [112].
- Rishi Kant, *An Analytical Method of Vector Diffraction for Focusing Optical Systems with Seidel Aberrations II: Astigmatism and Coma*, Journal of Modern Optics, 42:2, 299-320 [111].

Now, we will remark on the differences between [108, 112] and the integrals presented here. The derivation presented in [112] is too cumbersome compared to the derivation of Eq. (7.15). Here, the approach is straightforward, by doing a change of coordinates systems and variables, whereas [112] uses 52 equations. Additionally, the method here does not require additional so-called *optical coordinates* (as in equation 5 of [112]), making it simpler. Here, the method here is designed for freeform wavefronts, there is no need of concepts like surfaces of revolution, as described in equations 6 and 7 in [112].

One of the key aspects that simplify in the method here is Cartesian coordinates, while the method in [112] relies heavily on cylindrical (which are not optimal for describing freeform surfaces) and spherical coordinates. Ref. [112] strongly uses the concept of surfaces of revolution, which explains why the author ends with single integrals (equations 21, 22, and 23), even when dealing with wavefronts that have non-symmetrical aberrations. Here, the expressions can be reduced to a single integral *only* if the wavefront is a surface of revolution and the polarization angle α is constant. And we are going to study this in the next chapter.

Ref. [112] is limited to certain Seidel coefficients (see section 5), while the method here works directly with an explicit wavefront function $w(x, y)$. The computation method in [112] also involves several approximations that not considered in this chapter, such as those in equation 27, the expression after equation 28 (not numbered in the paper), the expression after equation 35 (also not numbered), and equations 51 and 52. In section 7 of [112], the method involves solving a single integral analytically, a result of using the surface of revolution concept—a concept that does not apply to freeform wavefronts. In contrast, our double integrals are solved numerically.

Another key difference is that Ref. [112] does not mention, and it is unclear, whether the propagation vector is assumed to be normal to the wavefront. This is a crucial step in the method here, where we explicitly assume that the propagation vector is normal to the wavefront.

Now, Lets address the differences between [111] and the method here. The derivation presented in [111] is based on [112], and like [112], it is too cumbersome in comparison with the method here. The method in [111] also relies mainly on cylindrical and spherical coordinates (see equation 25 in [111]), which are less suitable for freeform surfaces.

Ref. [111] assumes that the aberrations are small, as mentioned explicitly before equation 20. However, the term *small* is vague and can be ambiguous. Here the approach works directly with the wavefront shape $w(x, y)$, allowing for more flexibility. Ref. [111] also makes several approximations that we do not, such as those in equations 20, 21, 22, and 43. These approximations are not part of the framework presented in this chapter.

Similar to Ref. [112], Ref. [111] is limited to working with Seidel coefficients (see section 5), while the method here works directly with an explicit wavefront function $w(x, y)$. In section 5 of [111], the approach is based on solving an analytical single integral, whereas here the double integrals are presented due the lack of symmetry.

For the reasons and approximations implemented in [108], [112], and [111], the method here is significantly different from the paradigms presented in these works. It is more mathematically elegant and straightforward. It does not require any additional physical concept or approximation. It is merely based on Eq. (7.2). Eq. (7.2) is transformed to the Cartesian system in order to easterly address freeform wavefronts $w(x, y)$.

Due to limitations of space, we do not include here an example involving apertures with highly irregular geometries. However, from Eq. (7.23) it becomes immediately evident that the shape of the aperture can be fully encoded by an appropriate definition of the pupil function $P(x, y)$. In this case, the integration limits are set by the aperture boundaries, namely x_{min}, x_{max}, y_{min}, and y_{max}, which makes Eq. (7.23) particularly well-suited for describing systems with arbitrary physical apertures.

It is also worth contrasting this method with the numerical strategy presented in [113], which relies heavily on numerical approximations such as the homeomorphic Fourier transform. In sharp contrast, both Eq. (7.15) and Eq. (7.23) are obtained through a fully analytical derivation, without resorting to numerical approximations or iterative schemes. This closed-form nature constitutes a significant theoretical advantage, as it provides deeper physical insight and avoids convergence or discretization issues inherent in purely numerical approaches.

Although the examples presented in this chapter have focused primarily on quadratic wavefronts, there is no intrinsic limitation in Eqs. (7.15) and (7.23) regarding the functional form of the wavefront, provided it is a smooth and continuous real-valued function. Admittedly, the computational time increases with the complexity of the wavefront.

In the illustrative examples, in this chapter, set $q(x, y) = 1$ and $l_0(x, y) = 1$, as our primary interest lies in analyzing the intrinsic behavior of the integral rather than specific illumination patterns or apodization functions. Nevertheless, it should be stressed that both illumination and apodization can significantly influence the diffraction pattern. For example, two identical wavefronts endowed with different illumination patterns may yield drastically different diffraction outcomes. The same conclusion applies to apodization effects, which can strongly modulate the final intensity distribution.

Another key feature of our approach is that the sampling of $w(x, y)$ is not constrained by the numerical aperture of the optical system. Instead, the integrals directly operate on the polarization distribution $\hat{\mathbf{a}}(x, y)$ and the propagation vector $\mathbf{k}$. The numerical aperture can be readily computed from the marginal ray, i.e., from the propagation vector $\mathbf{k}$ at the edge of the aperture. Importantly, the structure of Eqs. (7.15) and (7.23) remains unchanged regardless of the marginal ray position. However, the accuracy of the results is directly proportional to the sampling density of $w(x, y)$. In other words, the more oscillatory or complex the wavefront (with multiple local maxima and minima), the finer the sampling required to achieve reliable results.

It is also important to clarify that Eqs. (7.15) and (7.23) are mathematically equivalent, differing only in their form of expression. Eq. (7.15) is more convenient when the polarization state is naturally described in terms of the angle $\alpha(x, y)$, as is the case for radially or azimuthally polarized beams. Conversely, Eq. (7.23) is better suited when the polarization is specified through Cartesian vector components $\mathbf{E}_{\text{inc}} = [e_x, e_y, 0]^T$, which is often the case in practical implementations.

7.10 Conclusion

In this chapter, the general diffraction integral has been introduced. It explicitly incorporates three fundamental characteristics of an optical system: the shape of the wavefront, denoted by $w(x, y)$; the polarization state, described by the parameter $\alpha(x, y)$; and the geometry of the aperture or pupil mask, represented by $P(x, y)$. The resulting formulation is mathematically expressed through Eq. (7.15) and its equivalent form Eq. (7.23).

Both integrals have been carefully tested in a variety of representative scenarios, and in every case the obtained results are consistent with theoretical expectations as well as with previously established formalisms, particularly those of Richards and Wolf [2]. For instance, the proposed equations reproduce the expected diffraction patterns when the incident wavefront is spherical and radially polarized, and similarly when the wavefront is spherical and linearly polarized. These results confirm the correctness of the formulation and its compatibility with classical diffraction theory.

Although Eq. (7.15) and Eq. (7.23) are mathematically equivalent, their practical usefulness differs depending on the manner in which the polarization state is expressed. Specifically, Eq. (7.15) is particularly convenient when dealing with polarization states that are naturally defined in terms of the angular parameter $\alpha(x, y)$, such as radially or azimuthally polarized beams. In contrast, Eq. (7.23) provides a more straightforward formulation when the input polarization is better expressed in Cartesian vector components, i.e., as $\mathbf{E}_{\text{inc}} = [e_x, e_y, 0]^T$. This dual representation provides flexibility for both theoretical analysis and numerical implementation.

It is crucial to highlight that the validity of both integrals relies on a key physical and mathematical condition: the wavefront $w(x, y)$ must be convergent and must constitute a real, smooth, and continuous function. Under these conditions, the integrals correctly describe the formation of the diffraction pattern without introducing non-physical artifacts.

Given its robustness and versatility, Eq. (7.15) can be rightfully regarded as a general diffraction integral capable of handling freeform wavefronts, arbitrary polarization states, and pupils of virtually any geometry.

Problems

Problem 1: Basic conceptual question Explain in your own words why the classical Richards–Wolf integral cannot be directly applied to strongly aberrated or freeform wavefronts. What fundamental physical assumption is violated in such cases?

Problem 2: Propagation vector Let the wavefront be given by

$$w(x, y) = 0.1\, x^2 + 0.15\, y^2 + 0.02\, xy.$$

(a) Compute the propagation vector $\mathbf{k}(x, y)$ (up to the overall factor k).

(b) At the point $(x, y) = (1, 1)$, calculate the direction cosines $\big(k_x/k, k_y/k, k_z/k\big)$.

(c) What is the angle (in degrees) between $\mathbf{k}$ and the optical axis ($\hat{\mathbf{z}}$) at this point?

Problem 3: Polarization basis vector Using the same wavefront as in Problem 2, compute the auxiliary polarization vector $\mathbf{a}(x, y)$ (before applying Rodrigues' rotation formula). Verify numerically at the point $(x, y) = (2, 0)$ that $\mathbf{a} \cdot \mathbf{k} = 0$ (orthogonality condition).

Problem 4: Jacobian Show that for a spherical wavefront,

$$w(x, y) = R - \sqrt{R^2 - x^2 - y^2},$$

the expression

$$\frac{\partial_x^2 w\, \partial_y^2 w - (\partial_{xy} w)^2}{(\partial_x w^2 + \partial_y w^2 + 1)^{3/2}},$$

simplifies to

$$\frac{1}{R\sqrt{R^2 - x^2 - y^2}}.$$

Give a brief physical interpretation of this result in the context of energy conservation and the apodization factor.

Problem 5: Radial, azimuthal and linear polarization via α
Prove the following statements for a *spherical* wavefront:

(a) Setting $\alpha(x, y) = 0$ corresponds to radial polarization.

(b) Setting $\alpha(x, y) = \pi/2$ corresponds to azimuthal polarization.

(c) Setting $\alpha(x, y) = -\varphi$ (where φ is the azimuthal angle) corresponds to uniform x-linear polarization.

Problem 6: Plane wave – why the formalism fails
Consider a strictly plane wavefront $w(x, y) = ax + by$ (tilted plane wave).

(a) Compute the Jacobian term $\dfrac{\partial_x^2 w\, \partial_y^2 w - (\partial_{xy} w)^2}{(\partial_x w^2 + \partial_y w^2 + 1)^{3/2}}$.

(b) Explain both physically and mathematically why the generalized integral (Eq. (7.15)) cannot be applied in this case.

Problem 7: Paraboloidal wavefront with uniform linear polarization
Consider the wavefront $w(x, y) = 0.08(x^2 + y^2)$ and a uniform x-polarized input field, i.e. $[e_x, e_y, 0]^T = [1, 0, 0]^T$. Using the simplified form (Eq. (7.23)), write down (without full algebraic simplification) the structure of the three Cartesian components of the integrand (ignore the overall constants and phase factor). Qualitatively discuss how the longitudinal field component E_z near focus is expected to behave compared to the case of radial polarization.

Problem 8: Energy conservation and the apodization factor $q(x, y)$
In many practical computations people set $q(x, y) = 1$ (flat apodization).

(a) What important physical effect is neglected by doing so?

(b) For which range of numerical apertures (NA) is this approximation usually acceptable?

(c) For which systems is neglecting the correct $q(x, y)$ clearly dangerous?

8

Vector diffraction of aspherical wavefronts

This chapter develops an advanced vectorial diffraction model for high-numerical-aperture systems with aspherical wavefronts, generalizing the classical Richards–Wolf integral beyond its spherical-wavefront assumption. Utilizing the Malus–Dupin theorem and angular spectrum representation, the approach directly incorporates true wavefront geometry to compute focal electric fields for linear, radial, and azimuthal polarization states. Closed-form, Bessel-function-based integrals are derived, enabling efficient PSF calculation for real microscope objectives described by Zernike polynomials. Examples reveal that objectives deemed diffraction-limited under scalar criteria exhibit noticeably different vectorial focal patterns and polarization-dependent resolution when even small aspherical aberrations are present, highlighting the necessity of this rigorous framework for accurate high-NA optical design and analysis.

8.1 Introduction

Building upon the foundations established in preceding chapters, this work addresses the evolution of diffraction theory from early scalar models to advanced vectorial frameworks capable of describing modern optical systems. Classical approaches, such as the Rayleigh–Sommerfeld and Kirchhoff theories, effectively describe paraxial propagation by treating light as a scalar quantity—most notably resulting in the characterization of the Airy disc [7, 1]. However, these models fail to capture the vectorial nature of light, which is essential for understanding the high-numerical-aperture systems used in high-resolution microscopy, optical trapping, and beam shaping [62, 66].

The Richards-Wolf integral, formulated in 1959, marked a pivotal advancement by incorporating polarization effects into the focal field calculation [2, 57]. By accounting for the vectorial components of the electromagnetic

DOI: 10.1201/9781003726494-12

field, this method enables the precise modeling of tightly focused light, showing how specific states—such as radial polarization—can reduce the focal spot size beyond the limits predicted by scalar theory [63, 65]. Despite its utility, the traditional Richards-Wolf approach relies on the idealization of an aplanatic system producing a perfectly spherical wavefront. This assumption limits its accuracy when dealing with real-world optical aberrations, aspherical surfaces, or the freeform wavefronts increasingly common in modern optical design [109].

To overcome these constraints, this chapter utilizes the novel generalization of the Richards-Wolf integral for converging freeform vectorial fields from Chapter 7 [71, 11]. Unlike previous models that treat aberrations as mere phase perturbations over a sphere [108], this approach leverages the Malus-Dupin theorem to directly account for the inherent geometry of the wavefront. This generalized framework provides a versatile tool for computing vectorial diffraction patterns across arbitrary pupil geometries and diverse polarization states.

In this chapter, we apply this advanced model to compute the PSF of microscope objectives characterized by aspherical wavefronts. By analyzing both linear, radial and azimuthal polarization states under non-ideal conditions, we explore the intricate relationship between aberrations and focal field distribution. This analysis aims to provide a more accurate understanding of microscope performance, ultimately paving the way for enhanced precision in the design of polarization-sensitive optical technologies.

8.2 Mathematical model

The model to compute the aberrated vector PSF employs the angular spectrum representation, a versatile formalism that describes optical fields in homogeneous media as a continuous superposition of plane and evanescent waves—both exact solutions of Maxwell's equations. Plane waves propagate along directions associated with geometric rays, providing an intuitive physical interpretation. Using this framework, the electric field $\mathbf{E}(r, \varphi, z)$ in cylindrical coordinates at the focal region can be conveniently expressed as follows,

$$\mathbf{E}(r, \varphi, z) = \frac{E_0}{2\pi} \int_{-\infty}^{\infty} \int_{-\infty}^{\infty} \frac{\mathbf{E}_\infty(k_x, k_y)}{k_z} \exp(-\mathrm{i}\mathbf{k} \cdot \mathbf{r}) \mathrm{d}k_x \mathrm{d}k_y, \tag{8.1}$$

E_0 denotes a constant amplitude that establishes both the far-field approximation and ensures consistency with the energy conservation theorem. The vector $\mathbf{k} \equiv [k_x, k_y, k_z]$ represents the wave vector, while $\mathbf{r} \equiv [r\cos\varphi,\ r\sin\varphi,\ z]$ defines the spatial position of the focal region in cylindrical coordinates. $\mathbf{E}_\infty(k_x, k_y)$ is the electric field at the exit pupil,

$$\mathbf{E}_\infty(k_x, k_y) \equiv q(k_x, k_y) l_0(k_x, k_y) \hat{\mathbf{a}}(k_x, k_y), \tag{8.2}$$

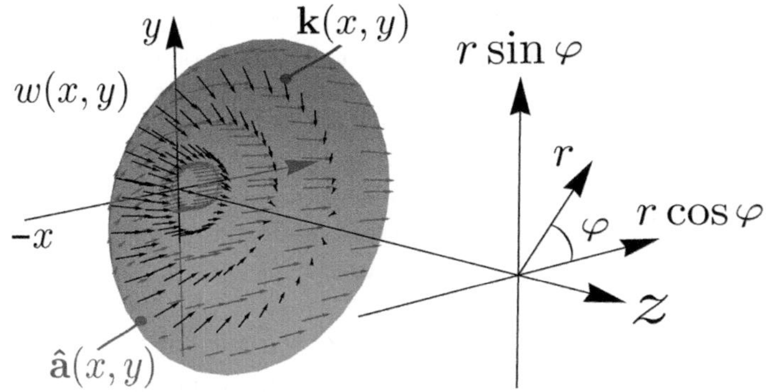

FIGURE 8.1
Diagram of the diffraction phenomena. The wavefront $w(x, y)$ is placed in the coordinate system (x, y). The vector $\mathbf{k}$ is normal to $w(x, y)$ and $\hat{\mathbf{a}}$ is normal to $\mathbf{k}$. The diffraction pattern is in the coordinate system $(r\cos\varphi, r\sin\varphi)$, and z is the axis of propagation.

where $q(k_x, k_y)$ represents the apodization factor derived from the principle of energy conservation, $l_0(k_x, k_y)$ denotes the complex beam amplitude distribution across the angular spectrum, and $\hat{\mathbf{a}}(k_x, k_y)$ is a unit vector describing the local polarization state in the vicinity of the focal region. In Fig. 8.1, the wavefront $w(x, y)$ is depicted as a gray surface with its corresponding normal vectors $\mathbf{k}(x, y)$ (black), while the polarization vectors $\hat{\mathbf{a}}(x, y)$ are highlighted in red for clarity. The image plane, expressed in cylindrical coordinates, is shown on the right-hand side. Eq. (8.1) can be further generalized for an arbitrary continuous wavefront $w(x, y)$ by performing a coordinate transformation from (k_x, k_y) to the Cartesian pair (x, y) [11]. This transformation, studied in detail in Chapter 7, ultimately leads to,

$$\begin{aligned}\mathbf{E}(r,\varphi,z) &= \frac{E_0 k}{2\pi}\int\int_{\Omega} P(x,y)q(x,y)l_0(x,y)\left[\frac{\partial_x^2 w\partial_y^2 w-(\partial_{xy}w)^2}{(\partial_x w^2+\partial_y w^2+1)^{3/2}}\right]\\ &\times\left[\frac{\frac{\cos\alpha\partial_x w}{\sqrt{\partial_x w^2+\partial_y w^2+1}}-\sin\alpha\partial_y w}{\sqrt{\partial_x w^2+\partial_y w^2}}\hat{\mathbf{x}}+\frac{\frac{\cos\alpha\partial_y w}{\sqrt{\partial_x w^2+\partial_y w^2+1}}+\sin\alpha\partial_x w}{\sqrt{\partial_x w^2+\partial_y w^2}}\hat{\mathbf{y}}\right.\\ &\left.+\cos\alpha\sqrt{1-\frac{1}{\partial_x w^2+\partial_y w^2+1}}\hat{\mathbf{z}}\right]\exp\left[\frac{-izk}{\sqrt{\partial_x w^2+\partial_y w^2+1}}\right]\\ &\times\exp\left[-\frac{ir\cos\varphi k\partial_y w}{\sqrt{\partial_x w^2+\partial_y w^2+1}}-\frac{ir\sin\varphi k\partial_y w}{\sqrt{\partial_x w^2+\partial_y w^2+1}}\right]dxdy. \end{aligned}\tag{8.3}$$

where $\partial_x w$ and $\partial_y w$ denote the partial derivatives of the continuous function $w(x, y)$, which defines the shape of the wavefront. The function $P(x, y)$ specifies the exit pupil aperture (equal to unity within and zero outside) bounded by $x_{\min}, x_{\max}$ and $y_{\min}, y_{\max}$. The parameter α determines the local polarization rotation in the plane orthogonal to the propagation vector $\mathbf{k}$.

Eq. (8.3) can be further simplified when $w(x, y)$ is assumed to represent an aspherical wavefront. The aspherical wavefront can be expressed conveniently with the coordinates $x = \rho \cos \varsigma$ and $y = \rho \sin \varsigma$. Since the optical field is on-axis and the system exhibits linear symmetry, the resulting aberration remains purely aspheric and can therefore be described using only the linearly symmetric Zernike terms [25], for instance,

$$w(\rho) = R_a - \sqrt{R_a^2 - \rho^2} + Z_1 + Z_4 + Z_{12} + Z_{24}, \tag{8.4}$$

where R_a denotes the radius of the reference sphere. The functions Z_1, Z_4, Z_{12}, and Z_{24} correspond to the first, fourth, eleventh, and twenty-second Zernike polynomial terms, respectively, representing piston, defocus, primary spherical aberration, and secondary spherical aberration.

$$Z_1 = z_1, \quad Z_4 = z_4\sqrt{3}\left(2\rho^2 - 1\right), \quad Z_{12} = z_{12}\sqrt{5}\left(6\rho^4 - 6\rho^2 + 1\right),$$
$$Z_{24} = z_{24}\sqrt{7}\left(20\rho^6 - 30\rho^4 + 12\rho^2 - 1\right), \tag{8.5}$$

where z_1, z_4, z_{12}, and z_{24} denote the corresponding expansion coefficients, which quantify the relative magnitude or strength of each individual aberration term. By differentiating Eq. (8.4) with respect to the spatial coordinates x and y, the local gradients of the wavefront can be obtained. The explicit analytical expressions for $\partial_x w$ and $\partial_y w$ are summarized in Eq. (8.6).

$$\partial_x w = \cos \varsigma \Xi(\rho) \quad \partial_y w = \sin \varsigma \Xi(\rho), \tag{8.6}$$

where,

$$\Xi(\rho) = \frac{1}{\sqrt{R_a^2 - \rho^2}} + 12\sqrt{5} z_{12}\left(2\rho^2 - 1\right) + 120\sqrt{7} z_{24}\left(\rho^2 - 1\right)\rho^2$$
$$+ 24\sqrt{7} z_{24} + 4\sqrt{3} z_4. \tag{8.7}$$

For the on-axis optical field, the partial derivatives of $w(x, y)$ can consistently be expressed in the analytical form of Eq. (8.6). Owing to the inherent circular symmetry of the system, there exists a scalar function $\Xi(\rho)$ that depends solely on the radial coordinate ρ, whose explicit form is uniquely determined by the specific set of Zernike terms considered in the expansion.

Now the differential must be transformed via the Jacobean,

$$\mathrm{d}x\mathrm{d}y = \mathbf{Jcb}(\rho, \varsigma)\mathrm{d}\varsigma\mathrm{d}\rho. \tag{8.8}$$

By applying the Jacobian, we obtain the following relations for the differentials,

$$\mathbf{Jcb}(\rho,\varsigma)=\begin{bmatrix}\dfrac{\partial x}{\partial \rho} & \dfrac{\partial x}{\partial \varsigma}\\ \dfrac{\partial y}{\partial \rho} & \dfrac{\partial y}{\partial \varsigma}\end{bmatrix}. \tag{8.9}$$

These derivatives encapsulate how the wave vector changes as the position in space varies, and they allow us to properly transform the integration variables. Computing the determinant of the Jacobian matrix,

$$\mathbf{Jcb}(\rho,\varsigma)=\frac{\frac{\partial}{\partial\varsigma}\left(-\frac{\sin(\varsigma)\Xi(\rho)}{\sqrt{\Xi(\rho)^2+1}}\right)\frac{\partial}{\partial\rho}\left(-\frac{\cos(\varsigma)\Xi(\rho)}{\sqrt{\Xi(\rho)^2+1}}\right)-\frac{\partial}{\partial\rho}\left(-\frac{\sin(\varsigma)\Xi(\rho)}{\sqrt{\Xi(\rho)^2+1}}\right)\frac{\partial}{\partial\varsigma}\left(-\frac{\cos(\varsigma)\Xi(\rho)}{\sqrt{\Xi(\rho)^2+1}}\right)}{\frac{1}{\sqrt{\Xi(\rho)^2+1}}}. \tag{8.10}$$

Computing the derivatives,

$$\mathbf{Jcb}(\rho,\varsigma)=\sqrt{\Xi(\rho)^2+1}\left(-\frac{\sin(\varsigma)\Xi(\rho)\left(\frac{\sin(\varsigma)\Xi(\rho)^2\Xi'(\rho)}{(\Xi(\rho)^2+1)^{3/2}}-\frac{\sin(\varsigma)\Xi'(\rho)}{\sqrt{\Xi(\rho)^2+1}}\right)}{\sqrt{\Xi(\rho)^2+1}}\right.$$
$$\left.-\frac{\cos(\varsigma)\Xi(\rho)\left(\frac{\cos(\varsigma)\Xi(\rho)^2\Xi'(\rho)}{(\Xi(\rho)^2+1)^{3/2}}-\frac{\cos(\varsigma)\Xi'(\rho)}{\sqrt{\Xi(\rho)^2+1}}\right)}{\sqrt{\Xi(\rho)^2+1}}\right), \tag{8.11}$$

where $\Xi'(\rho)$ denotes the derivative of $\Xi(\rho)$ with respect to the radial coordinate ρ. After a long manipulation the last Jacobian divided by k_z can expressed as,

$$\frac{\mathbf{Jcb}(\rho)}{k_z}=\frac{\Xi(\rho)\Xi'(\rho)}{\left(\Xi(\rho)^2+1\right)^{3/2}}. \tag{8.12}$$

The linear polarization state across the aspherical wavefront is obtained by setting $\alpha=-\varsigma$. As it is shown in Fig. where it shows several values of α correspoding several polarizations states. $\alpha=0$ corresponds to a purely radial polarization state, whereas $\alpha=\pi/2$ yields an azimuthal polarization configuration. See Fig. 8.2

8.2.1 Linear polarization

If the incident on-axis aspherical wavefront exhibits linear polarization $\alpha=-\varsigma$, one may directly substitute the expressions given in Eq. (8.6) into the

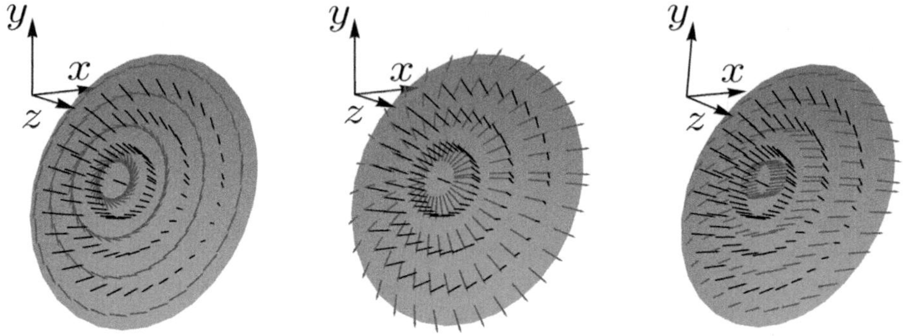

FIGURE 8.2
Polarization diagrams. *Left*: $\alpha = 0$ corresponds to a purely radial polarization state. *Center*: $\alpha = \pi/2$ yields an azimuthal polarization configuration. *Right*: $\alpha = -\varsigma$ gives the linear polarization.

integral formulation of Eq. (8.3),

$$\mathbf{E}(r,\varphi,z) = \frac{E_0}{2\pi}\int_0^{2\pi}\int_0^{\rho_{\max}} q(\rho)l_0(\rho)\left[\frac{\Xi(\rho)\Xi'(\rho)}{(\Xi(\rho)^2+1)^{3/2}}\right]$$
$$\times\left[\left(\frac{\cos^2\varsigma}{\sqrt{\Xi(\rho)^2+1}}+\sin^2\varsigma\right)\hat{\mathbf{x}}+\left(\frac{\cos\varsigma\sin\varsigma}{\sqrt{\Xi(\rho)^2+1}}-\cos\varsigma\sin\varsigma\right)\hat{\mathbf{y}}\right.$$
$$\left.+\cos\varsigma\sqrt{\frac{\Xi(\rho)^2}{\Xi(\rho)^2+1}}\hat{\mathbf{z}}\right]$$
$$\times\exp\left[\frac{-izk}{\sqrt{\Xi(\rho)^2+1}}-\frac{ir\cos(\varphi)k\cos(\varsigma)\Xi(\rho)}{\sqrt{\Xi(\rho)^2+1}}-\frac{ir\sin(\varphi)k\sin(\varsigma)\Xi(\rho)}{\sqrt{\Xi(\rho)^2+1}}\right]d\rho d\varsigma, \tag{8.13}$$

In Eq. (8.13), the pupil function $P(x,y)$ no longer appears explicitly, as it is assumed that the exit pupil is perfectly circular. The resulting integral may be evaluated efficiently by employing standard properties of Bessel functions, which naturally emerge in the context of circularly symmetric Fourier–Bessel expansions [16], providing a rigorous and computationally convenient framework for analyzing the field distribution. So let's start. From basic trigonometry, it is known that,

$$\sin\varphi\sin\varsigma+\cos\varphi\cos\varsigma=\cos(\varphi-\varsigma). \tag{8.14}$$

The $\hat{\mathbf{x}}$ component of Eq. (8.13) can be expressed as,

$$\sin^2\varsigma+\frac{\cos^2\varsigma}{\sqrt{\Xi(\rho)^2+1}}=\frac{\cos(2\varsigma)\left(1-\sqrt{\Xi(\rho)^2+1}\right)}{2\sqrt{\Xi(\rho)^2+1}}+\frac{1+\sqrt{\Xi(\rho)^2+1}}{2\sqrt{\Xi(\rho)^2+1}}. \tag{8.15}$$

The $\hat{\mathbf{y}}$ component of Eq. (8.13) can be reduced to,

$$\sin\varsigma\cos\varsigma = \frac{\sin(2\varsigma)}{2}. \tag{8.16}$$

Replacing Eqs. (8.6), (8.14), (8.15), and (8.16), in Eq. (8.13),

$$\begin{aligned}\mathbf{E}(r,\varphi,z) = \frac{E_0}{2\pi}\int_0^{2\pi}\int_0^{\rho_{\max}} & q(\rho)l_0(\rho)\left[\frac{\Xi(\rho)\Xi'(\rho)}{\left(\Xi(\rho)^2+1\right)^{3/2}}\right]\exp\left[\frac{-\mathrm{i}zk}{\sqrt{\Xi(\rho)^2+1}}\right] \\ & \times\left[\left(\frac{\cos(2\varsigma)\left(1-\sqrt{\Xi(\rho)^2+1}\right)}{2\sqrt{\Xi(\rho)^2+1}}+\frac{1+\sqrt{\Xi(\rho)^2+1}}{2\sqrt{\Xi(\rho)^2+1}}\right)\hat{\mathbf{x}}\right. \\ & \quad + \left(\frac{\sin(2\varsigma)}{2\sqrt{\Xi(\rho)^2+1}}-\frac{\sin(2\varsigma)}{2}\right)\hat{\mathbf{y}} \\ & \left. + \cos\varsigma\sqrt{\frac{\Xi(\rho)^2}{\Xi(\rho)^2+1}}\hat{\mathbf{z}}\right]\exp\left[-\frac{\mathrm{i}rk\cos(\varphi-\varsigma)\Xi(\rho)}{\sqrt{\Xi(\rho)^2+1}}\right]\mathrm{d}\rho\mathrm{d}\varsigma. \end{aligned} \tag{8.17}$$

From the last expression, the integral over ς can be obtained using the following well-known identities [16],

$$\int_0^{2\pi}\cos(m\varsigma)e^{\mathrm{i}u\cos(\varphi-\varsigma)}\mathrm{d}\varsigma = 2\pi(\mathrm{i}^m)\mathrm{J}_m(u)\cos(m\varphi), \tag{8.18}$$

and

$$\int_0^{2\pi}\sin(m\varsigma)e^{\mathrm{i}u\cos(\varphi-\varsigma)}\mathrm{d}\varsigma = 2\pi(\mathrm{i}^m)\mathrm{J}_m(u)\sin(m\varphi). \tag{8.19}$$

The result is Eq.(8.20).

$$\begin{aligned}\mathbf{E}(r,\varphi,z) = \frac{E_0}{2}\int_0^{\rho_{\max}} & q(\rho)l_0(\rho)\left[\frac{\Xi(\rho)\Xi'(\rho)}{\left(\Xi(\rho)^2+1\right)^{3/2}}\right]\exp\left[\frac{-\mathrm{i}zk}{\sqrt{\Xi(\rho)^2+1}}\right] \\ & \times\left[\left(\frac{\left(\sqrt{\Xi(\rho)^2+1}-1\right)}{\sqrt{\Xi(\rho)^2+1}}\cos(2\varphi)\mathrm{J}_2\left(\frac{kr\Xi(\rho)}{\sqrt{\Xi(\rho)^2+1}}\right)\right.\right. \\ & \quad \left. + \left(\frac{1+\sqrt{\Xi(\rho)^2+1}}{\sqrt{\Xi(\rho)^2+1}}\right)\mathrm{J}_0\left(\frac{kr\Xi(\rho)}{\sqrt{\Xi(\rho)^2+1}}\right)\right)\hat{\mathbf{x}} \\ & + \left(\frac{\left(\sqrt{\Xi(\rho)^2+1}-1\right)}{\sqrt{\Xi(\rho)^2+1}}\sin(2\varphi)\mathrm{J}_2\left(\frac{kr\Xi(\rho)}{\sqrt{\Xi(\rho)^2+1}}\right)\right)\hat{\mathbf{y}} \\ & \left. + 2\mathrm{i}\sqrt{\frac{\Xi(\rho)^2}{\Xi(\rho)^2+1}}\cos(\varphi)\mathrm{J}_1\left(\frac{kr\Xi(\rho)}{\sqrt{\Xi(\rho)^2+1}}\right)\hat{\mathbf{z}}\right]\mathrm{d}\rho, \end{aligned} \tag{8.20}$$

where J_m denotes the Bessel function of m-order. Eq. (8.20) is the central result of this manuscript, providing a compact expression for the diffraction pattern produced by an aspherical wavefront that is linearly polarized at the exit pupil of the optical system. The other electric field components are zero, so the intensity is directly given by the absolute value of Eq. (8.20). In the next section, we use Eq. (8.20) to compute the PSF of microscopes directly.

8.3 Radial polarization

If the input on-axis field is flat and radially polarized $\alpha = 0$, and replacing Eqs. (8.6) and (8.12) in Eq. (8.3), we get,

$$\mathbf{E}(r,\varphi,z) = \frac{E_0}{2\pi}\int_0^{2\pi}\int_0^{\rho_{\max}} q(\rho)l_0(\rho)\left[\frac{\Xi(\rho)\Xi'(\rho)}{\left(\Xi(\rho)^2+1\right)^{3/2}}\right] \times \left[\frac{\cos(\varsigma)}{\sqrt{\Xi(\rho)^2+1}}\hat{\mathbf{x}} + \frac{\sin(\varsigma)}{\sqrt{\Xi(\rho)^2+1}}\hat{\mathbf{y}} + \sqrt{1-\frac{1}{\Xi(\rho)^2+1}}\hat{\mathbf{z}}\right] \times \exp\left[\frac{-izk}{\sqrt{\Xi(\rho)^2+1}} - \frac{ir\cos(\varphi)k\cos(\varsigma)\Xi(\rho)}{\sqrt{\Xi(\rho)^2+1}} - \frac{ir\sin(\varphi)k\sin(\varsigma)\Xi(\rho)}{\sqrt{\Xi(\rho)^2+1}}\right] \mathrm{d}\rho\mathrm{d}\varsigma, \tag{8.21}$$

It is important to note that in Eq. (8.23) the pupil function $P(x, y)$ no longer appears explicitly.

Furthermore, the integration over the azimuthal angle ς in Eq. (8.23) can be evaluated by invoking well-known properties of Bessel functions, which naturally emerge in problems involving circular symmetry and Fourier–Bessel expansions [16], Just like the properties, we studied in Chapter 1 and applied in Chapter 4,

$$\mathbf{E}(r,z) = E_0\int_0^{\rho_{\max}} q(\rho)l_0(\rho)\left[\frac{\Xi(\rho)\Xi'(\rho)}{\left(\Xi(\rho)^2+1\right)^{3/2}}\right]\exp\left[\frac{-izk}{\sqrt{\Xi(\rho)^2+1}}\right] \times \left[\frac{i}{\sqrt{\Xi(\rho)^2+1}}J_1\left(\frac{kr\Xi(\rho)}{\sqrt{\Xi(\rho)^2+1}}\right)\hat{\mathbf{r}} + \sqrt{1-\frac{1}{\Xi(\rho)^2+1}}J_0\left(\frac{kr\Xi(\rho)}{\sqrt{\Xi(\rho)^2+1}}\right)\hat{\mathbf{z}}\right]\mathrm{d}\rho, \tag{8.22}$$

where J_0 and J_1 denote the Bessel functions of the zeroth and first order, respectively. Eq. (8.22) stands as the central result of this chapter, since it provides a concise formulation of the diffraction pattern generated by an aspherical wavefront that is radially polarized at the exit pupil of an optical system.

In the following section, Eq. (8.22) will be employed to directly compute the diffraction patterns produced by microscope objectives and to evaluate their imaging performance under realistic aberration conditions. From a theoretical perspective, Eq. (8.22) is notably more compact and elegant than Eq. (8.3), since it reduces the problem to a single independent radial parameter. From a computational standpoint, this simplification provides a significant advantage, as it minimizes the dimensionality of the integral and thereby improves both numerical efficiency and stability [11].

8.4 Azimuthal polarization

If the input on-axis field is flat and radially polarized $\alpha = \pi/2$ and replacing Eq. (8.6) in Eq. (8.3), we get,

$$\mathbf{E}(r,\varphi,z) = \frac{E_0}{2\pi}\int_0^{2\pi}\int_0^{\rho_{\max}} q(\rho)l_0(\rho)\left[\frac{\Xi(\rho)\Xi'(\rho)}{\left(\Xi(\rho)^2+1\right)^{3/2}}\right][\cos(\varsigma)\hat{\mathbf{x}} + \sin(\varsigma)\hat{\mathbf{y}}]$$
$$\times \exp\left[\frac{-\mathrm{i}zk}{\sqrt{\Xi(\rho)^2+1}} - \frac{\mathrm{i}r\cos(\varphi)k\cos(\varsigma)\Xi(\rho)}{\sqrt{\Xi(\rho)^2+1}} - \frac{\mathrm{i}r\sin(\varphi)k\sin(\varsigma)\Xi(\rho)}{\sqrt{\Xi(\rho)^2+1}}\right]\mathrm{d}\rho\mathrm{d}\varsigma, \tag{8.23}$$

where $\Xi'(\rho)$ denotes the derivative of $\Xi(\rho)$ with respect to the radial coordinate ρ. It is important to note that in Eq. (8.23) the pupil function $P(x,y)$ no longer appears explicitly.

Furthermore, the integration over the azimuthal angle ς in Eq. (8.23) can be evaluated by invoking well-known properties of Bessel functions, which naturally emerge in problems involving circular symmetry and Fourier–Bessel expansions [16], Just like the properties, we studied in Chapter 1 and applied in Chapter 4,

$$\mathbf{E}(r,z) = E_0\int_0^{\rho_{\max}} q(\rho)l_0(\rho)\left[\frac{\Xi(\rho)\Xi'(\rho)}{\left(\Xi(\rho)^2+1\right)^{3/2}}\right]\exp\left[\frac{-\mathrm{i}zk}{\sqrt{\Xi(\rho)^2+1}}\right]$$
$$\times\left[\mathrm{i}J_1\left(\frac{kr\Xi(\rho)}{\sqrt{\Xi(\rho)^2+1}}\right)\hat{\varphi}\right]\mathrm{d}\rho, \tag{8.24}$$

where J_0 and J_1 denote the Bessel functions of the zeroth and first order, respectively. Eq. (8.24) stands as the central result of this chapter, since it provides a concise formulation of the diffraction pattern generated by an aspherical wavefront that is radially polarized at the exit pupil of an optical system.

8.5 Illustrative examples

We present illustrative examples of diffraction patterns in microscope objectives for several polarization states. The diffraction patterns are computed directly using Eqs. (8.20), (8.22) and (8.24), for linear, radial and azimuthal polarizations, respectively. The wavefront at the exit pupil is represented through a Zernike polynomial expansion, as given in Eq. (8.4), with the corresponding expansion coefficients determined from `Quadoa Optical CAD`, a widely used commercial optical design software [114]. For the sake of simplicity and consistency across all examples, we assume uniform apodization and beam amplitude factors, setting $q = l_0 = 1$, and adopt a fixed wavelength of $\lambda = 550$ nm in all computations, allowing for direct comparison of the resulting diffraction patterns.

The diffraction patterns for linear polarization are illustrated in Fig. 8.7. The black curve corresponds to the diffraction pattern produced by a perfect spherical wavefront at the exit pupil, computed using the classical formulation of Richards–Wolf [57, 2]. The first example considers a diffraction-limited objective (NA = 0.749), whose exit-pupil wavefront is nearly spherical. The associated Zernike coefficients are $z_1 = -2.18\times10^{-08}$ mm, $z_4 = -2.89\times10^{-08}$ mm, $z_{12} = -6.90\times10^{-09}$ mm, and $z_{24} = -5.43\times10^{-09}$ mm. The diffraction pattern for this configuration, computed from Eq. (8.20), is shown in gray in Fig. 8.7,

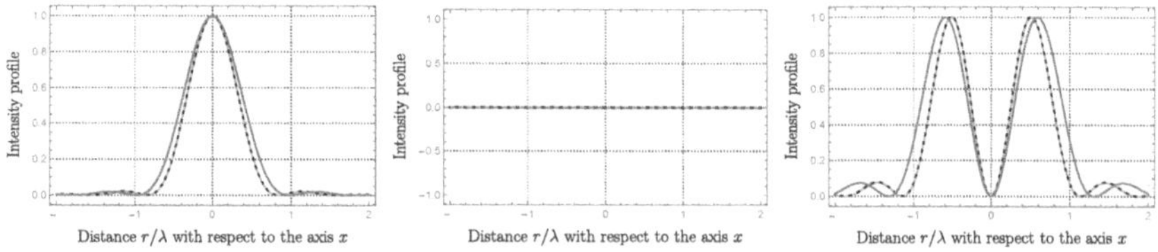

FIGURE 8.3
Normalized intensity profiles of several microscope objectives. In the *left* is the profile along z and in the *right* is a long r. The black line coincides with the diffraction pattern of a perfect spherical wavefront at the exit pupil computed with the classical formulation of Richards-Wolf [57, 2]. The dashed gray and light gray lines correspond to the results of the microscope objective in *left* and *right* of Table 8.1 and Table 8.2, respectively.

TABLE 8.1
Specifications of the microscope of Fig. 8.4.

Radius [mm]	Thickness [mm]	Index	Aperture [mm]
inf	5.0	Air	4.0
20.5533	1.5	1.73249	4.0
5.09748	3.0	1.69404	3.76865
12.265	2.0	Air	3.65133
-6.21604	2.0	1.45747	3.67833
-16.8403	6.0	Air	4.44219
100.259	3.0	1.45747	6.49604
-15.4398	1.0	Air	6.72081
24.3013	4.0	1.45747	6.88895
-11.3619	2.5	1.81235	6.84647
-57.9376	0.1	Air	7.27017
19.4423	3.5	1.54999	7.4213
10.3049	4.0	1.45747	6.95852
-87.3925	0.0	Air	6.90845
13.9012	3.5	1.73249	6.69251
-60.78	1.5	1.4882	6.27531
213.532	0.0	Air	5.59196
5.63103	3.5	1.75856	4.42499
9.21752	1.5	1.81235	2.93
3.91087	2.0	Air	1.77405

and it coincides almost perfectly with the Richards–Wolf prediction, as expected for an aberration-free optical system. The corresponding ray-tracing and stop diagrams are presented in Fig. 8.4, with design parameters summarized in Table 8.1. This configuration serves as a benchmark for assessing the impact of aberrations in more complex systems.

The second example examines another objective with the same numerical aperture (NA = 0.749), also designed to approach the diffraction limit. In this case, small but non-negligible aberrations are intentionally introduced, producing a slightly degraded diffraction pattern. The intensity distribution exhibits a broader central lobe, and the geometrical spot is noticeably enlarged, with FWHM values of 1.2λ in the radial direction and 0.701λ along the optical axis. This intensity distribution is plotted with light gray color in Fig. 8.7. These modifications are indicative of low-order aberrations, as quantified by the Zernike coefficients $z_1 = -3.12 \times 10^{-04}$, $z_4 = 3.20 \times 10^{-04}$, $z_{12} = -1.13 \times 10^{-05}$, and $z_{24} = -4.67 \times 10^{-05}$.

Although these deviations from ideal performance are minor, they clearly demonstrate the sensitivity of high-NA objectives to even small wavefront errors. The corresponding ray-tracing diagrams are shown in Fig. 8.5, providing a visual comparison of spot size and aberration behavior. Additional

TABLE 8.2
Specifications of the microscope of Fig. 8.5.

Radius [mm]	Thickness [mm]	Index	Aperture [mm]
inf	5.0	Air	4.0
21.2506	1.5	1.73249	4.0
5.13186	3.0	1.69404	3.77608
12.4164	2.0	Air	3.66356
-6.3095	2.0	1.45747	3.69292
-17.5105	6.0	Air	4.45573
96.969	3.0	1.45747	6.52555
-15.3181	1.0	Air	6.74518
24.0349	4.0	1.45747	6.90953
-11.3122	2.5	1.81235	6.86821
-57.7372	0.1	Air	7.29367
19.5144	3.5	1.54999	7.44342
10.5222	4.0	1.45747	6.98391
-88.3667	0.0	Air	6.92892
13.9311	3.5	1.73249	6.70706
-63.2024	1.5	1.4882	6.28521
199.153	0.0	Air	5.60167
5.63528	3.5	1.75856	4.43241
9.47958	1.5	1.81235	2.94718
3.94334	2.0	Air	1.78282

design parameters for this second example are provided in Table 8.2, ensuring completeness and reproducibility of the results. This analysis highlights how subtle aberrations can measurably affect the diffraction pattern, emphasizing the need for precise optical design and alignment in high-performance microscopy systems.

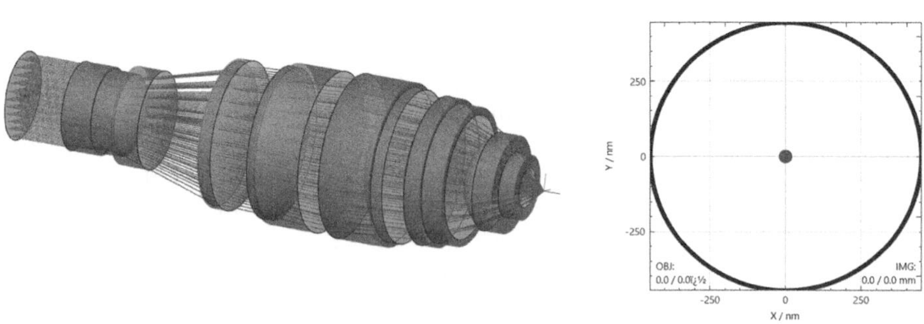

FIGURE 8.4
Microscope objective at *left* Table 8.1. On *left* is the ray tracing and on *right* is the spot diagram.

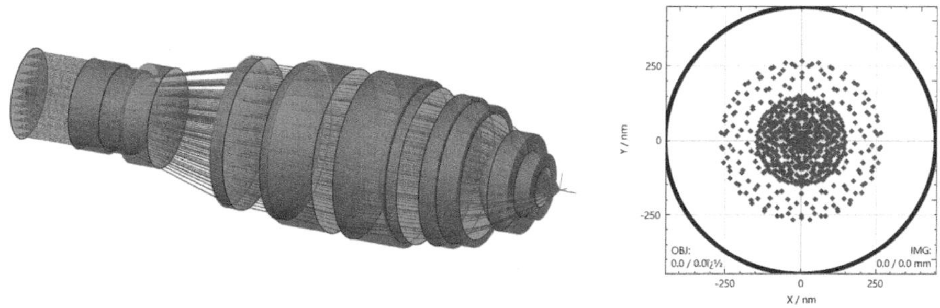

FIGURE 8.5
Microscope objective at *right* of Table 8.2. In *left* is the ray tracing and in *right* is the spot diagram.

The computed diffraction pattern for radial polarization, both along the optical axis (z-axis) and in the radial direction (r), is shown in Fig. 8.7. The results obtained from Eq. (8.22) are represented by the dashed gray curves, while the classical Richards–Wolf formulation [2, 57] is shown in dashed black. It can be observed that the dashed gray and dashed black curves overlap perfectly in both axial and radial directions. This agreement is expected, since the Richards–Wolf integral is valid for stigmatic systems, and in this case, the system is essentially aberration-free (design of Table 8.1). The RMS wavefront error is much smaller than the Airy disc size, which explains the nearly perfect overlap between the two formulations.

The second example (design of Table 8.2) is added to Fig. 8.7 and is shown in gray curves. Compared with the first case using Eq. (8.22), this objective exhibits slightly lower performance. Although still considered diffraction-limited, its geometrical spot size is larger than that of the first objective. The FWHM of the diffraction pattern is 1.2λ in the radial direction, and 0.701λ along the optical axis for a medium with positive refractive index.

The diffraction patterns for azimuthal polarization are shown in Fig. 8.7. The black line coincides with the diffraction pattern of a perfect spherical wavefront at the exit pupil computed with the classical formulation of Richards-Wolf [2, 57]. The diffraction pattern example of Table 8.1 is obtained from Eq. (8.24) and it is presented in dashed gray. It coincides with the Richards–Wolf formulation, as expected for an aberration-free system.

The intensity distribution of the second example of Table 8.2 is presented in Fig. 8.7, in light-gray curves. It exhibits a broader central lobe, and the geometrical spot is noticeably larger, with FWHM values of 1.2λ in the radial direction and 0.701λ along the optical axis.

The two microscope objectives demonstrate that systems with nearly equivalent numerical apertures, although both diffraction-limited under scalar theory, can display markedly different behavior when examined through vectorial analysis. This finding underscores the necessity of utilizing full vectorial

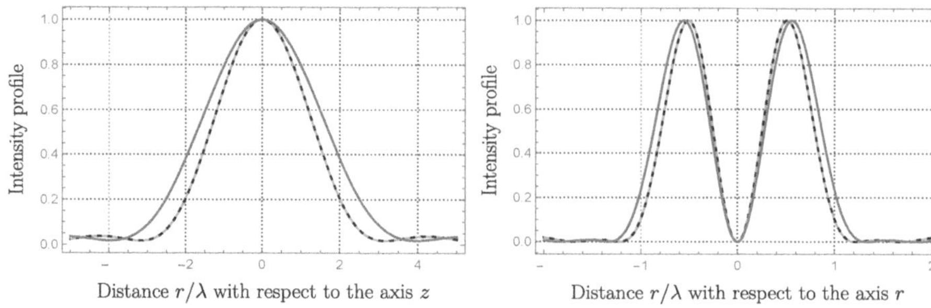

FIGURE 8.6
Normalized intensity profiles of several microscope objectives. In the *left* is the profile along z and in the *right* is a long r. The dashed black line coincides with the diffraction pattern of a perfect spherical wavefront at the exit pupil computed with the classical formulation of Richards-Wolf [2, 57]. The dashed gray and gray lines correspond to the results of the microscope objective in Tables 8.1 and 8.2, respectively.

diffraction models for high–NA optical instruments. In particular, scalar approximations are unable to capture subtle yet consequential polarization and phase effects that emerge due to the strong focusing inherent to high numerical apertures. As illustrated in the second example, even small deviations from an ideally spherical wavefront—quantified through low-order Zernike coefficients—can induce measurable alterations in the point spread function (PSF), focal spot dimensions, and energy distribution within the focal region.

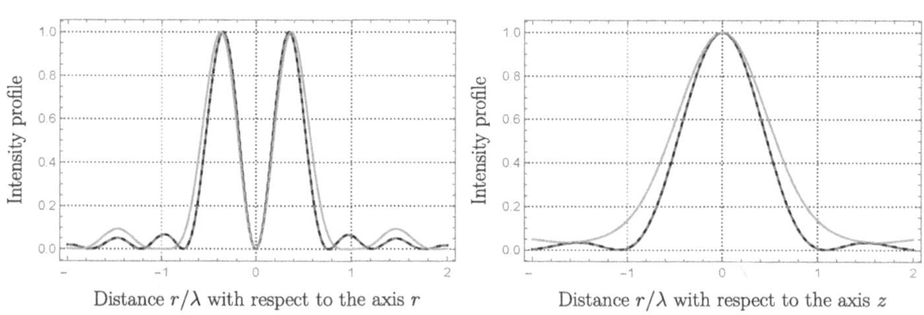

FIGURE 8.7
Normalized intensity profiles of several microscope objectives. In the *left* is the profile along z and in the *right* is a long r. The black line coincides with the diffraction pattern of a perfect spherical wavefront at the exit pupil computed with the classical formulation of Richards-Wolf [2, 57]. The dashed gray and light gray lines correspond to the results of the microscope objective in *left* and *right* of Table 8.1, respectively.

Vectorial formulations, such as the approach presented here, inherently account for the influence of polarization and aspherical aberrations, phenomena that become increasingly prominent as the numerical aperture increases. These effects are especially significant for high-resolution imaging applications [65, 62] and optical trapping [66], where precise manipulation of the focal field is critical [70]. Moreover, the effect of wavefront aberrations is not uniform across different polarization states: for example, radially or linearly polarized beams interact distinctly with system aberrations. These observations emphasize the vital importance of employing comprehensive vectorial approaches in both the design and evaluation of modern high-performance optical systems.

Traditional methods, particularly those proposed by Kant and colleagues [108, 112, 111], differ fundamentally from the present framework. Their approaches generally assume either aberration-free conditions or surfaces of revolution, depend on multiple auxiliary optical coordinates, and necessitate extensive sets of equations with numerous approximations. By contrast, our method is derived from a general freeform wavefront representation, operates directly on $w(x, y)$, and avoids such restrictive assumptions, offering a more adaptable and generalizable framework.

Similarly, Wang's numerical approach [115] contrasts with ours. While Wang's method relies on iterative approximations, Eq. (8.20) provides a closed-form analytical solution that reduces to a single integral under radial symmetry and uniform polarization. This simplification enhances both analytical clarity and computational efficiency, rendering the method widely applicable to a variety of high-NA optical systems without compromising accuracy.

Although the current examples focus on microscope objectives, the methodology is broadly applicable to linear, symmetric systems represented by Zernike wavefronts. Possible extensions, including the treatment of misalignment and off-axis fields involving polarization aberrations, are left for future investigation. These advancements will further broaden the utility of the vectorial framework, providing a versatile tool for the design and analysis of sophisticated optical systems.

8.6 Conclusions

In this chapter, we have established an advanced theoretical framework for computing vectorial diffraction patterns in high-numerical-aperture optical systems, overcoming the fundamental limitations of classical models. By deriving closed-form integrals for radial, azimuthal, and linear polarization, we validated the precision of this method, achieving exact agreement with Richards-Wolf theory for ideal spherical wavefronts. However, applying this model to real microscope objectives revealed that systems classified as *diffraction-limited* under scalar theories exhibit divergent behaviors when accounting for their full vectorial structure and aspherical aberrations. Our numerical simulations demonstrate that wavefront deviations induce significant focal asymmetries, polarization-dependent resolution degradation, and reduced Strehl ratios. Ultimately, this generalized approach, incorporating aberrations directly through Zernike expansions, provides a rigorous and efficient tool for the design and analysis of high-precision optical technologies, revealing subtle physical effects undetectable by traditional methods.

Problems

Problem 1: Derivation of the linear polarization case (main result) Start from the general expression Eq. (8.3) with linear polarization obtained by setting $\alpha = -\varsigma$. Using the coordinate change to polar coordinates in the pupil (ρ, ς) and the relations $\partial_x w = \cos\varsigma\, \Xi(\rho)$, $\partial_y w = \sin\varsigma\, \Xi(\rho)$, show step-by-step how one arrives at the intermediate form Eq. (8.13) and then, by applying the Bessel integral identities

$$\int_0^{2\pi} \cos(m\varsigma)\, \exp\bigl[iu\cos(\varphi - \varsigma)\bigr]\, d\varsigma = 2\pi\, (i^m) J_m(u) \cos(m\varphi),$$

$$\int_0^{2\pi} \sin(m\varsigma)\, \exp\bigl[iu\cos(\varphi - \varsigma)\bigr]\, d\varsigma = 2\pi\, (i^m) J_m(u) \sin(m\varphi),$$

derive the compact final expression Eq. (8.20). Briefly explain why the final field contains J_0, J_2 in the transverse components and J_1 in the longitudinal component.

Problem 2: Effect of primary spherical aberration on the longitudinal field Consider only primary spherical aberration: $z_{12} \neq 0$ and $z_1 = z_4 = z_{24} = 0$.

(a) Write the simplified expression for $\Xi(\rho)$.

(b) For small $z_{12} > 0$ (positive spherical aberration), explain qualitatively why the on-axis longitudinal component $E_z(r = 0, z)$ is reduced and shifted compared to the aberration-free case.

Problem 3: Comparison with the aberration-free Richards–Wolf case Show analytically that when the wavefront is perfectly spherical, i.e. $z_1 = z_4 = z_{12} = z_{24} = 0$, then

$$\Xi(\rho) = \frac{1}{\sqrt{R_a^2 - \rho^2}} = \frac{\sin\theta}{\cos\theta}, \quad \sqrt{\Xi^2 + 1} = \frac{1}{\cos\theta}, \quad u = kr\sin\theta,$$

and Eq. (8.20) reduces to the well-known Richards–Wolf result for linearly x-polarized light focused by an aplanatic system:

$$\mathbf{E}(r,\varphi,z) \propto \int_0^{\theta_{\max}} \sqrt{\cos\theta}\,\sin\theta\,\exp(-ikz\cos\theta) \times \Big[(1+\cos(2\varphi)\mathrm{J}_2(kr\sin\theta))\hat{\mathbf{x}} + \sin(2\varphi)\mathrm{J}_2(kr\sin\theta)\hat{\mathbf{y}} + 2\mathrm{i}\cos\varphi\mathrm{J}_1(kr\sin\theta)\hat{\mathbf{z}}\Big]\,\mathrm{d}\theta.$$

Indicate the change of variable used and any missing normalization factors.

Problem 4: Malus-Dupin vs. phase-screen aberration models
In conventional aberration theory, aberrations are usually added as an extra phase delay over an ideal spherical wavefront. The chapter instead uses the Malus–Dupin theorem to treat the aberrated wavefront directly. Conceptually, why can these two approaches give different vectorial diffraction patterns even when the optical path difference is mathematically identical?

Problem 5: Jacobian term physical meaning
The factor $\Xi(\rho)\Xi'(\rho)/(\Xi(\rho)^2+1)^{3/2}$ appears in the final Bessel-function integral as an energy-redistribution weight. Provide a clear physical interpretation of this term: why does it exist, and what happens to the diffraction pattern in the limit $\Xi'(\rho) \to 0$ (i.e., perfectly plane wavefront)?

Problem 6: Vectorial breakdown of scalar diffraction limit claims
Many papers claim that a microscope objective is *diffraction-limited* because its geometrical spot size lies inside the Airy disc and the RMS wavefront error is small. The chapter shows that two such objectives with the same NA can still produce noticeably different vectorial point-spread functions. Explain why scalar-based *diffraction-limited* criteria can be misleading when radial (or other non-paraxial) polarization is used.

Problem 7: Derivation of the radial case Start from the general vector diffraction integral given in Eq. (8.3) Assume an on-axis input field with purely azimuthal polarization ($\alpha = \pi/2$) and a radially symmetric wavefront $w(x,y) = w(\rho)$. Show step-by-step that the electric field in the focal region reduces to the compact form

$$\mathbf{E}(r,z) = E_0 \int_0^{\rho_{\max}} q(\rho)\,l_0(\rho)\,\frac{\Xi(\rho)\,\Xi'(\rho)}{(\Xi(\rho)^2+1)^{3/2}} \times \exp\Big(-\mathrm{i}kz/\sqrt{\Xi(\rho)^2+1}\Big)\,\mathrm{i}\,\mathrm{J}_1\Big(kr\Xi(\rho)/\sqrt{\Xi(\rho)^2+1}\Big)\,\hat{\boldsymbol{\varphi}}\,\mathrm{d}\rho$$

Explain the physical meaning of each major term.

Problem 8: Zernike aberration influence – defocus only Consider a wavefront containing only piston and defocus:

$$w(\rho) = z_1 + z_4\sqrt{3}(2\rho^2 - 1).$$

(a) Compute the explicit expression for the slope function $\Xi(\rho)$. (b) Discuss qualitatively how a small positive defocus coefficient $z_4 > 0$ affects the axial position of maximum intensity and the width of the focal spot compared to the aberration-free case ($z_4 = 0$). (c) If z_4 is large enough, what qualitative change in the on-axis intensity $|E_z(0,0)|^2$ do you expect?

Problem 9: Numerical sensitivity – small spherical aberration Use the expression for $\Xi(\rho)$ given in Eq. (8.7). Assume $z_{12} = 0.02\,\lambda$ (primary spherical aberration) and all other coefficients zero except piston (which can be ignored). (a) Plot or sketch $\Xi(\rho)$ for $\rho \in [0,1]$ and compare it with the aberration-free case $\Xi(\rho) = 1/\sqrt{R_a^2 - \rho^2}$. (b) Explain why even a small amount of primary spherical aberration ($z_{12} \sim 0.01\text{–}0.05\,\lambda$) can produce a noticeable broadening of the radial intensity profile $I(r,0)$ in a high-NA system (NA $\gtrsim 0.7$).

Problem 10: Polarization contrast Derive or argue why the final expression in Eq. (8.24) contains only the $\hat{\varphi}$ component (azimuthal electric field) when the input field is azimuthally polarized at the exit pupil. What would be the dominant polarization component in the focal region if the input polarization were purely radial instead ($\alpha = 0$)? Justify your answer using symmetry arguments.

Problem 11: Comparison with Richards–Wolf limit Show analytically that when the wavefront is perfectly spherical, i.e.,

$$\Xi(\rho) = \frac{1}{\sqrt{R_a^2 - \rho^2}} = \frac{\sin\theta}{\cos\theta} \qquad \text{with} \quad \sin\theta = \rho/R_a,$$

the generalized integral (Eq. (8.24)) reduces exactly to the classical Richards–Wolf result for azimuthally polarized light,

$$\mathbf{E}(r,z) \propto \mathrm{i}\int_0^{\theta_{\max}} \sin\theta q(\theta)\,\mathrm{J}_1(kr\sin\theta)\exp(-\mathrm{i}kz\cos\theta)\,\mathrm{d}\theta\,\hat{\varphi}.$$

Indicate clearly which change of variable you use.

9

Diffraction in off-axis freeform vector optical systems

The Richards–Wolf integral models vectorial diffraction assuming on-axis spherical wavefronts, limiting its use in off-axis freeform systems with polarization aberrations. This chapter extends a generalized freeform diffraction framework to compute vector PSFs for arbitrary wavefronts $w(x, y)$ and spatially varying polarization. Applied to a compact five-mirror freeform optical system, it reveals asymmetric, diffraction-limited PSFs with central intensity nulls from combined wavefront and polarization effects. The approach reduces to classical Richards–Wolf results for ideal cases, enabling accurate analysis of realistic off-axis polarized diffraction in freeform optical systems.

9.1 Introduction

The study of diffraction phenomena has long been at the forefront of optical research, with its roots tracing back to the 17th century when Francesco Grimaldi first observed the effects of light bending around obstacles. Since that time, a variety of theoretical frameworks have emerged, aiming to describe and quantify the intricacies of diffraction, with key developments emerging in both scalar and vector approaches. Scalar theories, such as the Rayleigh-Sommerfeld and Kirchhoff models, provide elegant solutions under the paraxial approximation, but their limitations become evident in more complex scenarios—especially when diffraction deviates from the paraxial regime. In contrast, vector theories offer a more comprehensive treatment of diffraction, accounting for the three-dimensional nature of the electric field and providing a more accurate description of diffraction effects beyond the paraxial regime. One of the most prominent vector-based models is the Richards and Wolf

DOI: 10.1201/9781003726494-13

theory, which is widely accepted for describing diffraction patterns resulting from spherical converging wavefronts.

Despite its strengths, the Richards-Wolf theory is limited in its scope, particularly in its treatment of on-axis, aberration-free fields. However, recent advancements have broadened the applicability of the theory. In Chapter 7, we introduced a generalization of the Richards-Wolf integral, extending its applicability to freeform wavefronts and accounting for a wide range of polarization states. This generalization opens up new avenues for exploring diffraction phenomena in systems where the wavefront shape is no longer constrained to simple spherical or planar geometries. Specifically, this manuscript explores the application of this generalized approach to diffraction patterns of off-axis, aberrated, and polarized fields in freeform optical systems.

Freeform optical systems, characterized by surfaces lacking rotational or translational symmetry, pose unique challenges in diffraction analysis. Unlike conventional optical systems, where wavefronts are often constrained to idealized shapes, freeform systems generate more complex, asymmetric wavefronts that can exhibit a variety of polarization distributions. These systems are particularly relevant in modern optics, including cutting-edge imaging systems, high-performance telescopes, and novel lithographic techniques. The ability to model and predict the diffraction patterns in such systems is crucial for optimizing their performance, especially in scenarios involving polarized sources, where the polarization state varies across the wavefront surface.

In this chapter, we use the generalization of the Richards-Wolf integral (Chapter 7) to address the diffraction characteristics of off-axis, aberrated, polarized fields in freeform optical systems. This framework allows for the computation of the point spread function (PSF) in three dimensions—x, y, and z—for any given optical system, even when the system produces a freeform wavefront with varying polarization states. The ability to study and quantify these diffraction patterns provides valuable insights into the quality of freeform optical systems, particularly in the context of polarized light.

The chapter is organized as follows: Section 2 recalls the generalized Richards-Wolf integral and its underlying mathematical framework. Section 3 presents an illustrative example of diffraction patterns resulting from off-axis aberrated polarized fields in a freeform optical system. Finally, the conclusions drawn from these analyses are discussed in the last section, highlighting the implications for the design and optimization of freeform optical systems in real-world applications.

9.2 Vector diffraction integral

In Ref. [11], a new vector diffraction integral was introduced, which can compute diffraction patterns for freeform wavefronts, accounting for any polarization state. We have studied this integral in Chapter 7, and special cases of it in chapters 8, 9, 10 for aspheric wavefronts.

The integral is grounded in the angular spectrum representation of optical fields, a framework derived from Fourier analysis. This approach represents optical fields in homogeneous media as a sum of plane waves and evanescent waves, both of which are solutions to Maxwell's equations. The plane waves in this representation contribute to the field, with their propagation directions aligning with those of the corresponding geometric optical rays. Taking these principles into account, the electric field $\mathbf{E}(r, \varphi, z)$, where (r, φ, z) are cylindrical coordinates, at focus can be expressed as,

$$\mathbf{E}(r,\varphi,z) = \frac{E_0 k}{2\pi}\int\int_\Omega P(x,y)q(x,y)l_0(x,y)\left[\frac{\partial_x^2 w\partial_y^2 w-(\partial_{xy}w)^2}{(\partial_x w^2+\partial_y w^2+1)^{3/2}}\right]$$

$$\times\left[\frac{\sqrt{1-\dfrac{1}{\partial_x w^2+\partial_y w^2+1}}}{(\partial_x w+e_y\partial_y w)\sqrt{\dfrac{(e_y\partial_x w-e_x\partial_y w)^2}{(\partial_x w^2+\partial_y w^2+1)(e_x\partial_x w+e_y\partial_y w)^2}+1}}(e_x\hat{\mathbf{x}}+e_y\hat{\mathbf{y}})\right.$$

$$\left.+\frac{\sqrt{1-\dfrac{1}{\partial_x w^2+\partial_y w^2+1}}}{\sqrt{\dfrac{(e_y\partial_x w-e_x\partial_y w)^2}{(\partial_x w^2+\partial_y w^2+1)(e_x\partial_x w+e_y\partial_y w)^2}+1}}\hat{\mathbf{z}}\right]\exp\left[\frac{-\mathrm{i}zk}{\sqrt{\partial_x w^2+\partial_y w^2+1}}\right]$$

$$\times\exp\left[-\frac{\mathrm{i}r\cos\varphi k\partial_y w}{\sqrt{\partial_x w^2+\partial_y w^2+1}}-\frac{\mathrm{i}r\sin\varphi k\partial_y w}{\sqrt{\partial_x w^2+\partial_y w^2+1}}\right]\mathrm{d}x\mathrm{d}y. \quad (9.1)$$

Here, E_0 is a constant amplitude that sets the scale of the field according to the far-field approximation and ensures consistency with the energy conservation theorem. The position vector of the focal point is given by $\mathbf{r} \equiv [r\cos\varphi,\ r\sin\varphi,\ z]$, expressed in cylindrical coordinates.

The function $q(x, y)$ is the *apodization factor*, derived from the energy conservation theorem, which accounts for the local ray density across the exit pupil. This factor plays a key role in modulating the amplitude of the field based on the ray distribution.

The term $l_0(x, y)$ represents the *amplitude distribution* of the incident beam at the exit pupil, while $w(x, y)$ denotes the *wavefront profile* of the optical field under consideration. The first-order partial derivatives of the wavefront with respect to x and y are written as $\partial_x w$ and $\partial_y w$, and describe the local slope of the wavefront. The second-order partial derivatives—$\partial_x^2 w$, $\partial_y^2 w$, and the mixed derivative $\partial_{xy} w$—describe the local curvature and are essential for evaluating the focusing behavior and aberrations in the system.

The function $P(x, y)$ defines the shape of the *exit pupil*, acting as an aperture function that equals one inside the pupil region and zero outside. The

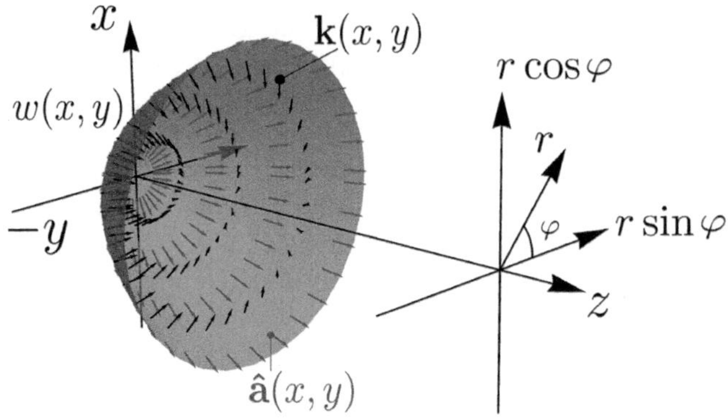

FIGURE 9.1
Diagram of the diffraction phenomena. The wavefront $w(x, y)$ is placed in the coordinate system (x, y). The vector $\mathbf{k}$ is normal to $w(x, y)$ and $\hat{\mathbf{a}}$ is normal to $\mathbf{k}$. The diffraction pattern is in the coordinate system $(r\cos\varphi, r\sin\varphi)$, and z is the axis of propagation.

pupil Ω is bounded in the x and y directions by the limits x_{min} to x_{max}, and y_{min} to y_{max}, respectively. These bounds define the integration domain and the region of the optical system that contributes to the diffraction integral.

See Fig. 9.1, in which it is represented the wavefront $w(x, y)$ is represented as a gray surface. The image plane in cylindrical coordinates is to the right of the figure.

The vector $[e_x(x, y), e_y(x, y), 0]^T$ represents the *input polarization vector* of the optical field, where $e_x(x, y)$ and $e_y(x, y)$ are the components of the electric field polarization in the x and y directions at each point (x, y) on the wavefront. This spatial variation in the polarization vector is crucial because it allows for the modeling of more complex wavefronts, where the polarization is not uniform but varies across the surface. This flexibility is what enables Eq. (9.1) to compute the diffraction pattern for freeform wavefronts with spatially varying polarization, providing a more accurate and comprehensive representation of real-world optical systems.

If the input field is linearly polarized along the x-axis, the polarization vector would be $\mathbf{E}_{\text{inc}} = [1, 0, 0]^T$. In this case, the expression for the field simplifies by substituting $\mathbf{E}_{\text{inc}} = [1, 0, 0]^T$ for the polarization vector in the relevant equations. However, it is important to note that the polarization vector can also vary as a function of position on the wavefront. This means that the polarization state can change from point to point across the wavefront, depending on the local optical conditions. Thus, the polarization vector becomes a spatially dependent function: $\mathbf{E}_{\text{inc}} = [e_x(x, y), e_y(x, y), 0]^T$, where both e_x and e_y vary with respect to x and y.

9.3 Vector diffraction in freeform optical systems

The illustrative example is an optical system consisting of five freeform mirrors, as previously detailed in [116]. The mathematical model behind the example presented in this chapter has been developed with a set of papers on stigmatic systems, which are the following references [117, 118, 119, 120, 121, 122, 123, 124, 125, 126, 127, 128, 129, 130, 131, 132, 133, 134, 135, 136, 137, 138]. This system incorporates a sensor with a diagonal size of 7.2 mm, specifically designed to capture the diffraction patterns generated by the freeform wavefronts. The positions, orientations, and specifications of the freeform mirrors, along with the sensor dimensions, are all presented in Table 9.1 for clarity.

The optical components in this example feature elliptical apertures, with the respective half-axes along the x- and y-directions of the ellipse specified in the last column of Table 9.1. The use of elliptical apertures provides several advantages, including enhanced control over the system's field of view and diffraction properties, particularly in optical systems where space is limited and the light path is highly constrained. This configuration allows for better management of the incoming light and helps mitigate common optical aberrations.

An important aspect of this optical system is its compact form factor, which makes it particularly well-suited for integration into space-constrained devices, such as tablets, smartphones, and other portable imaging systems. Despite its small size, the system is engineered to maintain excellent optical performance. The optical stack height along the Z-axis is tightly constrained to just 10 mm, ensuring minimal thickness for integration into compact devices. In contrast, the X and Y dimensions, which are parallel to the device screen, allow for greater flexibility in their tolerances. This design choice enables a larger lateral footprint, which can help accommodate more optical elements and improve system performance without affecting the device's overall thickness. This balance of compactness and performance makes the system

TABLE 9.1
Global coordinates of the element in the illustrative example. X, Y, and Z are in millimeters. Rot X, Rot Y, and Rot Z are in degrees.

Element	X	Y	Z	Rot X	Rot Y	Rot Z	Half axis X/Y
Stop	0	0	0	0	0	0	6/6
Mirror 1	0	0	10	45	0	0	5/7
Mirror 2	0	21.8	10	90	44	3.4	10/7
Mirror 3	15	21.3	10	-90	48	180	10/7
Mirror 4	13.8	5.1	10	-90	-38	-180	10.5/7.5
Image	-10	8.9	10	-90	-81	180	7.2

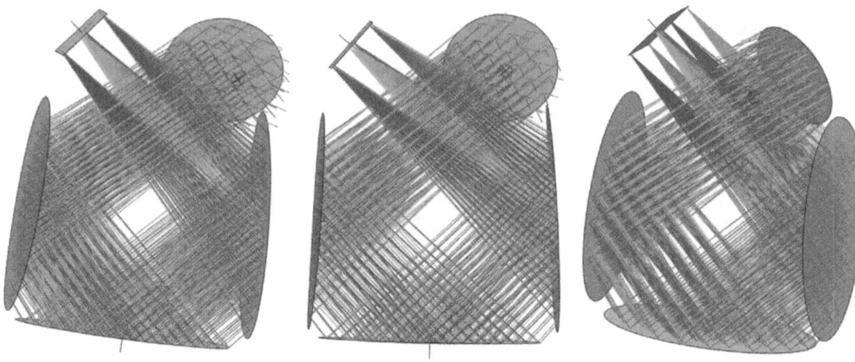

FIGURE 9.2
Different angles of the ray-tracing of illustrative example.

ideal for modern portable and mobile applications, where both size and optical quality are critical.

The field of view of the system is $8.6^\circ \times 6.5^\circ$, which defines the angular extent of the scene that can be captured by the optical system. These values correspond to the horizontal and vertical angles subtended by the observed scene, ensuring that the system is capable of imaging a sufficiently wide area while maintaining high resolution. The effective focal length of the system is $f = 34$ mm, with an f-number of 3.3, indicating a relatively large aperture for the given focal length, which helps in gathering more light and improving low-light performance. Notably, the effective focal length is the same in both the x- and y-directions, ensuring that the system maintains symmetry in its optical characteristics across both axes.

For the following computations, the wavelength is set to 550 nm, which is a typical value for visible light. Ray-tracing of the illustrative example is shown in Fig. 9.2. The ray-tracing simulation provides a detailed visualization of the light paths through the optical system, allowing for an analysis of the system's focusing characteristics. By following the rays through the five freeform mirrors, we can gain insights into the behavior of the optical system.

The system has been designed using the method outlined in [116] and implemented with the help of `Quadoa-Optical CAD`, a commercial optical design software [40].

The profile of the surfaces of the illustrative example is displayed in Fig. 9.3. In this figure, the aperture of each surface is also shown, providing a clear visualization of the spatial distribution of the optical elements. The freeform surfaces are described by xy polynomials of order 8, which offer a high degree of flexibility in modeling complex surface shapes. These polynomials allow for precise control over the curvature and slope of each surface, ensuring that the system can meet the desired optical performance.

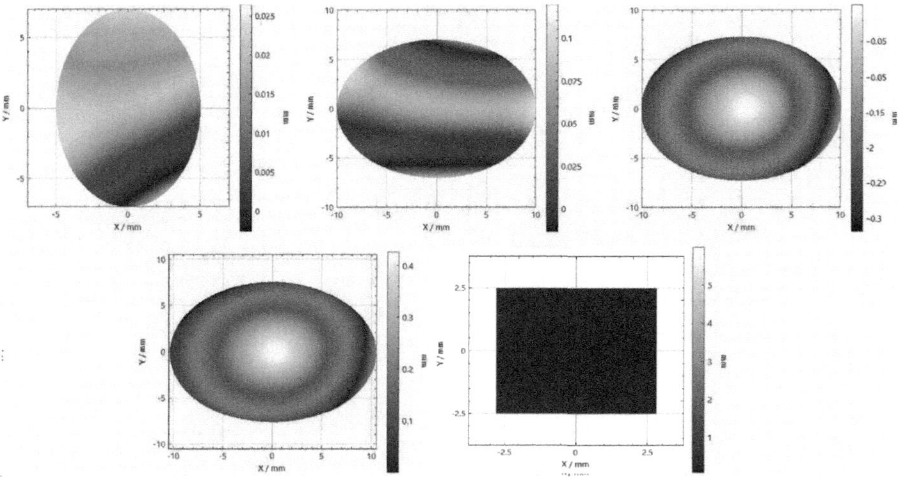

FIGURE 9.3
Profile of the surfaces of the illustrative freeform optical system. *At-top from left to right* are the first three surfaces. *At-bottom* are the fourth surface and sensor.

From the figure, it is evident that surfaces 3 and 4 exhibit slight symmetry, indicating that their shapes are similar but not identical. This symmetry may contribute to more predictable diffraction behavior. On the other hand, the first and second surfaces are completely unsymmetrical, which introduces a more complex optical behavior. These asymmetries are intentionally designed to achieve specific imaging characteristics, such as improved field of view, reduced aberrations, or enhanced light gathering, depending on the requirements of the optical system.

This combination of symmetric and asymmetric surfaces highlights the flexibility and power of freeform optical designs, where both the shape and arrangement of surfaces can be optimized for specific system objectives. By using higher-order polynomials to describe these surfaces, the design can better correct for aberrations and improve overall imaging performance.

The input field for the illustrative example is linearly polarized, and the polarization distributions both at the input and output along the pupil are presented in Fig. 4. The input field is perfectly linearly polarized, ensuring a uniform polarization state across the entire wavefront.

In contrast, the polarization states in the output field are defined point-by-point due to the polarization aberrations introduced by the optical system [53]. These aberrations lead to a spatially varying polarization state across the field, which is a key feature of the system's behavior.

It's important to note that the field selected for this example is not an on-axis field, but rather an off-axis field with a field of view. This off-axis configuration introduces additional complexities in terms of polarization and

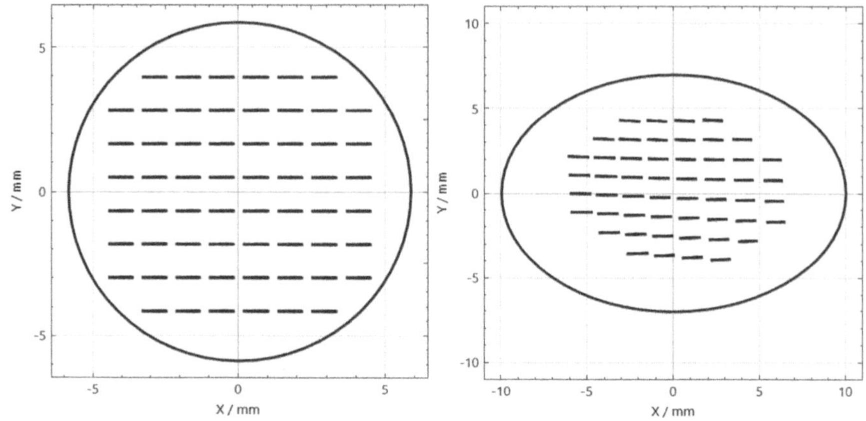

FIGURE 9.4
Polarization map of the input field *left* and output field *right.*

diffraction, making it a more representative scenario for practical optical systems.

To compute the vector point spread function (PSF) of the system shown in Fig. 9.2, the direct evaluation of Eq. (9.1) is required. In this calculation, the aberration coefficients at the output wavefront are used to characterize $w(x, y)$, following a similar approach to the one proposed in [14]. The data for the output polarization, which includes the components $\mathbf{E}_{\text{inc}} = [e_x(x, y), e_y(x, y), 0]^T$, is exported from `Quadoa-Optical CAD` to `Wolfram Mathematica`, where Eq. (9.1) is numerically integrated.

In the case of `Quadoa-Optical CAD`, the polarization is provided in the form of the vector $\mathbf{E}_{\text{inc}} = [e_x, e_y, 0]^T$. This vector is then adapted to fit the requirements of Eq. (4) and integrated within the framework of Eq. (9.1). The distribution of rays shown in Fig. 9.4 corresponds to the apodization function, which, as detailed in [14], is already taken into account in the numerical integration of Eq. (9.1). Thus, the apodization function is inherently considered during the computation, which helps in simulating the effects of the system's optical characteristics more accurately.

The normalized intensity of the diffraction produced by the output wavefront $w(x, y)$ for an off-axis field at $8.6° \times 6.5°$ is presented in Fig. 9.5. The diffraction patterns along the $r \cos \varphi$ and $r \sin \varphi$ axes are shown for the x-, y-, and z-components of the electric field in the *left*, *center*, and *right* panels, respectively. As anticipated, the diffraction patterns along the x- and y-axes differ due to the aberrations suffered by the wavefront. Notably, both diffraction patterns exhibit a central null in the electric field, which is a signature of the polarization and wavefront aberrations. When the aberrations are removed,

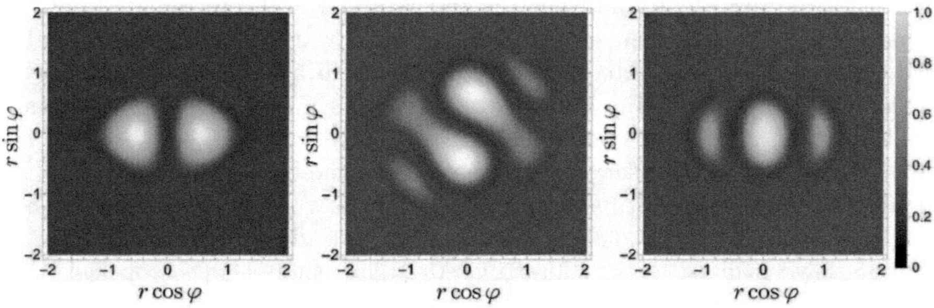

FIGURE 9.5
Normalized intensity of the diffraction pattern design of Fig. 2. *left*, *center*, and *right* are the intensity profiles concerning the x, y, and z axes, respectively in the axis of the image plane.

the resulting diffraction pattern aligns with the expected results described in classical diffraction theory, confirming the validity of the approach.

The results presented in Fig. 9.5 demonstrate that the system's point spread function is largely diffraction-limited, though its shape is distorted due to the combined effects of the freeform wavefront aberrations and polarization aberrations. This distortion is typical of systems with freeform optics, where the complexity of the optical surfaces introduces non-trivial variations in the diffraction pattern. The effect of stronger freeform aberrations on the diffraction pattern has been explored by the author in [11], where the resulting patterns were shown to deviate from the ideal case.

If the system were radially symmetric and the field were on-axis, Eq. (9.1) could be significantly simplified. This condition, which assumes no off-axis effects and no wavefront distortions beyond radial symmetry, has been utilized in previous works by the author, such as in [14]. Under these idealized conditions, the analysis becomes more tractable, providing insights into the diffraction behavior of simpler systems.

The scope of this chapter is focused exclusively on theoretical computations and simulations. While the results presented provide valuable insights into the behavior of freeform wavefront diffraction, the natural progression of this research involves experimental validation, which is left as a subject for future work. Such experiments would help confirm the accuracy and applicability of the models in practical optical systems.

Both Richards and Wolf's integral and Eq. (9.1) are formulated under the far-field approximation framework [11]. Traditionally, Richards and Wolf's integral assumes a perfect spherical wavefront (as seen in chapter 4) and its associated normal vectors. In contrast, our approach generalizes this by considering any continuous wavefront surface with its corresponding normal vectors, allowing for the analysis of complex freeform surfaces.

Eq. (9.1) represents the Richards and Wolf integral reformulated in Cartesian coordinates (as seen in Chapter 7). This transformation is particularly advantageous because Cartesian coordinates facilitate a more straightforward calculation of normal vectors for arbitrary continuous freeform surfaces, as well as their associated vector fields.

Obtaining the necessary information about the freeform wavefront $w(x, y)$ and the input polarization vector is straightforward with the use of modern commercial optical design software. For the illustrative example in this chapter, we utilized `Quadoa Optical-CAD` [40], though other optical design platforms are equally capable of generating the required data. This accessibility makes our method practical and readily implementable for a wide range of optical system designs involving freeform optics.

9.4 Conclusions

In this chapter, we have employed a generalized integral to compute the diffraction pattern of off-axis fields in freeform optical systems, specifically considering the polarization aberrations introduced by the system. The integral, expressed in Eq. (9.1), incorporates both the aberration coefficients of the wavefront $w(x, y)$ and the polarization components at the exit pupil of the system, represented by the vector $\mathbf{E}_{\text{inc}} = [e_x(x, y), e_y(x, y), 0]^T$. These parameters are computed using the commercial optical design software `Quadoa Optical-CAD`, which facilitates the accurate modeling of both wavefront aberrations and polarization states across the optical system.

The wavefront aberrations are characterized by the deviations from an ideal spherical wavefront, which are quantified through the aberration coefficients. These coefficients play a crucial role in determining the diffraction characteristics of the system, particularly when considering off-axis fields. Similarly, the polarization vector $\mathbf{E}_{\text{inc}} = [e_x(x, y), e_y(x, y), 0]^T$ captures the spatially varying polarization state across the pupil, which significantly influences the diffraction pattern, especially in systems that handle polarized light.

When the aberrations are set to zero, and the polarization is perfectly linear—i.e., the polarization vector becomes constant across the wavefront—Eq. (9.1) reduces to the classical formulation of the Richards-Wolf integral [57]. This case corresponds to ideal conditions where the wavefront is free of distortions, and the polarization remains uniform across the optical field. Under these assumptions, the diffraction pattern generated by the system follows the well-established predictions of the Richards-Wolf theory, which serves as a baseline for comparing more complex, real-world systems.

The results obtained from our study align with the theoretical predictions, confirming the validity of the approach for systems involving freeform surfaces and varying polarization states. These findings highlight the importance of considering both wavefront aberrations and polarization effects in the design and analysis of modern optical systems, where such complexities are common, especially in high-performance or miniaturized applications. This manuscript serves as a foundation for further exploration and development in the field of diffraction modeling for freeform optics, with a particular focus on polarization aberrations and off-axis field effects.

Problems

Problem 1: Physical interpretation and limiting cases of the generalized integral The core equation of the chapter is the generalized vector diffraction integral in Cartesian coordinates (Eq. (9.1)).

(a) Identify and briefly explain the physical meaning of the following key parts of the integrand:
i. The prefactor $\dfrac{\partial_x^2 w\, \partial_y^2 w - (\partial_{xy} w)^2}{(\partial_x w^2 + \partial_y w^2 + 1)^{3/2}}$
ii. The complicated polarization projection terms involving, e_x, e_y, $\partial_x w$, $\partial_y w$
iii. The phase factors,
$\exp\left[-\mathrm{i}kz/\sqrt{\ldots} - \mathrm{i}kr\cos\varphi\, \partial_y w/\sqrt{\ldots} - \mathrm{i}kr\sin\varphi\, \partial_y w/\sqrt{\ldots}\right]$

(b) Explain why the integral is formulated in Cartesian coordinates (x, y) rather than polar or spherical coordinates, especially for off-axis freeform systems.

(c) Show qualitatively (or argue) that when the wavefront is perfectly spherical ($w(x, y) = R_a - \sqrt{R_a^2 - x^2 - y^2}$), the polarization is constant and uniform (e.g. $[e_x, e_y] = [1, 0]$), and the observation point is on-axis ($r = 0$), the integral reduces in structure to the classical Richards–Wolf expression.

Bibliography

[1] J. Braat and P. Török, *Imaging Optics.* Cambridge University Press, 2019.

[2] B. Richards and E. Wolf, "Electromagnetic diffraction in optical systems, ii. structure of the image field in an aplanatic system," *Proceedings of the Royal Society of London. Series A. Mathematical and Physical Sciences*, vol. 253, no. 1274, pp. 358–379, 1959.

[3] M. L. Boas, *Mathematical methods in the physical sciences.* John Wiley & Sons, 2006.

[4] O. Rodrigues, "Des lois géométriques qui régissent les déplacements d'un système solide dans l'espace, et de la variation des coordonnées provenant de ces déplacements considérés indépendamment des causes qui peuvent les produire," *J. Math. Pures Appl*, vol. 5, no. 380-400, p. 5, 1840.

[5] G. B. Arfken, H. J. Weber, and F. E. Harris, *Mathematical methods for physicists: a comprehensive guide.* Academic press, 2011.

[6] K. B. Wolf, "Construction and properties of canonical transforms," in *Integral Transforms in Science and Engineering*, pp. 381–416, Springer, 1979.

[7] M. Born and E. Wolf, *Principles of Optics: Electromagnetic Theory of Propagation, Interference and Diffraction of Light.* Elsevier, 2013.

[8] J. W. Goodman, *Introduction to Fourier optics.* Roberts and Company publishers, 2005.

[9] S. J. Orfanidis, *Introduction to signal processing.* Prentice-Hall, Inc., 1995.

[10] M. Rezai and J. A. Salehi, "Fundamentals of quantum Fourier optics," *IEEE Transactions on Quantum Engineering*, vol. 4, pp. 1–22, 2022.

[11] R. G. González-Acuña and S. Thibault, "General diffraction integral for converging freeform vectorial fields," *Physica Scripta*, vol. 98, no. 11, p. 115503, 2023.

[12] J. M. Boone, T. Yu, and J. A. Seibert, "Sinusoidal modulation analysis for optical system MTF measurements," *Medical Physics*, vol. 23, no. 12, pp. 1955–1963, 1996.

[13] P. J. de Groot and X. Colonna de Lega, "Fourier optics modeling of interference microscopes," *Journal of the Optical Society of America A*, vol. 37, no. 9, pp. B1–B10, 2020.

[14] R. G. González-Acuña and S. Thibault, "Focusing of an aberrated radially polarized beam by microscope objectives," *Optical Engineering*, vol. 63, no. 3, pp. 035108–035108, 2024.

[15] R. N. Bracewell and R. N. Bracewell, *The Fourier transform and its applications*, vol. 31999. McGraw-Hill, 1986.

[16] F. Bowman, *Introduction to Bessel functions.* Courier Corporation, 2012.

[17] D. G. Duffy, *Advanced engineering mathematics with MATLAB®*. Chapman and Hall/CRC, 2016.

[18] T. L. Heath *et al.*, *The Thirteen Books of Euclid's Elements.* Courier Corporation, 1956.

[19] I. Mueller, *Philosophy of Mathematics and Deductive Structure of Euclid's" Elements".* Dover Publications, 1981.

[20] G. De Young, "The Arabic textual traditions of Euclid's elements," *Historia Mathematica*, vol. 11, no. 2, pp. 147–160, 1984.

[21] M. Herzberger, *Modern Geometrical Optics.* Interscience, 1958.

[22] R. G. González-Acuña, H. A. Chaparro-Romo, and J. C. Gutiérrez-Vega, *Analytical Lens Design.* IOP Publishing, 2020.

[23] R. G. González-Acuña and H. A. Chaparro-Romo, *Stigmatic Optics.* Institute of Physics Publishing, 2020.

[24] D. Malacara-Hernández and Z. Malacara-Hernández, *Handbook of Optical Design.* CRC Press, 2016.

[25] R. Kingslake and R. B. Johnson, *Lens Design Fundamentals.* Academic Press, 2009.

[26] H. Sun, *Lens Design: A Practical Guide.* CRC Press, 2016.

[27] H. A. Buchdahl, *An Introduction to Hamiltonian Optics.* Courier Corporation, 1993.

[28] R. S. Westfall, *Never at Rest: A Biography of Isaac Newton.* Cambridge University Press, 1983.

[29] R. S. Westfall, *The Life of Isaac Newton.* Cambridge University Press, 1994.

[30] M. S. Mahoney, *The Mathematical Career of Pierre de Fermat, 1601-1665.* Princeton University Press, 1994.

[31] R. K. Luneburg and M. Herzberger, *Mathematical Theory of Optics.* University of California Press, 1964.

[32] H. Gross, W. Singer, M. Totzeck, F. Blechinger, and B. Achtner, *Handbook of Optical Systems*, vol. 1. Wiley Online Library, 2005.

[33] H. Gross, F. Blechinger, and B. Achtner, *Handbook of Optical Systems, Survey of Optical Instruments*, vol. 4. Wiley-VCH, 2008.

[34] H. Gross, H. Zügge, M. Peschka, and F. Blechinger, *Handbook of Optical Systems, Aberration Theory and Correction of Optical Systems*, vol. 3. Wiley-VCH, 2007.

[35] M. Laikin, *Lens Design.* CRC Press, 2018.

[36] D. C. Dilworth, *Lens Design: Automatic and Quasi-Autonomous Computational Methods and Techniques.* IOP Publishing, 2020.

[37] A. E. Conrady, *Applied Optics and Optical Design, Part One.* Courier Corporation, 2013.

[38] D. Malacara, *Handbook of Optical Engineering.* CRC Press, 2001.

[39] M. C. Rocha, *Aplanatic Lens Correctors for Imaging Optics.* PhD thesis, National University of Ireland Galway, 2022.

[40] R. G. González-Acuña, *CAD-Based Optical Design with Quadoa.* CRC Press, 2025.

[41] O. N. Stavroudis, *The Mathematics of Geometrical and Physical Optics: the K-Function and its Ramifications.* John Wiley & Sons, 2006.

[42] O. Stavroudis, *The Optics of Rays, Wavefronts, and Caustics*, vol. 38. Elsevier, 2012.

[43] R. G. González-Acuña and H. A. Chaparro-Romo, "General formula for bi-aspheric singlet lens design free of spherical aberration," *Applied Optics*, vol. 57, no. 31, pp. 9341–9345, 2018.

[44] K. Niu and C. Tian, "Zernike polynomials and their applications," *Journal of Optics*, vol. 24, no. 12, p. 123001, 2022.

[45] C. McAlinden, M. McCartney, and J. Moore, "Mathematics of Zernike polynomials: a review," *Clinical & Experimental Ophthalmology*, vol. 39, no. 8, pp. 820–827, 2011.

[46] V. Lakshminarayanan and A. Fleck, "Zernike polynomials: a guide," *Journal of Modern Optics*, vol. 58, no. 7, pp. 545–561, 2011.

[47] J. Borne, D. Panneton, M. Piché, and S. Thibault, "Analytical inversion of the focusing of high-numerical-aperture aplanatic systems," *JOSA A*, vol. 36, no. 10, pp. 1642–1647, 2019.

[48] R. G. González-Acuña, J. Borne, and S. Thibault, "On the diffraction of a high-na aplanatic and stigmatic singlet," *JOSA A*, vol. 38, no. 9, pp. 1332–1338, 2021.

[49] C. Huygens, *Traité de la lumière*. 1690.

[50] C. S. Williams and O. A. Becklund, *Introduction to the Optical Transfer Function*, vol. 112. SPIE Press, 2002.

[51] A. Vijayakumar and S. Bhattacharya, *Design and Fabrication of Diffractive Optical Elements with MATLAB®*. SPIE Press, 2017.

[52] D. H. Goldstein, *Polarized Light*. CRC Press, 2017.

[53] R. A. Chipman, W.-S. T. Lam, and G. Young, *Polarized light and optical systems*. CRC Press, 2018.

[54] L. Mandel and E. Wolf, *Optical Coherence and Quantum Optics*. Cambridge University Press, 1995.

[55] L. Novotny and B. Hecht, *Principles of Nano-Optics*. Cambridge University Press, 2 ed., 2012.

[56] E. Wolf, "A scalar representation of electromagnetic fields: II," *Proceedings of the Physical Society (1958-1967)*, vol. 74, no. 3, p. 269, 1959.

[57] E. Wolf, "Electromagnetic diffraction in optical systems-i. an integral representation of the image field," *Proceedings of the Royal Society of London. Series A. Mathematical and Physical Sciences*, vol. 253, no. 1274, pp. 349–357, 1959.

[58] R. G. González-Acuña, "Stigmatic singlet with a user-defined apodization pupil function," *JOSA A*, vol. 39, no. 2, pp. 213–218, 2022.

[59] S. George H, *Optical Design of Microscopes*, vol. 1. SPIE Press, 2010.

[60] D. Panneton, G. St-Onge, M. Piché, and S. Thibault, "Needles of light produced with a spherical mirror," *Optics Letters*, vol. 40, no. 3, pp. 419–422, 2015.

[61] H. Dehez, A. April, and M. Piché, "Needles of longitudinally polarized light: guidelines for minimum spot size and tunable axial extent," *Optics Express*, vol. 20, no. 14, pp. 14891–14905, 2012.

[62] G. Thériault, Y. De Koninck, and N. McCarthy, "Extended depth of field microscopy for rapid volumetric two-photon imaging," *Optics express*, vol. 21, no. 8, pp. 10095–10104, 2013.

[63] J. Liu, J. Tan, T. Wilson, and C. Zhong, "Rigorous theory on elliptical mirror focusing for point scanning microscopy," *Opt. Express*, vol. 20, pp. 6175–6184, Mar 2012.

[64] J. Liu, ed., *Elliptical Mirrors.* 2053-2563, IOP Publishing, 2018.

[65] F. O. Fahrbach, V. Gurchenkov, K. Alessandri, P. Nassoy, and A. Rohrbach, "Light-sheet microscopy in thick media using scanned Bessel beams and two-photon fluorescence excitation," *Optics express*, vol. 21, no. 11, pp. 13824–13839, 2013.

[66] Q. Zhan, "Trapping metallic Rayleigh particles with radial polarization," *Optics express*, vol. 12, no. 15, pp. 3377–3382, 2004.

[67] A. April and M. Piché, "4π Focusing of TM_{01} beams under nonparaxial conditions," *Optics Express*, vol. 18, no. 21, p. 22128, 2010.

[68] C. Varin, M. Piché, and M. A. Porras, "Acceleration of electrons from rest to GEV energies by ultrashort transverse magnetic laser pulses in free space," *Physical review E*, vol. 71, no. 2, p. 026603, 2005.

[69] K. Jahn and N. Bokor, "Solving the inverse problem of high numerical aperture focusing using vector Slepian harmonics and vector Slepian multipole fields," *Optics Communications*, vol. 288, pp. 13–16, 2013.

[70] T. Liu, J. Tan, J. Liu, and H. Wang, "Modulation of a super-gaussian optical needle with high-na fresnel zone plate," *Optics letters*, vol. 38, no. 15, pp. 2742–2745, 2013.

[71] R. G. González-Acuña and S. Thibault, "Exact vectorial model for nonparaxial focusing of freeform wavefronts," *Optics Express*, vol. 30, no. 13, pp. 23656–23663, 2022.

[72] T. A. Nieminen, N. R. Heckenberg, and H. Rubinsztein-Dunlop, "Forces in optical tweezers with radially and azimuthally polarized trapping beams," *Optics letters*, vol. 33, no. 2, pp. 122–124, 2008.

[73] H. Kawauchi, K. Yonezawa, Y. Kozawa, and S. Sato, "Calculation of optical trapping forces on a dielectric sphere in the ray optics regime produced by a radially polarized laser beam," *Optics letters*, vol. 32, no. 13, pp. 1839–1841, 2007.

[74] Q. Zhan, "Cylindrical vector beams: from mathematical concepts to applications," *Advances in Optics and Photonics*, vol. 1, no. 1, pp. 1–57, 2009.

[75] K. S. Youngworth and T. G. Brown, "Focusing of high numerical aperture cylindrical-vector beams," *Optics express*, vol. 7, no. 2, pp. 77–87, 2000.

[76] I. Moshe, S. Jackel, and A. Meir, "Correction of spherical and azimuthal aberrations in radially polarized beams from strongly pumped laser rods," *Applied optics*, vol. 44, no. 36, pp. 7823–7827, 2005.

[77] M. Liu, Y. Lei, L. Yu, X. Fang, Y. Ma, L. Liu, J. Zheng, and P. Gao, "Super-resolution optical microscopy using cylindrical vector beams," *Nanophotonics*, vol. 11, no. 15, pp. 3395–3420, 2022.

[78] D. Panneton, *Focalisation extrême par des éléments optiques non-paraxiaux.* Thèse de doctorat, Université Laval, Québec, Canada, 2018. Doctorat en Physique.

[79] B. Richards and E. Wolf, "Electromagnetic diffraction in optical systems. ii. structure of the image field in an aplanatic system," *Proceedings of the Royal Society of London. Series A, Mathematical and Physical Sciences (1934-1990)*, vol. 253, no. 1274, pp. 358–379, 1959.

[80] D. Michaelis, P. Schreiber, and A. Bräuer, "Cartesian oval representation of freeform optics in illumination systems," *Optics Letters*, vol. 36, no. 6, pp. 918–920, 2011.

[81] U. Geyer, A. Hellwig, T. Heßling, and M. C. Hübner, "Application of the multiple Cartesian oval method for glass lens design," in *Illumination Optics V* (T. E. Kidger and S. David, eds.), vol. 10693, pp. 46 – 51, International Society for Optics and Photonics, SPIE, 2018.

[82] J. P. Sutter and L. Alianelli, "Ideal Cartesian oval lens shape for refocusing an already convergent beam," in *AIP Conference Proceedings*, vol. 2054, p. 030007, AIP Publishing LLC, 2019.

[83] M. W. McDowell and H. W. Klee, "Use Of A Cartesian Oval Fibre-Optic 'Lens' As A Linear To Angular Position Converter," in *1985 International Lens Design Conference* (D. T. Moore and W. H. Taylor, eds.), vol. 0554, pp. 260 – 264, International Society for Optics and Photonics, SPIE, 1986.

[84] R. Rashed, "A pioneer in anaclastics: Ibn Sahl on burning mirrors and lenses," *Isis*, vol. 81, no. 3, pp. 464–491, 1990.

[85] R. G. González-Acuña, H. A. Chaparro-Romo, and J. C. Gutiérrez-Vega, *Analytical Lens Design.* Institute of Physics Publishing, 2020.

[86] C.-C. Hsueh, T. Elazhary, M. Nakano, and J. Sasian, "Closed-form sag solutions for Cartesian oval surfaces," *Journal of Optics*, vol. 40, no. 4, pp. 168–175, 2011.

[87] J. Valencia-Estrada and H. Bedoya-Calle, "Lentes asféricas ovales," *Mexican patent application. Instituto mexicano de la propiedad industrial (IMPI)*, 2012.

[88] J. C. V. Estrada, Á. H. B. Calle, and D. M. Hernández, "Explicit representations of all refractive optical interfaces without spherical aberration," *J. Opt. Soc. AM. A*, vol. 30, no. 9, pp. 1814–1824, 2013.

[89] C. E. Gutiérrez and Q. Huang, "The near field refractor," in *Annales de l'Institut Henri Poincare (C) Non Linear Analysis*, vol. 31, pp. 655–684, Elsevier, 2014.

[90] S. Cho, "Explicit superconic curves," *J. Opt. Soc. Am. A*, vol. 33, no. 9, pp. 1822–1830, 2016.

[91] A. Silva-Lora and R. Torres, "Explicit Cartesian oval as a superconic surface for stigmatic imaging optical systems with real or virtual source or image," *Proceedings of the Royal Society A*, vol. 476, no. 2235, p. 20190894, 2020.

[92] A. Silva-Lora and R. Torres, "Aplanatism in stigmatic optical systems," *Optics Letters*, vol. 45, no. 23, pp. 6390–6393, 2020.

[93] A. Silva-Lora and R. Torres, "Rigorously aplanatic Descartes ovoids," *JOSA A*, vol. 38, no. 8, pp. 1160–1169, 2021.

[94] A. Silva-Lora and R. Torres, "Superconical aplanatic ovoid singlet lenses," *JOSA A*, vol. 37, no. 7, pp. 1155–1165, 2020.

[95] R. G. González-Acuña, H. A. Chaparro-Romo, and J. C. Gutiérrez-Vega, "General stigmatic surfaces," *J. Opt. Soc. Am. A*, vol. 38, pp. 298–302, Mar 2021.

[96] D. Panneton, G. St-Onge, M. Piché, and S. Thibault, "Exact vectorial model for nonparaxial focusing by arbitrary axisymmetric surfaces," *JOSA A*, vol. 33, no. 5, pp. 801–810, 2016.

[97] R. G. González-Acuña, H. A. Chaparro-Romo, and J. C. Gutiérrez-Vega, "General formula to design freeform singlet free of spherical aberration and astigmatism," *Applied Optics*, vol. 58, no. 4, pp. 1010–1015, 2019.

[98] R. G. González-Acuña and H. A. Chaparro-Romo, *Optical Path Theory: Fundamentals to Freeform Adaptive Optics*. IOP Publishing, 2022.

[99] R. G. González-Acuña, H. A. Chaparro-Romo, and J. C. Gutiérrez-Vega, "Analytic solution of the eikonal for a stigmatic singlet lens," *Physica Scripta*, 2020.

[100] S. Quabis, R. Dorn, M. Eberler, O. Glöckl, and G. Leuchs, "The focus of light–theoretical calculation and experimental tomographic reconstruction," *Applied Physics B*, vol. 72, no. 1, pp. 109–113, 2001.

[101] W. v. Ignatowsky, "Diffraktion und reflexion, abgeleitet aus den maxwellschen gleichungen," *Annalen der Physik*, vol. 328, no. 10, pp. 875–904, 1907.

[102] W. v. Ignatowsky, "Diffraktion und reflexion, abgeleitet aus den maxwellschen gleichungen," *Annalen der Physik*, vol. 330, no. 1, pp. 99–117, 1908.

[103] F. Kottler, "Zur theorie der beugung an schwarzen schirmen," *Annalen der Physik*, vol. 375, no. 6, pp. 405–456, 1923.

[104] J. A. Stratton and L. Chu, "Diffraction theory of electromagnetic waves," *Physical Review*, vol. 56, no. 1, p. 99, 1939.

[105] H. Green and E. Wolf, "A scalar representation of electromagnetic fields," *Proceedings of the Physical Society. Section A*, vol. 66, no. 12, p. 1129, 1953.

[106] J. Dumont, F. Fillion-Gourdeau, C. Lefebvre, D. Gagnon, and S. MacLean, "Efficiently parallelized modeling of tightly focused, large bandwidth laser pulses," *Journal of Optics*, vol. 19, no. 2, p. 025604, 2017.

[107] S. Guha and G. D. Gillen, "Vector diffraction theory of refraction of light by a spherical surface," *JOSA B*, vol. 24, no. 1, pp. 1–8, 2007.

[108] R. Kant, "An analytical solution of vector diffraction for focusing optical systems," *Journal of Modern Optics*, vol. 40, no. 2, pp. 337–347, 1993.

[109] A. April, P. Bilodeau, and M. Piché, "Focusing a tm 01 beam with a slightly tilted parabolic mirror," *Optics express*, vol. 19, no. 10, pp. 9201–9212, 2011.

[110] X. Zeng and X. Chen, "Characterization of tightly focused vector fields formed by off-axis parabolic mirror," *Opt. Express*, vol. 27, pp. 1179–1198, Jan 2019.

[111] R. Kant, "An analytical method of vector diffraction for focusing optical systems with seidel aberrations ii: Astigmatism and coma," *Journal of Modern Optics*, vol. 42, no. 2, pp. 299–320, 1995.

[112] R. Kant, "An analytical solution of vector diffraction for focusing optical systems with seidel aberrations: I. spherical aberration, curvature of field, and distortion," *Journal of Modern Optics*, vol. 40, no. 11, pp. 2293–2310, 1993.

[113] Z. Wang, O. Baladron-Zorita, C. Hellmann, and F. Wyrowski, "Generalized debye integral," *Optics Express*, vol. 28, no. 17, pp. 24459–24470, 2020.

[114] *Quadoa Optical CAD Software Manual v.* Quadoa Optical Systems, 2022.

[115] Z. Wang, O. Baladron-Zorita, C. Hellmann, and F. Wyrowski, "Generalized debye integral," *Opt. Express*, vol. 28, pp. 24459–24470, Aug 2020.

[116] R. G. G.-A. na, "Freeform aplanatic reflective imaging systems with several stigmatic pairs," *Appl. Opt.*, vol. 64, pp. 721–727, Jan 2025.

[117] R. G. González-Acuña, "Aplanatic freeform-mirror-based optical systems," *Appl. Opt.*, vol. 62, pp. 5260–5266, Jul 2023.

[118] R. G. González-Acuña, "Power set of stigmatic freeform catadioptric systems," *Appl. Opt.*, vol. 62, pp. 5427–5432, Jul 2023.

[119] R. G. González-Acuña and H. A. Chaparro-Romo, *Optical Path Theory: Fundamentals to freeform adaptive optics.* IOP Publishing, 2022.

[120] R. G. González-Acuña, H. A. Chaparro-Romo, and J. C. Gutiérrez-Vega, "Uniqueness of stigmatic solutions," *JOSA A*, vol. 37, no. 11, pp. 1832–1838, 2020.

[121] R. G. González-Acuña, "Equations to design an aplanatic catadioptric freeform optical system," *Applied Optics*, vol. 62, no. 27, pp. 7226–7232, 2023.

[122] R. G. González-Acuña, "Design of freeform-mirror-based optical systems with several stigmatic pairs," *Applied Optics*, vol. 63, no. 31, pp. 8068–8073, 2024.

[123] R. G. González-Acuña, "Exact equations to design aplanatic sequential optical systems," *Applied Optics*, vol. 60, no. 30, pp. 9263–9268, 2021.

[124] R. G. González-Acuña, "Analytical equations for a nonconfocal stigmatic three-freeform-mirror system," *Applied Optics*, vol. 61, no. 28, pp. 8317–8322, 2022.

[125] R. G. González-Acuña, "Power set of mirror-based non-symmetric stigmatic optical systems," *Applied Optics*, vol. 62, no. 3, pp. 536–540, 2023.

[126] R. G. González-Acuña and J. C. Gutiérrez-Vega, "Exact equations to design a stigmatic singlet that meets the Herschel's condition," *Optics Communications*, vol. 485, p. 126727, 2021.

[127] R. G. González-Acuña and J. C. Gutiérrez-Vega, "General formula to eliminate spherical aberration produced by an arbitrary number of lenses," *Optical Engineering*, vol. 58, no. 8, p. 085106, 2019.

[128] R. G. González-Acuña and J. C. Gutiérrez-Vega, "Analytic formulation of a refractive-reflective telescope free of spherical aberration," *Optical Engineering*, vol. 58, no. 8, p. 085105, 2019.

[129] R. G. González-Acuña and J. C. Guitiérrez-Vega, "Generalization of the axicon shape: the gaxicon," *Journal of the Optical Society of America*, vol. 35, no. 11, pp. 1915–1918, 2018.

[130] R. G. González-Acuña, M. Avendaño-Alejo, and J. C. Gutiérrez-Vega, "Singlet lens for generating aberration-free patterns on deformed surfaces," *J. Opt. Soc. Am. A*, vol. 36, no. 5, pp. 925–929, 2019.

[131] R. G. González-Acuña and J. C. Gutiérrez-Vega, "General formula of the refractive telescope design free spherical aberration," in *Novel Optical Systems, Methods, and Applications XXII*, vol. 11105, p. 111050Z, International Society for Optics and Photonics, 2019.

[132] R. G. González-Acuña and J. C. Gutiérrez-Vega, "General formula for aspheric collimator lens design free of spherical aberration," in *Current Developments in Lens Design and Optical Engineering XX*, vol. 11104, p. 111040P, International Society for Optics and Photonics, 2019.

[133] R. G. González-Acuña, H. A. Chaparro-Romo, and J. C. Gutiérrez-Vega, "Exact equations for stigmatic singlet design meeting the Abbe sine condition," *Opt. Commun.*, p. 126415, 2020.

[134] R. G. González-Acuña and J. C. Gutiérrez-Vega, "Analytic formulation of a refractive-reflective telescope free of spherical aberration," *Opt. Eng.*, vol. 58, no. 8, p. 085105, 2019.

[135] R. G. González-Acuña and J. C. Gutiérrez-Vega, "Analytic design of a spherochromatic singlet," *J. Opt. Soc. Am. A*, vol. 37, no. 1, pp. 149–153, 2020.

[136] R. G. González-Acuña, "Design of a pair of aplanatic mirrors," *Applied Optics*, vol. 61, no. 8, pp. 1982–1986, 2022.

[137] R. G. González-Acuña, "Two-mirror system for tunable apodization," *Applied Optics*, vol. 60, no. 34, pp. 10756–10760, 2021.

[138] R. G. González-Acuña, "Set of all possible stigmatic pairs of mirrors," *Applied Optics*, vol. 61, no. 10, pp. 2513–2517, 2022.

Index

Note: Page numbers followed by f, t, and n indicate figures, tables, and notes, respectively.

For Product Safety Concerns and Information please contact our EU representative GPSR@taylorandfrancis.com Taylor & Francis Verlag GmbH, Kaufingerstraße 24, 80331 München, Germany

Batch number: 10725802

Printed by Printforce, the Netherlands